DEUTSCHE AKADEMIE DER WISSENSCHAFTEN ZU BERLIN
INSTITUT FÜR ORIENTFORSCHUNG
Veröffentlichung Nr. 54

J. R. HARRIS

LEXICOGRAPHICAL STUDIES IN ANCIENT EGYPTIAN MINERALS

AKADEMIE-VERLAG · BERLIN
1961

Erschienen im Akademie-Verlag GmbH, Berlin W 8, Leipziger Straße 3-4
Lizenz-Nr. 202 · 100/43/61
Satz und Druck: IV/2/14 · VEB Werkdruck Gräfenhainichen · 1328
Bestellnummer: 2013/54
Preis: DM 89,50
Printed in Germany
ES 7 L

TO MY PARENTS

In Sincere Gratitude and Affection

ACKNOWLEDGEMENTS

In preparing the present thesis I have been fortunate in being able to work in the Griffith Institute Oxford, and to make use of its excellent library. I am grateful to the Librarian, Mr. R. F. Ovenell, and to his staff for their unfailing helpfulness, and in particular to Miss B. M. Sewell who has so readily assisted me in consulting the collections of unpublished material preserved in the Institute. To Miss R. L. B. Moss, Mrs. E. W. Burney, and Miss H. V. Murray I am indebted for much invaluable bibliographical information, and for their kindness in answering my many enquiries, while for access to the WÖRTERBUCH Zetteln and to Sethe's copies of unpublished inscriptions, I owe my thanks to Prof. H. Grapow and his colleagues in the Berlin Academy. Prof. H. W. Fairman and Mr. I. E. S. Edwards, to whom the thesis was submitted, have very kindly communicated their views on a number of important points, the former also supplying me with innumerable references from his indices of Ptolemaic texts. My greatest debt is, however, to my supervisor, Prof. J. Černý, to whom the inception of this work was due, and without whose help and encouragement its completion would hardly have been possible. He has allowed me to search at will among his notebooks, and to make use of his many transcriptions of unpublished papyri and ostraca, and I have had the inestimable advantage of being able to discuss problems with him and profit from his acute observations. Finally, I must express my sincere gratitude to the Governing Body of Christ Church, Oxford, without whose sympathetic financial assistance I should not have been able to undertake this research, and to the Institut für Orientforschung of the Deutsche Akademie der Wissenschaften zu Berlin who have so generously undertaken its publication.

In order to reduce printing costs, which have made publication of this work in Oxford impossible, the use of hieroglyphs has been reduced to a minimum. Except in certain instances, the orthography of words has not been given, and for this reference should be made to the WÖRTERBUCH, and to H. VON DEINES und H. GRAPOW, Wörterbuch der ägyptischen Drogennamen, where the variant writings of words occurring in the medical texts are indicated in full.

J. R. Harris
Christ Church, Oxford

CONTENTS

REFERENCES

References within angular brackets ⟨ ⟩ are to unpublished texts included in the Wörterbuch Belegstellen, those within square brackets [ ] are to unpublished texts given in the autographed pages at the end of this volume.

Classical authors are quoted by book and section. Unless otherwise stated, references to Theophrastus indicate De Lapidibus, and to Dioscorides De Materia Medica (ed. Wellmann).

NB. The medical texts are now available in Grapow, Grundriss der Medizin der alten Ägypter vol. V, with translation and commentary in vol. IV pts. 1 & 2.

Several of the Philae texts quoted from Dümichen, Geographische Inschriften and from the Philae Photos are also more readily accessible in Junker, Der große Pylon des Tempels der Isis in Philä, for which the following concordance should be noted.

Düm. G. I. IV, 162–163 (Philae Phot. 210–211) = Große Pylon Abb. 52–53; pp. 99f.

Düm. G. I. IV, 164 (Philae Phot. 217–218) = Große Pylon Abb. 44–45; pp. 84f.

Düm. G. I. IV, 186–188 (Philae Phot. 265–266, 241, 245) = Große Pylon Abb. 156–158; pp. 265f.

Philae Phot. 524–519 (text of Filling the Sound Eye) = Große Pylon Abb. 54–56; pp. 104f.

INTRODUCTION

'We are still far from having reached the stage when the meaning of an Egyptian word can be proclaimed ex cathedra, and every learner who has spent more than a couple of years over the subject ought to be put in a position to question, if he considers himself to have good cause, the inherited lexicographical beliefs of his teachers.' GARDINER, Onomastica, p. XIII.

The completion of the Berlin WÖRTERBUCH nearly thirty years after the appearance of the first volume was an event of no small importance in the history of Egyptology. With the reprinting of Vol. I of the Belegstellen, and the addition of the unpublished material, scholarship is at last in possession of the evidence which the editors deemed significant in establishing the probable meanings of words. It is at once an end and a beginning. An essential tool is placed in the hands of Egyptologists, but, more important, the way along which future lexicographical research should move is pointed out.

By a proper scientific conservatism, none but the most evident conclusions have been drawn from the mass of material originally collected. In cases susceptible of the slightest doubt, only the basic and general meaning of a word is given, and many variants are allowed separate existence in order to avoid the risk of telescoping distinct but superficially similar terms[1].

That a more detailed and specialised study of words and groups of words is capable of yielding more precise results cannot be doubted, and has, indeed, been demonstrated by GARDINER in his Onomastica, the most important contribution to Egyptian lexicography since the five main volumes of the WÖRTERBUCH appeared. In his introduction[2], GARDINER rightly stresses that future progress is most likely to be made by the exhaustive investigation of groups of related words, advocating in particular the compilation of separate dictionaries for certain texts and collections of texts. Progress in this direction has already been made by the Berlin Academy itself, whose researches into the medical papyri[3] will place the lexicographical, grammatical, and therapeutic study of these documents on an entirely new basis.

But although work upon a group of similar texts may be of the greatest general value, embracing grammar, style, and content, as well as vocabulary, it is not, from a purely lexicographical standpoint, as satisfactory as the systematic study of groups of words. However restricted and characteristic the vocabulary of a particular set of texts may appear to be, it is always possible that the most significant example for determining the meaning of a word will be found elsewhere. In fact, there is much truth in IVERSEN's observation[4] that the more out of place a word is in a given context, the more revealing it is likely to be. Moreover, the final delimitation of the meaning of a term may depend to a considerable

[1] Eg. *ʿmʿt* (I. 185. 12), *ʿmʿt* (I. 185. 17), and *ʿmʿmt* (I. 186. 12); cf. Onom. 1 p. 10.

[2] Onom. 1 p. xix f.

[3] Grundriß 9 Vols., in particular Vol. VI, Wörterbuch der ägyptischen Drogennamen, and Vol. VII, Wörterbuch der medizinischen Sprache.

[4] Paints and Pigments p. 6.

extent on the meaning accorded to others of similar general connotation, which may not appear in the same class of text. The value of such research into closely related groups of words has not, however, been overlooked, and successful work along these lines has been done by SETHE[1], IVERSEN[2] and others[3].

The necessity for greater lexicographical precision is nowhere more apparent than in the case of Egyptian materia technica, the accurate identification of which is the concern not only of the linguist, but also of the archaeologist. Inability to adduce reliable inscriptional evidence, and an involuntary reliance upon contradictory and often confused translations are the greatest defects of LUCAS' otherwise admirable work[4]. In the absence of any uniform and reasoned rendering of the various substances occurring in Egyptian tribute-lists and elsewhere, any author not himself a lexicographer must perforce draw upon the work of others, without the ability to distinguish between what is reliable and what is not. Thus LUCAS himself gives equal credence to LORET's excellent article proving the meaning turquoise for *mfkꜣt*[5], and NEWBERRY's note suggesting that *šsmt* is malachite[6], a hypothesis at best. Conversely, although WIEDEMANN[7] long ago showed that *msdmt* was galena, not antimony sulphide, and the same conclusion was reached by HOFMANN[8] and FLORENCE and LORET[9], the point has been ignored for over half a century until its reiteration by LUCAS[10] and JONCKHEERE[11].

The uncritical endorsement of unfounded translations in standard archaeological works by reputable authorities can only be of ultimate disservice to lexicographical research, although, faute de mieux, it is to be excused. More misleading, however, are the attempts of technologists to solve lexicographical problems with insufficient, if indeed any first-hand knowledge of the language. In a recent work[12], FORBES has compiled six tables showing the ancient nomenclature of certain pigments, the Egyptian names being, it is stated, taken from the WÖRTERBUCH, though several are not[13]. Of a total of eighteen identifications, only seven are probably correct, four are doubtful, and seven are almost certainly wrong. In addition, three substances of which the names are in fact known are left without translations. Elsewhere[14], the same author completely misunderstands and misinterprets the inscription of Harwerrēᶜ (Sinai 90), in spite of the extensive literature that has been devoted to it.

Of the materials used by the ancient Egyptians, and mentioned by them, it is the mineral products that are likely to be the most susceptible of intelligent study with any profitable results. The richness of Egypt in minerals of all kinds, and their relative freedom from organic change due to the passage of time, make criteria of provenance, colour and the like fairly reliable. Whereas vegetable or animal products may deteriorate and change beyond recognition over 3,500 years, it is only rarely that a mineral substance presents an altered appearance, for example the rusting of iron, or the rare conversion of yellow ochre to red by the heat from fires lighted in tombs, and other colour changes due to smoke. As a result, it is possible to make significant comparisons between textual and archaeological data, which may be of the greatest assistance in establishing the meaning of certain words.

---

[1] Bau- und Denkmalsteine. [2] Paints and Pigments.
[3] Ebbell in Act. Or. 17 and ZÄS. 64; Dawson in JEA. 18ff. [4] Materials.
[5] Kemi 1 p. 99. [6] Griff. Stud. p. 230.
[7] Die Augenschminke mestem (Aeg. Stud.); Varieties of Ancient Kohl (Medum p. 41–4).
[8] Über Mesdem. [9] In De Morgan, Dahchour 1894 p. 153f. [10] Materials p. 100f.
[11] La Mesdemet, in Histoire de la Médecine 1952 No. 7, p. 2f. [12] Studies 3 p. 202f.
[13] Eg. *ṯms*, said to be red ochre, though Wb. (V. 369.7) makes it clear that it is merely a colour nuance and not a pigment. [14] Metallurgy p. 338.

This aspect of lexicography has, however, been considerably neglected, and LEPSIUS' essay[1], published over eighty years ago, is still the only comprehensive work on metals and semi-precious stones. At the time, it gave rise to great interest, and a number of useful contributory articles by DÜMICHEN[2] and others[3] appeared in ZÄS., but little advance has been made since. BRUGSCH accepted many of the conclusions in the Ergänzungsband of his Wörterbuch, and a few years later WENDEL published a thesis[4] dealing with stones of all kinds, though without resolving any of the major problems. From that time, the interest shown in materials has been small, and the contributions to knowledge, with a few notable exceptions, negligible, because unsystematic.

SETHE's important paper[5] has elucidated the various problems connected with the major building and monumental stones, though without exhausting the list, and BALCZ[6] has discussed the stones used for vessels in the early dynasties. EBBELL and DAWSON[7] have studied materia medica, and, more recently, IVERSEN has produced an interesting and important essay on pigments[8], which, however, suffers from its omission of the less common pigments, and a failure to take into account three of the four most important lists[9].

There have, of course, been numerous other articles dealing with individual words, but except in one or two instances, they do not carry conviction through their treating of a material in isolation, and not in relation to others of the same kind or similar employment. LORET's article[10] re-establishing and proving the meaning turquoise for *mfkꜣt*, and the detailed re-assessment of the evidence, with much additional material, by ČERNÝ[11], together with the various contributions by VARILLE[12], LUCAS and ROWE[13], BRUNTON[14], and SHIAH[15] concerning *bḫn* stone, are of prime importance, but the majority of articles and brief incidental discussions, such as those of JÉQUIER[16], have fallen prey to the same shortcomings, which the systematic study of groups of words can avoid.

In compiling the present thesis, the intention has been, as far as possible, to summarise all articles and studies of any value, though GARDINER's advice[17] has generally been followed in eschewing the unprofitable, except in cases where a wildly aberrant translation has gained undeserved acceptance, or has given rise to serious misconceptions.

The author of an article who proposes to identify a given word with a particular material, is not usually concerned to find a suitable equivalent either for the substance by which the word has previously been translated, or for the word until then connected with the material. Hence, in attempting to provide a solution to one question, he may very easily pose two more. Thus SETHE[18], while suggesting that *ỉbhty* is to be translated amethyst, does not attempt to reconcile previous identifications of the stone as imperial porphyry[19] or a species of marble[20], nor does he pause to consider an alternative meaning for the word *ḥmꜣgt*, widely held to represent amethyst[21]. Similarly IVERSEN[22], in his excellent treatment of *mnšt*, omits to

[1] Die Metalle in den ägyptischen Inschriften (1872), translated and augmented as Les Métaux dans les Inscriptions égyptiennes (1877), to which all citations in the present work refer.

[2] ZÄS. 10 p. 42; p. 98; 11 p. 46.

[3] Renouf, ZÄS. 11 p. 119; Chabas, ZÄS. 12 p. 1 etc.

[4] Bau- und Edelsteine.

[5] op. cit. p. 10 n. 1.

[6] Mitt. Kairo 3 p. 64.

[7] loc. cit. p. 10 n. 3.

[8] op. cit. p. 10 n. 2.

[9] Ostr. Tor. A. 11, of which Iversen was aware, but which he was unable to consult; Ostr. Cairo 25594; P. Ch. B. V rt. 8, 12–14.

[10] Kemi 1 p. 99.

[11] Sinai 2 p. 7 f.

[12] BIFAO. 34 p. 93.

[13] Ann. Serv. 38 p. 127; p. 677; 41 p. 347.

[14] Ann. Serv. 40 p. 617.

[15] Ann. Serv. 41 p. 189.

[16] BIFAO. 19 passim.

[17] Onom. 1 p. xxi.

[18] Bau- und Denkmalsteine p. 50.

[19] Eg. Wendel, Bau- und Edelsteine p. 77.

[20] Brugsch, Wb. Erg. p. 438.

[21] Eg. Wendel, Bau- und Edelsteine p. 107; Jéquier, BIFAO. 19 p. 121.

[22] Paints and Pigments p. 28.

compare his interpretation as red ochre with Dawson's view[1] that that substance was called *dỉdỉ*, and does not discuss possible Egyptian names for the pigments with which *mnšt* has been equated by others[2].

When, however, dealing with materials in groups, such vagueness and inconsistency is impossible, since all the words within a given group must be identified with equal plausibility. The existence of two or even more words apparently referring to the same substance is not, of course, an impossibility, but it seems likely from the long lists of minerals preserved in the Chester Beatty papyrus[3], at Edfu[4], Dendera[5] and elsewhere, that as a rule the Egyptians connected one substance with each name. Apparent anomalies must therefore be examined, and possible explanations sought for the duplication of terms or the representation of more than one material by the same name. In some cases minerals were subdivided and distinguished in ways which modern science does not recognise, in others, two or more apparently but not in fact related substances were grouped under one head because of their colour or some similar criterion. A new loan-word might replace an older term or exist parallel to it, eg. *ḳḏ*, *ỉbnw*, and in texts of the Graeco-Roman period artificial designations were created as doublets for more usual words without, however, any perceptible difference of meaning, eg. *ḥb*, *ḥkn*.

The various mineral substances are therefore to be dealt with in separate groups – metals, building and monumental stones, miscellaneous stones, semi-precious stones, pigments, miscellaneous minerals, salts, and earths, with, in addition, discussion of certain words the mineral nature of which is questionable. Within each group it seems most logical to treat the words in their Egyptian alphabetical order[6], except in the cases of gold, silver, and the major building stones, where all the various designations are more satisfactorily grouped together and discussed in relation to one another. The range of mineral substances known to the Egyptians and to which therefore references may be expected, can roughly be gathered from Lucas' work, though it must be kept in mind that archaeological evidence is not wholly conclusive, and that certain other materials are known from other indications, such as references in the Amarna letters, to have been accessible to them.

In treating of the minerals in this way, it is well to remember that the Egyptian classification may not necessarily coincide with our own. Indeed, to speak of classification at all is perhaps improper, since the Egyptians, at least before the Graeco-Roman period, had not even the rudimentary scientific knowledge of the Assyrians. Until comparatively recently, there was no precise scientific classification of minerals, and Theophrastus[7], using criteria of colour, hardness and weight, divided minerals into metals, stones, including gems, and earths. Little or no chemistry or crystallography was known to Pliny, and colour remained the most important characteristic of mineral substances until the end of the 18th century. Quist (1768), Bergmann (1792) and Hauy (1817) were the first to use more scientific methods founded on weight, crystal form and hardness[8].

But although the Egyptians had no rational classification of their minerals, there is nevertheless clear evidence of the way in which they thought of their materials, from the frequent recurrence of the same lists of words in approximately the same order[9]. For this reason it may be necessary to displace certain substances from their position in relation to others of the same scientific grouping, and treat them in connection with those with

---

[1] JRAS. 1927 p. 497.
[2] Eg. Brugsch, ZÄS. 18 p. 6; Dawson, JEA. 20 p. 188.
[3] P. Ch. B. IV vs. 7, 4–8, 6.
[4] Edfou VI, 201 f.; cf. Dend. Chass. IV, 159.
[5] Dend. Mar. IV, 36, 49–50; 39, 140–2.
[6] The practice of the Wörterbuch is abandoned in one particular, namely in combining 𓊃 and 𓋴. For justification of this deviation, the convenience of which cannot be gainsaid, cf. Onom. 1 p. xxi.
[7] περὶ λίθων § 1–7.
[8] Cf. Ball, Roman Book p. 13; p. 16.
[9] Cf. Lepsius, Métaux p. 3.

which the Egyptians associated them. Such is the case with, for example, the pigments. Similarly, it will be appropriate to discuss a small number of non-mineral substances, which were either regarded as mineral by the Egyptians, or which, though recognised as vegetable or animal products, nevertheless occur in close connection with a mineral group. An example of the former is *inr n r(ꜣ)*, read *inr n spt (rꜣ) mw* by EBBELL[1], which seems to refer to marine shells[2]. Among the latter may be mentioned *ḳmyt*, which occurs in some lists of painting materials, and is evidently the gum employed as a vehicle for the powdered pigments. Finally, there are a few words which must be considered, if only to be rejected, because, though vegetable products or even less likely things, cf. for example *iḳḥw*, they have previously been thought to be minerals by certain Egyptologists.

A disadvantage of this system of treatment is that some few words will inevitably come under discussion more than once under different heads. For example, certain of the semi-precious stones that are also pigments, such as *ḫsbḏ* and *wꜣḏ*, will have to be considered in both contexts as appropriate. On the other hand, it has not seemed desirable, or indeed altogether feasible, to separate the various words for faience, glass, frit etc. from the semi-precious stones. It is not a distinction of which the Egyptians themselves were ever acutely conscious, and it is evident that the name of almost any stone could equally well be applied to its imitation in faience or glass, it being quite impossible in many cases to tell which is meant. In the same way it is hardly practicable to maintain a strict distinction between the stones and minerals in, for example, the Chester Beatty list, and the groups of miscellaneous stones and miscellaneous minerals, chapters 4 and 7, include many items which might be classified with greater precision were more known concerning them.

In the absence of any Egyptian documents of a similar nature to the Assyrian syllabaries, evidence of provenance, colour and use becomes the most reliable guide to any identification. Provenance is, of course, only useful in as far as the names of the regions mentioned in inscriptions can be identified with more or less certainty, and is even then dependent on the continued existence of the same minerals in those regions. BALL[3] has pointed out that the ancients obtained their gemstones from four sources: 1. deposits still known to us and in some instances still being operated, 2. deposits worked by them and since exhausted, 3. deposits known to them but with which we are unacquainted, and 4. deposits producing too poor a product to be operated today. The same can be said of most minerals, and where the ancient deposits are either worked out or lost, not even quarry inscriptions in situ are of any significance.

Colour is of greater importance, since it was one of the chief criteria by which the Egyptians themselves, in common with other peoples of the ancient world, differentiated between semi-precious and other stones. It may be established from the colour of heaps of named materials shown in wall-paintings, or by the colour given to objects known to be of a certain substance. In both cases, allowances must be made for the limitations of the Egyptian palette and for regular colour conventions, which may tend to confuse the issue[4]. We do not, for instance, know how the Egyptians represented the distinctive colour of amethyst, though it is probable that they saw it as a nuance of red. More rarely, colour can be ascertained by precise statements or similes in literary texts, eg. 'redder than *ḫnmt*'[5], 'colour of *nšmt*'[6], of *mfkꜣt*[7], etc. Here again caution is necessary, and account must be taken of literary extravagance. Thus *ḫsbḏ* is sometimes used of the colour of vegetation, for example a moringa[8],

---

[1] ZÄS. 64 p. 49.

[2] Cf. a somewhat similar mistake in Pliny, HN. 37, 60, where Hammonis cornu, fossil ammonite, is regarded as a stone.

[3] Roman Book p. 48; cf. p. 78.

[4] Cf. Appendix v.

[5] Love. Tur. 2, 3.

[6] ib.

[7] Totb. Leps. 109, 3.

[8] Nauri 11.

where no literal comparison can be intended, unless we assume visual peculiarities in the ancient Egyptians[1].

Use is, naturally, only a general guide, except in cases where, for example, a type of amulet is prescribed to be made of a certain material[2], and the majority of amulets of that type are in fact found in one particular stone.

The information to be derived from mere lists of materials is small in itself, but the order in which the items occur may be of considerable interest, as, for example, the precedence of the various grades of gold and silver, and the uniformly high position accorded to *ḫsbḏ* and *mfkꜣt*. But too much cannot be argued from the exact order of a word, or from the type of substances with which it occurs. Lists such as those at Dendera[3] display the most disturbing inconsistency, and others like the Edfu mineral list[4] and that of the Chester Beatty papyrus[5] are quite unsystematically compiled, as are other scribes' exercises containing outlandish words. But as the number of unknown substances in a list is reduced, it is sometimes possible by comparison of one with another to deduce the approximate nature of a lesser known material which takes the place of one more common. This is the case with the pigment lists[6], and the Dendera tribute lists[7], where the texts accompanying figurès bearing named mineral products include variant words, which may represent simply alternative names, or substances of similar use and appearance.

In this connection, the nature of evidence from the later periods, particularly Graeco-Roman times, must be considered very carefully. It is, for instance, clear that contact with the outside world led to the introduction of loan-words to represent distinctions not before observed, eg. *ktmt* for particularly pure gold, and *ʿrḳwr* for silver. At the same time, some terms such as *ḏʿm* became obsolete and acquired purely 'literary' meanings. The development of vocabulary must always be taken into account, since it is evident that certain words which, in the early stages of the language, had a wide meaning, eg. *nbw, msdmt, biꜣ, wꜣḏ, bsn*, gradually became more specialised and had their significance modified or made more precise. With the creation or borrowing of new terms for materials lately distinguished, shades of meaning were subtracted from the wider designations, and a word, the translation of which is based on late evidence alone may have an extended range in the best period. In some instances, only the later terms are known, and the original words cannot certainly be determined. Thus there are New Kingdom terms for alum and related astringent materials (*ỉbnw*), gypsum (*ḳḏ*), ? haematite (*ḳʿ*), and particular forms of black eye-paint (*ḥtm, gsfn*). For the last *msdmt* was undoubtedly the original all-embracing designation, but for the others no earlier word is at once apparent, thought the substances were in use.

More direct evidence of the materials represented by some Egyptian words is also forthcoming, but must be treated with reserve and correctly interpreted. The most significant is that drawn from quarry inscriptions, where it is possible to arrive at a fairly reliable conclusion as to the material named. In this way, *ḥsmn* has been established as amethyst from inscriptions at Wadi el Hudi[8], and *mntt* as diorite-gneiss (anorthosite-gneiss) from stelae

---

1 Cf. Ransom-Williams, Per-Nēb p. 57.
2 Eg. Berl. 20600; Dend. Mar. IV, 87; Edfou VI, 299f.
3 Dend. Mar. IV, 36, 49–50; 39, 140–2.
4 Edfou VI, 201f.; cf. Dend. Chass. IV, 159.
5 P. Ch. B. IV vs. 7, 4–8, 6.
6 Ostr. Stras. H. 41; Ostr. Tor. A. 11; Ostr. Cairo 25594; P. Ch. B. V rt. 8, 12–14.
7 Dend. Mar. I, 70; I, 71; II, 12; cf. also Düm. G. I. II, 63; 66; 69 etc.
8 Ann. Serv. 41 p. 51.

in the quarries near Tushka[1], though whether it also refers to the commoner diorite remains uncertain[2]. But in some instances, the existing workings do not afford any clue as to the material sought, as in the case of *mḫn(m)t*, the second mineral brought from Tushka. This may be due to its having been mined at a lower level than that visible today, though it is also possible that the deposit has been worked out or that the stones were collected on the surface in the form of pebbles, all of which have long ago been removed[3].

An additional source of direct evidence are monuments themselves, which in many instances bear inscriptions mentioning the material from which they are made. Clues of this kind must not, however, be treated too literally. The various naoi and other monuments stated to be of *bḫn* are not all of the same stone, the majority, it is true, being of the Wadi Hammamat greywacke, but with some exceptions. Such exceptions, however, do not necessarily imply that the word *bḫn* covered a number of similar stones, since an Egyptian sculptor carving a more or less conventional text on a naos was not particularly concerned with the real nature of the material, or if he was, the statement that it was *bḫn* stone sufficed to convince the god and to make it such for all practical purposes. This is more especially the case with amulets, when, for example, a heart scarab of *nmḥf* might be of any one of a variety of green stones, but the recitation of the relevant spell of the 'Book of the Dead' was enough to establish it as true *nmḥf*, probably serpentine or green jasper.

There are, however, a number of instances in which texts inscribed on objects are quite conclusive in establishing the name of a certain material, for example the lumps of gypsum found at Amarna, bearing the date and the description *ḳḏ*[4].

The too literal interpretation of statements of this kind has, however, led to considerable confusion, and in no case has the misunderstanding been greater than in dealing with the Berlin amulet board[5]. On the basis of that document, *ḥnmt* has been rendered as carnelian by Jéquier[6] and others, and *nšmt* as 'weißblauer Feldspat' by the Wörterbuch[7]. But it is clear that the names of materials carved on the board do not necessarily refer to the materials of the inserted amulets, but rather to the materials of which the particular amulets should ideally be made. In order therefore to arrive at any valid conclusion based on the evidence of the board, the substance in which a given amulet is most frequently found must be ascertained, and for this Petrie's Amulets is a useful guide, though some of its terminology is unfortunate, eg. durite, basalt, paste, glaze etc., and one may doubt the confident citation of certain rare stones. Such an investigation makes it clear that the *tít* amulet of *ḥnmt* is most usually of red jasper, and the *wꜣḏ* sceptre of *nšmt* of green feldspar, the carnelian and blue felspar of the pieces in the Berlin board being one a mistake or a substitute, the other a near approximation.

In short, it must be realised that a number of apparent inconsistencies in the occurrence and use of materials and their names is inevitable. These are not, however, to be taken too seriously, since they arise very largely from the peculiar psychological attitude of the Egyptians[8]. For them, words and names had a real significance which it is perhaps difficult to

---

1 Ann. Serv. 33 p. 65; p. 75; 38 p. 395; p. 682.

2 Cf. below p. 87f.

3 Cf. Ball, Roman Book p. 48; p. 78.

4 ZÄS. 58 p. 51; Cit. of Akh. 3 pl. 97–8 nos. 331–5.

5 Berl. 20600. Cat. Kennard Coll. 1912 no. 530, pl. 11; Amtliche Berichte 34, no. 2 (Nov. 1912) col. 24, Äg. Inschr. Berl. 2 p. 312.

6 BIFAO. 19 p. 121. 7 II. 339.19.

8 For some valuable observations on the problems of Egyptian thought and logic cf. Iversen, Fragments of a Hieroglyphic Dictionary p. 9f.

comprehend at the present day, and any object could for magico-religious purposes become whatever it was stated to be. A portrait might not be an exact likeness of the person represented, but if it bore his name it automatically became such, and the mere changing of the name effected a complete usurpation. Similarly, an object might be made of a substance other than that prescribed, but a statement that it was of the correct material was sufficient to rectify the mistake. In the same way, artificial substances were often used as inferior substitutes for the real ones, without differentiation, though the word *mꜣꜥ*, 'true', may be used to distinguish the genuine product. It is worth noting, however, that the Egyptians never seem to have felt any great distinction between the value of semi-precious stones and that of glass or faience imitations[1].

The nature and value of evidence dating from the late period has already been alluded to, and the possibility of less precise translations for earlier phases of the language suggested. A similar caution must be exercised in assessing the worth of translations based on apparent parallels with Akkadian, Hebrew, Greek and other sources. In the first place it is by no means certain that an Egyptian loan-word will have precisely the same meaning as the original from which it is borrowed, but further, the original may well prove to have a wider range of meaning than that generally assigned to it, and this appears to be the case in a few instances.

ỉsmr, later *smr*, 'emery', is so translated because of its relation to Sumerian ᵃAŠ.MUR, Hebrew *šāmîr*, and Greek σμύρις, σμιρίτης λίθος, but all the archaeological evidence is against such a rendering[2]. It is, however, clear that Sumerian ᵃAŠ.MUR has a fairly wide range of meaning[3], since it is used, inter alia, as an eye salve, for which any form of corundum would be quite unsuited, and similarly, the Hebrew and Greek evidence does not appear to warrant any very precise translation. The most that can be said is that σμύρις is a powder used by lapidaries[4], and this might well be nothing more than ferruginous sand or powdered quartz, meanings quite consistent with the appearance of *ỉsmr* in Egyptian contexts. Another example is that of *ỉbnw*, usually translated 'alum', though it is quite clear that both Coptic **OBN̄** and Greek στυπτηρία, from which this meaning is derived, refer to any one of a number of astringent substances[5].

Evidence from Coptic and Demotic is of varying significance. In some cases, eg. **OBN̄** cited above, the meaning of a word may be readily apparent, and so provide a translation for the Egyptian term from which it originated. On the other hand, the use of Greek loanwords in both Coptic and Demotic may suggest that there was no native Egyptian term for the material in question, or that it was of some rarity. But this is not always the case, as is shown by the use of ἀρσενικόν and σανδαράχη, Egyptian *ḳnỉt* and *ꜣwt-ỉb*, and of certain other words for which the Egyptian equivalent is known.

An added source of difficulty when treating of Egyptian words in relation to classical and other parallels is that the latter have often undergone considerable change of meaning, in many cases unrecognised. *ꜥš*, for example, is clearly the classical 'cedar', but the latter is equally clearly not the same as modern cedar[6]. *bḫn* is evidently PLINY's 'basalt', but that word is itself a misnomer, and of no real scientific significance[7]. In the same way, many of

---

[1] Cf. Ransom-Williams, Jewelry p. 6.
[2] Cf. Lucas, Materials p. 294f.
[3] Cf. Campbell Thompson, Dictionary p. 52; p. 133; p. 167.
[4] Dioscorides 5, 147; Hesychius sv. σμύρις.
[5] Cf. Liddell & Scott sv. στυπτηρία.
[6] Cf. Loret, Ann. Serv. 16 p. 33; Glanville, ZÄS. 68 p. 8; Gardiner, Onom. 1 p. 8 n. 1; Sethe, Bau- und Denkmalsteine p. 3.
[7] Cf. Sethe, op. cit. p. 33.

the precious stones mentioned by THEOPHRASTUS[1] and PLINY[2] are other than those which a modern lapidary knows by the same names. For this reason, the citation of too many classical and other parallels has, as far as possible, been avoided in this thesis, particularly in cases where the meaning of an Egyptian word can satisfactorily be established without. Where, however, the range of meaning of a classical term may be significant in delimiting the application of the Egyptian equivalent, eg. στυπτηρία mentioned above, it has been discussed in full.

The problem of terminology as a whole is one of considerable delicacy, since in many instances inadequate or incorrect terms are well established in Egyptological literature, and it would seem almost inevitable that they should be used as translations. But a movement towards greater accuracy and precision is necessary in all archaeological matters, and in proposing an exact translation for any word, the maximum possible accuracy consistent with intelligibility should be aimed at. Thus 'basalt' is to be rejected as too vague a word, diorite is to be amplified to diorite-gneiss as a rendering of *mntt*, and so on. Certain inaccurate and confusing terms should now be eradicated altogether, such as 'nitre' for natron, 'lime' for gypsum, and expressions like 'glaze', 'paste', 'enamel' and 'pâte de verre' used in reference to faience or glass[3].

Bearing in mind all the possible pitfalls, the minerals are to be treated systematically, in relation to others of the same nature, all the available indications from whatever source being considered, and the relative value of each type of evidence carefully weighed in cases of apparent inconsistency. By the gradual establishment of equivalents for the more clearly defined items within a group, the possible translations for the remainder are restricted, and criteria of colour and provenance assume greater importance. LUCAS' work is useful as a guide in determining the range of possibilities, providing as it does a fair conspectus of information regarding the occurrence and use of actual substances. But it can be no more than a guide, since it does not exhaust all the reliable literature, and there is evidence elsewhere of materials not mentioned by LUCAS. Further, it is always possible that a particular substance employed in small quantities has not yet been discovered in use. This seems to be the case with realgar, which IVERSEN has recently suggested[4] was in all probability known under the name *ꜣwt-ib*, and also with alum, *ibnw*, the use of which in ancient Egypt can be inferred, but not proved[5].

That the Egyptians must have known a number of materials of which no archaeological evidence has yet been found is clear from, for example, the Amarna letters, where a variety of minerals, and other materials, are recorded as being sent to Egypt. Of these, many are not easily identified, eg. *ᵃmušširru*, which occurs with lapis lazuli, but of the few that are, *ᵃdušu*, crystal glass, *ᵃḫulâlu*,? white lead, and *ᵃabašmu*,? amber, are worthy of note. Neither of the latter is known from Egypt, and yet, since they were imported in the New Kingdom, it is not unlikely that there may be mention of them in inscriptions[6]. Instances such as these demonstrate the danger of confining attention to those substances the knowledge and use of which can be proved on purely archaeological grounds, for there may be words which will never be translated because they represent materials of which no trace survives. More important still, if purely philological evidence suggests a particular rendering, it cannot be discarded out of hand solely through lack of concrete confirmation.

---

[1] περὶ λίθων; cf. Caley & Richards, Theophrastus.

[2] HN. Bks. 2, 9, 19, 31, 33–7; cf. Bailey, Elder Pliny.

[3] Cf. Lucas, JEA. 10 p. 128 and Materials, passim.

[4] Paints and Pigments p. 38.

[5] Cf. Lucas, Materials p. 291 f.

[6] The meaning of *ḫulâlu* is doubtful, cf. below p. 178.

Any advance made along these lines which increased our knowledge of the Egyptian materia technica as mentioned on the monuments and in texts would be of the greatest practical importance. Not only would it make possible a more precise and intelligent appraisal of the tribute lists, and hence lead to a clearer conception of at least one aspect of the ancient Egyptian economy, but it would also prepare the way for more accurate and illuminating translations of the texts accompanying technological scenes. From such translations our knowledge of the capabilities of the ancient craftsmen and the techniques they employed would be considerably increased, and in general the textual evidence corroborating archaeological facts would be more firmly grounded.

CHAPTER 1

GENERAL TERMS

In attempting to understand the lexicography of ancient Egyptian minerals, it is first necessary to consider how far the minerals as a whole were differentiated from other substances, and to what extent there was recognised any distinction between one class of minerals and another. To this end it is convenient to discuss a number of words which, while not the names of individual materials, are none the less significant in helping to throw some light on the Egyptian conception of mineralogy. Most important among these are the few general terms for stone and the like, examination of which reveals several interesting facts, and enables some tentative conclusions to be drawn.

ỉnr (Wb. I. 97.12f.), Coptic ^{S}ⲱⲛⲉ, ^{B}ⲱⲛⲓ, was recognised by WENDEL[1] as the generic word for stone, in particular the building and monumental stones, and not apparently including any other class of minerals. This definition is substantially correct, and, in fact, three broad divisions of meaning may be distinguished; 1. stone (as a mineral), 2. stone as a building material, and 3. stone-block, though this latter meaning is also extended to blocks of other materials.

As a simple term for stone, *ỉnr* forms the basic element in the names of several of the building stones, eg. *ỉnr ḥḏ nfr n ꜥỉn*, *ỉnr ḥḏ nfr n rwḏt*, and *ỉnr km*, while other stones such as *mꜣṯ*, *bỉꜣt*, and *bḫn* may equally well be referred to as *ỉnr mꜣṯ*, *ỉnr bỉꜣt*, and *ỉnr bḫn*. Similarly, terms are compounded of *ỉnr* and a geographical location, as *ỉnr ꜣbw*[2], *ỉnr n ḥwt-nbw*, and *ỉnr n ḏw dšr*, and *ỉnr* also occurs in the names of semi-precious stones, for example *ỉnr n wḏḥ* and *ỉnr n ḏbꜣ*. In addition to such examples where it appears as part of a compound name, *ỉnr* is also found in certain set phrases to do with stone-working, in the title *ms-ỉnr* (Wb. II. 135.22), and linked with the verbs *wbꜣ*, *wḫꜣ*, *šꜥd* etc. (Wb. I. 290.1; I. 346.15; IV. 422.3), and in these cases too it would seem to have a quite general meaning.

More commonly, however, *ỉnr* is applied to stone used in building[3], particularly in opposition to *ḏbt*, brick[4], and all kinds of constructions are said to be executed in *ỉnr*. Various parts of buildings are also specifically referred to as being of stone, including walls[5], pylons[6], columns[7], doors[8], door posts[9], shrines[10], and stelae[11], and in all instances of this kind, a translation (building)-stone is required.

---

[1] Bau- und Edelsteine p. 42.

[2] For an instance of *ỉnr ꜣbw* with a rare writing if *ỉnr*, cf. Shoshenk Stela 1.9, JEA. 27 pl. X; p. 89 n. 39.

[3] Eg. Amada 12; Urk. IV, 183; Palermo 5, 2; Abyd. Mar. I, 42b, 4 etc.

[4] Eg. Urk. IV, 765; Petos. 81, 77; Mem. Miss. XV, 10, 1.

[5] LD. III, 152a; Hier. Ostr. 22, 1 rt. 4 etc.

[6] Harr. 45, 6.

[7] Urk. IV, 819; P. Salt 124, 2, 6.

[8] Berl. 13272; Dend. Mar. II, 55b; Düm. K. I. 60b.

[9] Harr. 58, 10; Urk. I, 38.

[10] Karn. Mar. 15, 19.

[11] Amarna V, 28; Canop. 37.

As a term for block, *ỉnr* frequently refers to those used in building construction or the like, the material not being specified[1], though occasionally some further description such as *ỉnr n ḳd*[2], *ỉnr n snt*[3], or *ỉnr bꜣk m pꜣ ḏw*[4] is given, or the dimensions of the block are listed[5]. In one instance[6], a slab of stone actually bears an inscription describing it as *ỉnr*. But blocks are also mentioned for purposes other than building or the carving of monuments, for example as weights[7], whetstones[8] and writing grounds[9], and in several instances *ỉnr* merely indicates the shape of a particular mass, the material being specifically stated as *ꜥꜣt ḥḏt*[10], *ḫsbḏ*[11], *mḫn(m)t*[12], and even *ḥḏ*[13]. Of peculiar interest is the expression *m ỉnr wꜥ*[14], literally 'of one block', which must, as the Wörterbuch suggests, be the Egyptian term for monolithic, and it may also be noted that in one or two cases *ỉnr* refers not to a regular block, but to a boulder or the like[15].

In none of the uses mentioned above does *ỉnr* have any meaning other than stone or block, and in no way does it conflict with the term *ꜥꜣt* to be discussed below. There are, however, some few examples of *ỉnr* where it does appear to do so, and these require careful consideration. It has already been mentioned that *ỉnr* is found as the formative element in the names of at least two precious stones, *ỉnr n wḏḥ* and *ỉnr n ḏbꜣ*, where the use of *ꜥꜣt* might have been expected. Similarly, *ỉnr* occurs in the phrase *mḥ m ỉnr*[16], with obvious reference to the use of precious minerals for inlay, and here again there is some inconsistency. In certain other instances too, *ỉnr* appears where *ꜥꜣt* would be usual, notably in the phrase *ỉnr nb n mꜣꜥ*[17] for the more usual *ꜥꜣt nb(t) n mꜣꜥ*, and in the Sehel inscription[18] where the minerals are referred to as *ỉnr* rather than *ꜥꜣt*. A further source of difficulty is the expression *ỉnr (n) mꜣꜥ*[19], the exact significance of which cannot be determined, though it would most naturally refer to some semi-precious stone.

It is thus apparent that whereas *ỉnr* has, in the majority of cases, the meaning 'stone' or 'block', and is applied in particular to the building and monumental stones, there are some few instances where it is used with the wider reference of *ꜥꜣt*, from which it is normally distinguished.

ỉnr rwḏ (Wb. II. 410.13) is noted by Sethe[20] as a general designation of hard stone, and the same view is taken by the Wörterbuch. This is undoubtedly correct, though the expression is not to be confused with *ỉnr (n) rwḏt* and *ỉnr nfr (n) rwḏt*, which refer to sandstone, and is, in fact, somewhat rare. Sethe quotes only two examples[21], and it seems doubtful whether a third given by the Wörterbuch[22] is relevant, since it is probably *ỉnr nfr (n) rwḏ(t)*. It is tempting to regard the *ỉnr rwḏ* occurring with *mnšt* in a magical text[23] as this term, though

[1] Eg. Giza J. VIII fig. 62; P. B. M. 10052, 5, 20; 10403, 1, 19; Amonhymn. Leid. 1, 10; 3, 10; LD. II, 138c, 4; 149d; P. Kah. 14, 5 etc.

[2] Hier. Ostr. 17, 1, 10.

[3] P. Berl. 10621, 3; Bersh. I, 14.

[4] P. Berl. 10615; cf. Ostr. DM. 389 vs. 1–2.

[5] Ostr. UCL. ⟨2⟩.

[6] Ed. Soc. Ant. 912.

[7] Hier. Ostr. 67, 2 rt. 2, 4, 5 etc.

[8] Lansing 5, 6.

[9] P. Ch. B. IV vs. 2, 10.

[10] Urk. IV, 701.

[11] Urk. IV, 668; 732; Harr. 14a, 2–3; 33a, 12–13.

[12] LD. III, 258a b, 8 = Bub. Port. 22; Urk. IV, 638.

[13] Coff. Texts I, 261f.; Flor. Sch. 1617; Emb. Rit. 10, 18; Rec. Trav. 27, 225.

[14] Urk. IV, 932; LD. III, 24d; Louvre Wb. ⟨224⟩ (referring to metal).

[15] Urk. IV, 894; Hier. Ostr. 20, 4, 2.

[16] Harr. 6, 1; 45, 5–6.

[17] Ombos I, 168, 216b, 4.

[18] Sehel 29.

[19] Berl. 17271; Düm. Baug. 29; P. Tur. unpublished [1].

[20] Bau- und Denkmalsteine p. 14.

[21] Urk. IV, 879; P. Berl. 11292.

[22] LD. Text IV, 98, 7.

[23] P. Bib. Nat. 237, Cart. 2,6 [2].

identification cannot be certain. In the Berlin ostracon[1], *ỉnr rwḏ* is distinguished from *ẖꜣt* and ordinary *ỉnr*, and may therefore refer to some specific kind of stone, perhaps a particularly hard outcrop in the quarry. No more precise translation than 'hard stone' is desirable, and it is worth noting that the few certain examples of *ỉnr rwḏ* can scarcely be differentiated from similar references to *ꜥꜣt rwḏt*.

ꜥꜣt (Wb. I. 165.13f.) is unquestionably a term of much wider reference than *ỉnr*, as was recognised by Brugsch[2], who points out that it included both metals and semi-precious stones, and also, on occasion, other materials. The same view was taken by Gensler[3], who regards *ꜥꜣt šps* as a general expression for 'mineral', and by Wendel[4], who recapitulates Brugsch's observations. The Wörterbuch has 'kostbarer Stein', which is too restricted, as is shown by Varille[5], who quotes a Louvre stela[6] where *ꜥꜣt* includes everything from silver and gold to ivory and ebony, and also refers to the dubious *ꜥꜣt ꜥš* of Medum[7], proposing to translate *ꜥꜣt* as 'une matière dure'. Balcz[8], on the other hand, reads *ꜥꜣt ꜥnw* in the latter case, and so takes *ꜥꜣt* as 'wertvolle Steine', ignoring other instances where *ꜥꜣt* has a broader meaning.

Now it is evident that *ꜥꜣt* does in fact cover a very wide range of materials, largely minerals, but also including other substances, and it is difficult to see what it was that led the Egyptians to class them together. Most frequently, *ꜥꜣt* refers to the noble metals and precious stones, such as gold, silver, lapis lazuli, turquoise, red jasper, green felspar, carnelian and *ṯḥnt*[9], *ḥsmn*, *mnw ḥḏ*, *ꜥꜣt ḥḏt* and *mnw* (*?km*)[10], *mḫn(m)t*[11], and the like. In other instances, however, such as the lists of the 24 *ꜥꜣt* at Dendera[12], and the Edfu mineral list[13], it covers a considerable variety of minerals, including *wꜣḏ*, *msdmt*, *bḳs-ꜥnḫ*, *ṯmḥy*, *snn*, and *ḏrwy*[14], the base metals, *sšw*, and *ỉsmr*[15], and even *ḥmꜣt* and *nṯry*[16]. Further, *ꜥꜣt* is applied to certain monumental stones, *bỉꜣt*, *mꜣt*, and *ỉnr km*[17], while *ꜥꜣt n mꜣt*[18], *ꜥꜣt bỉꜣt*[19], and *ꜥꜣt rwḏt*[20] are specifically mentioned. Lastly, in the passage cited by Varille[21], *ꜥꜣt* includes all art materials from silver and gold down to ivory and ebony.

And yet it is clear that *ꜥꜣt* had some definite meaning, as is shown by such expressions as *ꜥꜣt nb(t)*, *ꜥꜣt nbt špst*[22], *ꜥꜣt nbt mnḫt*[23], *ꜥꜣt nb n mꜣꜥt*[24] etc., and in the use of *ꜥꜣt* alone as designation of the material of certain objects. Among the latter are vessels[25], statues[26], inlay[27] and other decoration[28], and in every case *ꜥꜣt* probably refers to a stone of some description.

That the common quality of all *ꜥꜣt* is hardness as Varille suggests is most unlikely, since although the semi-precious stones and the few monumental stones classed as *ꜥꜣt* were undoubtedly hard, many of the other minerals mentioned among *ꜥꜣt* are friable and even powdery. More feasible is that *ꜥꜣt* refers primarily to value, though again there are objections, since neither granite nor natron was particularly valuable.

---

1 P. Berl. 11292.
2 Wb. p. 159.
3 ZÄS. 8 p. 141.
4 Bau- und Edelsteine p. 43.
5 Ann. Serv. 33 p. 90.
6 Louvre C 14.
7 Medum 13.
8 Mitt. Kairo 3 p. 70.
9 Edfou II, 215, 4–5; VI, 165, 15–16.
10 Urk. IV, 722.
11 LD. III, 127b; Karn. I, 27 = Urk. IV, 1668.
12 Dend. Mar. IV, 36, 49–50; 39, 140–2.
13 Edfou VI, 201f.
14 Dend. Mar. IV, 36, 49–50; 39, 140–2.
15 Urk. IV, 731.
16 Edfou VI, 203, 8.
17 Ann. Serv. 38, 220f.
18 Brugsch, Wb. p. 159.
19 LD. Text III, 144; Six Temp. 12, 4 etc.
20 Cf. below p. 22.
21 Louvre C 14.
22 Harr. 73, 13; Urk. IV, 183; Kenam. 17–18 etc.
23 Ann. Serv. 2, 283; Urk. IV, 537; 913.
24 Petos. 81, 43; cf. Edfou VI, 299; Harr. 4, 6.
25 Urk. IV, 183; 641, 665 etc.
26 Urk. IV, 537; 913; Goshen 6, 5; 6, 6 etc.
27 Urk. IV, 705; Harr. 6, 12; Berl. 20600 etc.
28 Urk. IV, 913 etc.

Alternatively, the conception of ꜥꜣt may be based on the form in which the material was obtained, or the use to which it was destined to be put. Thus, ꜥꜣt were brought from the mountains[1], and probably therefore mined, and were generally in small lumps as opposed to the larger blocks of building stones, *inr*[2]. In addition, nearly all the examples of the use of ꜥꜣt are for sculpture, vessels, inlay, and other quasi-artistic purposes, and this may perhaps be significant.

However, none of these explanations of ꜥꜣt is wholly satisfactory, and it seems likely that GENSLER was nearer the truth in taking it as, in essence, the Egyptian word for mineral. This would be apposite to every known example of its use, save that in the Louvre stela, where ivory and ebony are included. In this case there is obviously some confusion, due perhaps to nuances of value or rarity being also present in the meaning of ꜥꜣt, but since it is an isolated instance, it may be of no very great importance. Moreover, if ꜥꜣt is a general term for mineral substances, *ꜥꜣt špst*, *ꜥꜣt mnḫt*, and perhaps even *ꜥꜣt n mꜣꜥt* are readily explained as referring specifically to the more valuable ones, ie. the noble metals and semi-precious stones, though the last may also imply genuine minerals as distinct from imitations. In this way, ꜥꜣt would be parallel to the Akkadian *abnu*, which is used as determinative for most minerals, including glass, hailstones, coal etc., and is often added to metals[3].

If then ꜥꜣt is to be regarded as a word for mineral, with perhaps certain implications of value and rarity, the distinction between ꜥꜣt and *inr* becomes clear, since the latter refers principally to stones which were quarried in large quantities. That the distinction was not always observed has been noted above, where various examples of *inr* have been shown to bear the same general meaning as ꜥꜣt, or at least to refer to semi-precious stones, normally classed as ꜥꜣt. The same inconsistency can be exemplified in reverse, as, for example, in the title *ms-ꜥꜣt* (Wb. II. 138.19), which is simply a word for stone-mason, and has no reference to minerals in general[4]. It is interesting, also, that in Demotic there is no differentiation between the two words, which have practically become merged into one. Apart from two occurrences of *ꜥꜣy*, in a text parallel to hieratic[5], only *ꜣny*, ie. *inr*, is used, where hieroglyphic would have either *inr* or ꜥꜣt. Thus *wty n ꜣny*[6] is clearly for *wḏyt n inr*, while in other instances[7] *ꜣny* is clearly ꜥꜣt, and in some, eg. *mḥ m ꜣny*[8] and *ꜣny ḏry*[9], it may be for either[10]. In Coptic, ꜥꜣt has disappeared altogether.

Why this should be is not easily appreciated, unless it is that ꜥꜣt had gradually become so general a term that it had no real meaning, while *inr* could without difficulty be extended to cover all stones of whatever nature.

ꜥꜣt rwḏt (Wb. I. 165.19; II. 410.14; II. 412.18) is taken by SETHE[11] as an alternative to *inr rwḏ* as a general designation of hard stone, and this seems to be essentially correct. The WÖRTERBUCH reference to 'schwarzer Granit' is evidently derived from the description of a stela as *nty ꜥꜣt rwḏt*, Greek ἐκ στερεοῦ λίθου[12], the actual monument being usually described

[1] Dend. Mar. II, 28; Hatshepsut obelisk, unpublished [3].

[2] Hatnub Gr. 9; cf. however Urk. IV, 638, where *inr* are irregular lumps.

[3] Campbell Thompson, Dictionary p. xxxvi.

[4] Not to be confused with *ms-ꜥꜣt* (Wb. II. 135.22), which is quite distinct from the parallel *ms-inr* (Wb. II. 135.22); cf. Gardiner, Onom. 1 p. 68; p. 72.

[5] P. Rhind I, 8d, 2; 8d, 3.

[6] Priesterdek. A. 20; Mythus 6, 33; cf. Sius. 6, 17–18; P. Ryl. IX, 11, 9.

[7] Eg. P. Dem. Mag. vs. 3; Mythus 6, 16; 7, 4; Philensis I, 6; II, 5.

[8] Priesterdek. A. 16; B. 58.

[9] Priesterdek. R. 32.

[10] Cf. Lexa, Grammaire p. 975.

[11] Bau- und Denkmalsteine p. 14.

[12] Urk. II, 197.

as black basalt, which it probably is not. However, this is not to be taken literally, since another version of the same text in Cairo is inscribed on limestone, and *ꜥꜣt rwḏt* also occurs on a statue of 'calcaire compact[1]'. Further, references to *ꜥꜣt nbt rwḏt*[2], would show that *ꜥꜣt rwḏt* has a wide meaning, and cannot be confined to one specific stone. Other examples are not numerous[3], and in none of them is there any apparent distinction between the use of *ꜥꜣt* and the similar occurrence of *inr* mentioned above.

rwḏt (Wb. II. 412.14f.) is in some instances a word for sandstone, developed from the more common *inr ḥḏ nfr n rwḏt*, as SETHE[4] points out. On the other hand, it has also a more general meaning, corresponding to the 'festes Gestein', 'harter Stein', of the WÖRTERBUCH, and a very clear example of this is the reference to *rwḏt* on an ostracon from the tomb of Senmut[5], where it seems to mean, as HAYES points out[6], a particularly tough, flinty portion of the hill. Equally certain is the description of a statue of alabaster as of *inr n rwḏt*[7], which can only mean 'a block of hard stone', though why alabaster should be called hard stone it is difficult to imagine[8]. A general meaning for *rwḏt* may also be inferred from its occurrence in a title, 'chief of the hard-stone workers of His Majesty'[9], and from the mention of 16 statues of Amenemhet III *n rwḏt*[10], which are unlikely to have been of sandstone, which is only rarely used for statuary, and scarcely at all before the New Kingdom. Possibly a writing of *rwḏt* is *rwḏ* (Wb. II. 413.6), which is found once in Ebers[11], though the identification cannot be certain. At all events, there can be little doubt that *rwḏt* is a term for hard stone in general, though which stones would come into this category it is difficult to say, especially in view of the reference to alabaster as *rwḏt*.

ẖꜣt (Wb. III. 360.11–15; III. 359.6) is primarily, and in the majority of examples, a word for quarry[12], but by an extension of meaning it also comes to refer to the thing quarried[13], and in the Greek period is sometimes used as a synonym for *ꜥꜣt*[14]. It is thus, etymologically, the nearest equivalent to the English 'mineral', ie. a thing mined, but is of too limited occurrence to be of any real significance.

Among the general terms of this kind must be mentioned one which probably existed at some stage in the language, though no example of its use can now be quoted. *nšd**, regarded by MONTET[15] as a lost word for precious stone, is known only from a number of related words of which it constitutes the basic element, and from which its original existence may be inferred. Most significant is undoubtedly *ms-nšd* (Wb. II. 146.11), formed on the same pattern as the well known *ms-ꜥꜣt* (Wb. II. 138.9), 'stonemason', and apparently a word for jeweller or gem-cutter, as is shown by the occupation of the workers it describes, namely the boring of carnelian, *wbꜣ ḥrst*[16]. To this may be compared a similar, and perhaps even identical term *nšd-msw* or *nšdw* (Wb. II. 342.19)[17], and the more common *nšdy* (Wb. II.

---

[1] Rec. Trav. 4, 98.
[2] Sall. II, 5, 1; Cairo 583, 16; Sehel 11 etc.
[3] Eg. Urk. II, 55; Goshen 6, 6; Düm. G. I. II. 90; Rec. Trav. 36, 99.
[4] Bau- und Denkmalsteine p. 13; p. 15.
[5] Ostr. Senmut 74, 3.
[6] Ostraca etc. p. 37.
[7] Bersh. I, 14.
[8] Note also Urk. I, 107, 17, though the reading is doubtful, cf. Montet, Kemi 3 p. 44.
[9] Münch. Glypt. 35, 7.
[10] B. M. 569.
[11] Eb. 90, 11.
[12] Eg. LD. III, 140b, 2; Harr. 78, 2; 78, 3; Dend. Mar. II, 77e etc.
[13] Eg. P. Berl. 11292, 10.
[14] Eg. Philae ⟨2753⟩, Phot. 995.
[15] Scènes p. 288.
[16] D. el Geb. I, 13; I, 14; Caillaud, Arts et Métiers pl. 17; Wilkinson MSS. III, 24.
[17] Cairo 1445; 1731; LD. II, 96.

342.18), which GARDINER[1] translates specifically as 'worker in precious stones'. This meaning is proved by three examples in particular, where the *nšdy* is shown using a bow-drill[2], receives beads etc. of *mḫn(m)t*[3], and is engaged in adorning a temple[4].

From these titles, with which *nšdw* (Wb. II. 343.1), 'Art Schmucksband', may also be connected, it is perhaps reasonable to deduce that there was in Egyptian a word *nšd**, meaning 'precious stone' or the like. That no example can now be traced may be due purely to chance, though it is also possible that the word early became absorbed into *ꜥꜣt*.

Considering these generic terms in relation to one another, it is possible to arrive at some significant conclusions regarding the ancient Egyptian attitude to raw materials as a whole, and to mineral substances in particular, though certain points must of necessity remain obscure. In general there seems to have been a fairly clear distinction drawn between those natural resources[5] which were mineral, and those which were of vegetable or animal origin, the former being referred to as *ꜥꜣt*, the latter as *šmw* (Wb. IV. 481.1) or *rdt nbt* (Wb. II. 462.20; II. 463.8), and this distinction is normally observed in lists of tribute or the like, where the minerals are enumerated first, and the other substances follow after[6]. Moreover, there are some few lists in which practically all the items are mineral[7], which suggests that, with a number of exceptions, the Egyptians had a reasonably accurate knowledge of the nature of the many materials used by them.

That the minerals were classed together as *ꜥꜣt* is altogether probable, though the word appears in many instances to refer more specifically to the noble metals and semi-precious stones, particularly when qualified as *ꜥꜣt špst*. Furthermore, there are perhaps suggestions of value and rarity implicit in *ꜥꜣt*, which tend to modify the meaning, so that on one occasion at least it includes two other costly products, ivory and ebony, which are not mineral.

More restricted terms for the different classes of mineral substances are almost entirely lacking, and it is quite evident that there was little or no differentiation between the metals and semi-precious stones, perhaps because in the first place metals were found native and worked cold in the same way as the latter[8]. Both were *ꜥꜣt* or *ꜥꜣt špst*, and as far as can be ascertained, no specific word for metal or precious stone existed in the language. Whether there was at some stage a term *nšd**, meaning gem-stone, it is impossible to say with certainty, though there are indications which seem to point in that direction. However, the exact reference of such a word cannot be determined, and it must in any event have been rare.

For the building and monumental stones there was the generic term *inr*, which also applied to large blocks of more or less any description. Normally *inr* was quite distinct from *ꜥꜣt*, though in some few examples, discussed above, there appears to be an interchange of meaning, and it is, moreover, impossible to discover any obvious difference between the expressions *inr rwḏ* and *ꜥꜣt rwḏt*, both of which refer simply to hard stone.

Other groups of minerals, such as salts and earths, were not referred to by any distinctive name, and although some individual terms seem to have a fairly wide range of meaning, covering one primary substance and several others thought to be related to it, there is no evidence of any very clear classification.

---

[1] Onom. 1 p. 67; cf. JEA. 22 p. 177. 'Faience-maker' is apparently *bꜥbꜥ* (Wb. I. 447.5), cf. Onom. 1 p. 68. [2] Atlas I, 73a. [3] P. Boul. 19. [4] Anast. IV vs. C, 6.

[5] *mꜣꜥw* (Wb. II. 23.10–13) or *ḳmꜣt* (Wb. V. 36.8).

[6] Such apparent exceptions as Koller 3, 7f. are of small importance, since they merely represent exercises in the writing of little-known words.

[7] Eg. P. Ch. B. IV vs. 7, 4f.; Edfou VI, 201f.

[8] In historical times, however, most metals were obtained by smelting ores, and were worked molten.

In short, whatever criteria the Egyptians may have adopted to differentiate between separate classes of mineral substances, they were not sufficiently definite to be reflected in the language, and on lexicographical evidence alone there is no reason for a conventional division of metals from stones, or stones from other minerals.

In relation to the present discussion, some account must be given of a number of miscellaneous terms which refer not to mineral substances as such, but to the form in which they were found or employed. Often of no particular interest in themselves, words of this kind may sometimes help to define the meaning of other terms to which they refer, or may otherwise contribute to what is known of Egyptian minerals.

ifd (Wb. I. 71.10–11) is one of a number of similarly written words which derive their meaning from *fdw* (Wb. I. 582.13), 'four'[1]. Brugsch[2] regards it as a term for 'Sockel, Sockel-Würfel', but the context in which it occurs, describing a statue as *ifd m pr m ḏw ꜥꜣ*[3] seems to demand some such interpretation as that of Sethe[4] who says '*ifd* wird den rohen, blockartigen Zustand des Denkmals bezeichnen, das später durch die feinere Ausarbeitung in den Händen der Bildhauer bedeutend an Masse und Gewicht (*šꜣw*) verlieren sollte'. This is supported by the description *pr m ḏw ꜥꜣ*, which shows *ifd* to be the form in which the stone was quarried, namely a cuboid block. A similar word in the Rhind mathematical papyrus[5] may perhaps be related, as the Wörterbuch assumes, though it refers simply to a cube of 10 units.

inr wbꜣ (Wb. I. 291.16) seems to occur only in the Harris papyrus[6], where it appears to be the name of some object, since the quantity is not indicated by weight, but simply by a number. *wbꜣ* is presumably connected with the verb *wbꜣ* (Wb. I. 290.1), 'to bore', and *inr wbꜣ* would therefore seem to mean either 'bored stone', or 'stone-borer', the former perhaps referring to a kind of bead, the latter to a sharp stone used in drilling or graving. Either interpretation is possible in the context.

inr spdw (Wb. I. 97.16) occurs in a number of medical contexts[7], from which, however, little can be deduced. It would most naturally refer to some species of stone, but the example of *inr n spt mw*, discussed below, shows that such an assumption is not necessarily justified. On the other hand, there may be some connection with a word *spdw* which is found in one of the Kahun papyri[8], and is regarded by Griffith[9] as a kind of pebble, or in two instances[10] more specifically as scraping stones for cleaning papyrus or leather. This seems appropriate, since new papyrus was quite possibly rubbed smooth with an abrasive, perhaps pumice or siliceous sand, and there is evidence to suggest that a kind of 'erazer' of abrasive stone was used to correct mistakes[11]. Alternatively, *spdw* could equally well refer to a sharp stone used for splitting the pith into slices, though sharp stones would be unsuitable in the medical contexts, and 'abrasive stone' or the like is perhaps a preferable interpretation.

---

[1] Cf. Blackman & Fairman, JEA. 32 p. 77 n. 7 for *ifd* meaning 'girdle-wall', 'four sides of a building' etc.
[2] Wb. p. 61; Erg. p. 61.
[3] Bersh. I, 14, 6.
[4] Lesest. Erl. p. 119.
[5] P. Rhind 44, 1.
[6] Harr. 41b, 4; 56a, 11; 73, 15.
[7] Eb. 25, 12; 25, 13; 64, 17; Hearst 11, 8; P. Med. Berl. 6, 3.
[8] P. Kah. 19, 25; 19, 53; 20, 28; 2, 14.
[9] Kah. Pap. p. 49.
[10] P. Kah. 19, 25; 20, 28.
[11] Ann. Serv. 41 p. 91. The actual specimen published is of 'grès grisâtre'.

ʿnw (Wb. I. 189.18) may conceivably be a writing of *ʿin*, 'limestone'[1], but an interpretation as 'pebble' appears altogether more likely. The word seems to occur only three times, once as a product of the Wadi Natrun[2], once in a medical context, parallel to *šʿy*[3], and once in a passage where it evidently refers to sling-stones[4]. The last is the most significant, and suggests pebbles, which would be equally applicable in either of the other two instances, and the word would thus be analogous to the somewhat more common *ʿr*.

ʿnḫ (Wb. I. 204.10) is described by the Wörterbuch as 'Bez. des Alabasters oder Bez. von etwas aus diesem Hergestellten', the first of which it is certainly not. Anthes[5] rightly takes the phrase *ʿnḫ n pr nsw*[6] as an unity, and regards *ʿnḫ* as an inscribed monument, or an expression for 'die für den König bestimmte Lieferung'. No very clear conclusion can be drawn from the three examples known, though the occurrence of both □ and ▬ as determinatives might suggest that the former interpretation is the more likely.

ʿr (Wb. I. 208.11), Coptic ⲗⲗ, is correctly regarded by Lepsius[7] and Wendel[8] as a word for small stone or pebble, the latter noting that precious stones are more often depicted in small lumps than in large, or are packed in bags. Burchardt[9] similarly renders *ʿr* as 'Kiesel', but thinks the word is a vox peregrina. Certainly *ʿr* first occurs in Late Egyptian texts, and the wide range of writings may point to an attempt to represent a foreign word in Egyptian. On the other hand, group-writing is common enough, and affects the orthography of many native words, so that in default of any obvious Semitic original, no definite conclusion is possible.

There can be no doubt that the meaning 'small stone' or 'pebble' is approximately correct, though in one or two instances, *ʿr* seems to refer rather to small rocks or the like. Thus in two similar passages in Anastasi I[10], the way through a ravine is blocked with *dḫwt* (det. □ ꞁꞁꞁ) and *ʿrw* (det. ° ꞁꞁꞁ), the one apparently meaning boulders, the other smaller rocks. Likewise, the *ʿrw ḥr tbn n pꜣ is*[11], to which Caminos[12] compares *inr ḥr ḏꜣḏꜣ*[13], must surely refer to small boulders, perhaps forming a cairn over the top of a tomb, and there are also other instances in which something larger than pebbles seems to be implied[14].

But *ʿr* can only mean pebble or small stone in the expressions *mfkꜣt n ʿr*[15], *wꜣḏ n ʿr*[16], and *shrt n ʿr*[17], which seem to distinguish the stones picked up at random from those mined, and may indicate some difference in quality. Moreover, references to *ʿrw ḥr ḫꜣst* as something of proverbial worthlessness[18] would best be translated by pebble. In at least one case[19], however, *ʿr* appears to have a somewhat specific meaning, occurring in a list of precious stones after *ḫsbḏ*, *ḥnmt*, *mfkꜣt*, *bks*, and *ḥts*, though not connected with the last. It might almost be a term for all gem-stones collected in the form of pebbles, and somewhat analogous to *gmw n ḫꜣst* discussed below.

[1] Eg. Nauri 10; Beni Has. I, 26. [2] Peas. R. 17. [3] Hearst 1, 6.
[4] Pianchi 32. [5] Hatnub p. 51. [6] Hatnub Gr. 22; 28; 31.
[7] Métaux p. 40. [8] Bau- und Edelsteine p. 44.
[9] Fremdworte II p. 15 nos. 270, 274. [10] Anast. I, 23, 3; 24, 2.
[11] P. Tur. 1880 rt. 4, 5–6 = RAD. 57, 11. [12] LEM. p. 179.
[13] P. Salt 124 rt. 2, 5; cf. also Ostr. DM. 389 vs. 1–2.
[14] Ed. Soc. Ant. 912, 3; 912, 7; P. Tur. PR. 47, 5.
[15] Dend. Mar. IV, 36, 49–50; 39, 140–1. [16] Dend. Mar. IV, 39, 140–1.
[17] ib. [18] P. Mag. Harr. 4, 7; Flor. 1101; Flor. Wb. ⟨30⟩; Mett. 118.
[19] Philae ⟨1991⟩ Phot. 241.

ʿr is also found in two obscure expressions[1] discussed by PEET[2] and GARDINER[3], neither of which is here relevant, and a possible example of *ʿr* occurs in the Pyramid texts[4] (Wb. I. 208.14), though the latter was understood as the word for goat in the Middle Kingdom[5]. However, it is just feasible that 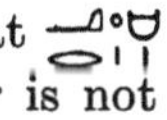 is to be read *ʿr bỉꜣ*, and equated with Coptic ⲁⲗⲙ̄ⲡⲉ, 'hail', though the fact that *ʿr* is not otherwise known before the New Kingdom must be against such an interpretation.

nws (Wb. II. 224,13–14), translated 'Klumpen o. ä.' by the WÖRTERBUCH, and regarded by PIEHL as 'poids en forme d'anneau', is evidently some standard measurement of metals etc. In the Harris papyrus[6] it appears to be a weight, and the only other reference suggests a specific form in which lead was transported, parallel to *ḏbt*, 'bricks'[7]. It may perhaps refer to the characteristic ingots[8] frequently depicted in scenes of tribute, and if these were of standard size or weight, a similar interpretation would be possible for the Harris examples.

nmst (Wb. II. 269.9) is evidently another form in which metals were traded[9], and also, like *nws*, a unit of weight[10]. CAMINOS[11] suggests that it is a bar or ingot of some kind, and this seems altogether probable, since *nmst* is parallel to *ḏbt*[12]. Again there is every reason to suppose that the bars were of some standard size, and perhaps distinctive shape, since otherwise they would scarcely be used as a unit of weight. Connection with the well known *nmst* vase (Wb. II. 269.7) appears unlikely.

nḏꜣ (Wb. II. 377.7–8) is clearly a general word for 'chip', 'splinter' or the like, and is probably connected with *nḏr* (Wb. II. 382.11) and Coptic ⲛⲟⲩϫⲉ as HAYES[13] suggests. In the story of Anubis and Bata, *nḏꜣ* refers to a chip of persea wood, but other examples are undoubtedly stone. Thus HAYES takes the mention of *nḏꜣ* on ostraca from the tomb of Senmut[14] as indicating the flakes of limestone on which the inscriptions are written, while in one of the Kahun papyri[15], *nḏꜣ* of stone are employed for building, and must be sizeable fragments, perhaps used for core masonry.

ḥnk (Wb. III. 120.6) seems to occur only once in a caption attached to a representation of the Pharaoh offering a heap of rectangular blocks, apparently bricks, in connection with a foundation ceremony[16]. No precise interpretation is possible, but the word may well be a technical term for an offering of this kind, or refer specifically to foundation deposits.

ḥrtt (Wb. III. 149.3), translated 'Klumpen o. ä.' by the WÖRTERBUCH, seems to indicate one form in which semi-precious stones were traded, namely as lumps of raw material. From

---

[1] P. B. M. 10052, 14, 4; Onom. 12, 14.
[2] Tomb Robberies p. 167.
[3] Onom. 2 p. 230.
[4] Pyr. 771c.
[5] Sethe, Kommentar III p. 421.
[6] Harr. 55a, 8–9; 55a, 11; 40b, 10; 41a, 8; 73, 12.
[7] Urk. IV, 708.
[8] Shaped ⊂⊃, but called *ḏbt* in some instances, eg. Urk. IV, 637.
[9] Anast. IV, 17, 7; VIII, 1, 3.
[10] Joppa 2, 3; Hor. & Seth 5, 2; Ostr. DM. 450, 1.
[11] LEM. p. 218.
[12] Anast. IV, 17, 7.
[13] Ostraca etc. p. 37.
[14] Ostr. Senmut 78, 2; 78, 3.
[15] P. Kah. 16, 3.
[16] Düm. Baug. 52.

the two examples quoted[1] it might appear that the term was appled specifically to lapis lazuli, but it is clear from another context[2] that it could refer to other stones as well.

ḥṯ (Wb. III. 204.7) is known only from an ostracon in the collection of University College London[3], which is 'a list of work done in dragging stone (*inr*) from the quarry (*ḫ3t*)', and includes four items. The first is ⟨*inr*⟩ *dp*, the second ⟨*inr*⟩ *ḥṯ*, and the third and fourth *inr ḏbt*, the dimensions of each being added. Thus *ḥṯ* appears to be the name of a block of stone of a particular shape or size[4], and distinct from either *dp* or *ḏbt*, the former being apparently larger, the latter somewhat smaller. A connection with the verb *ḥṯ* (Wb. III. 204.1) seems probable, and the meaning would then be 'facing-block' or the like.

ḫšḫš (Wb. III. 339.7) is apparently confined to two examples, both in the Leopold-Amherst papyrus[5], where Gardiner[6] comments that it "seems clearly to refer to something fortuitous that concealed the plaster facing of the burial-place. This something can only have been 'rubble' or 'sand'; and since *šꜥ* is the word for sand, the meaning of *ḫšḫš* must surely be 'rubble'..." There can be little doubt that this interpretation is correct, and the form of the word suggests something broken up, cf. *krkr* below. Other terms for rubble or the like, eg. *ḫmꜥ*, and perhaps *ššnyt*, are also known, and no more precise translation is possible.

s3t (Wb. III. 412.14) is evidently the name of a specific kind of stone block, and may refer either to the shape or to the destination for which the blocks were intended. The word occurs twice in an account of stones extracted from the quarry[7], where it is distinguished from *spt* and *inr n t3 ḫ3t*, though the total of all three is given simply as the last. But the word is not confined to these examples as Gunn[8] suggests, and is found also in at least two other contexts[9] which seem to point to a meaning 'paving-block' or 'column-base'. The former is the translation accepted by the Wörterbuch, and is perhaps more likely, particularly if *spt* is to be given the latter significance.

spt (Wb. IV. 100.8) is likewise the name of a particular kind of block, and is found in the same ostracon as *s3t*, where[10] again Gunn[11] takes it to be unique. However, the word also occurs in other contexts, several times referring to the bases of columns[12], and once to the base of a large stela[13], and from these examples it is clear that a translation 'base-block' is necessary for the ostracon reference. Thus three types of block are mentioned, one unspecified, one for the bases of columns etc., and one for paving or the like, and this conforms to the known Egyptian practice of dressing blocks to their approximate shape at the quarry before transport.

---

[1] Urk. IV, 638; 668.
[2] Junker, Große Pylon p. 88, 8.
[3] Ostr. UCL. ⟨1⟩, probably from the cenotaph of Seti I at Abydos.
[4] 2 cubits 3 spans × 2 cubits × 1 cubit 3 spans.
[5] P. Leo. Am. 2, 9; 2, 11.
[6] JEA. 22 p. 179.
[7] Ostr. Cen. Seti 1, 3; 1, 5 (Cen. Seti I pl. 90).
[8] Cen. Seti I, p. 93, reading *s3st*.
[9] LD. II, 118g; Q. Saqq. III p. 161, 3.
[10] Ostr. Cen. Seti 1 3 (Cen. Seti 1 pl. 90).
[11] Cen. Seti I p. 93, reading *sprt*.
[12] LD. V, 1b; Düm. H. I. II, 56; Karn. R. III 23, A 18; Ann. Serv. 44, 204; cf. Yoyotte, Chronique 28 p. 37.
[13] Harr. 7, 1.

snt (Wb. IV. 152.7), like the two terms just discussed, also refers to the form of a block of stone. BRUGSCH[1] compares Coptic ^{S}ⲤⲈⲚⲦⲈ, ^{B}ⲤⲈⲚⲦ and regards it as a word for 'basis, fundamentum', but SETHE[2] is probably more correct in maintaining that '*n snt* wird die monolithe Natur des Denkmals oder eine Folge davon aussprechen'. The word seems to occur twice, once in a text concerning an alabaster colossus[3], and once in an ostracon dealing with the extraction of stone from a quarry[4]. In both cases the reference is to *ỉnr n snt*, and *snt* therefore describes the block and must indicate either size or shape, probably the former, since the stone is described as *štꜣ* in the Bersheh passage.

snt (Wb. IV. 152.8) is apparently confined to two occurrences of the phrase *ḥwỉ m snt*[5] as the description of a scene depicting craftsmen polishing an object. MONTET[6] connects this word with *snt* discussed above, and comments, 'il est possible que le même mot *snt*, désigne une pierre quand il est déterminé par ▭ et, quand il est suivi des trois ∘∘∘, la poudre obtenue avec cette pierre'. Such an interpretation is inconsistent with the representation in the tomb of Ti, where the polishing is apparently being done with a small block held in the hand. It is not unlikely that the Egyptians used sandstone for smoothing wood[7], and *snt* may then represent a rubbing block of that material.

sẖt (Wb. IV. 264.4) is probably connected with the verb *sẖt* (Wb. IV. 263.6) as used of striking bricks, and may indicate the shape of blocks of stone. That it is to stone that it refers is clear from the expression *ỉnr sẖt*[8], and from the context in which the one example of *sẖt* alone occurs[9]. Parallel instances of *ỉnr ḏbt* may be quoted[10], in which *ḏbt* specifies the size and shape of the block, and *sẖt* may perhaps have a similar meaning, though no very satisfactory conclusion is possible.

sšnyt (Wb. IV. 294.1), which seems to occur only once[11], is obviously derived from the verb *sšn* (Wb. IV. 293.17), and the WÖRTERBUCH translation 'Abbruch (als Baumaterial für einen Neubau)' is almost certainly correct. *sšnyt* may refer either to rubble, as *ẖmꜥ* and *ẖšẖš*, or, more specifically, to blocks from another building re-used as core masonry or otherwise usurped, a common Egyptian practice.

ḳbyt (no Wb. ref.) is evidently the description of some kind of stone, probably foundation or paving blocks. The word occurs twice at Sinai[12] in contexts which ČERNÝ[13] translates '. . . this hall as a stretch of *ḳbyt* stones', and '. . . stretch of *ḳbyt* stones which the King of Upper and Lower Egypt . . . made . . .', and once elsewhere[14], where it appears to have a meaning 'foundation' or the like. It is impossible to determine the exact reference of *ḳbyt*, though a connection with foundations or flooring seems likely.

---

1 Wb. Erg. p. 1069. But Coptic ⲤⲚⲦⲈ is from *snṯt* (Wb. IV. 179.9).
2 Lesest. Erl. p. 119.
3 Bersh. I, 14, 2.
4 P. Berl. 10621, 3 – probably from the cenotaph of Seti I at Abydos.
5 Ti pl. 132; D. el Geb. I, 14.
6 Scènes p. 307.
7 Cf. Lucas, Materials p. 510.
8 Ostr. Cairo 25677 rt. 25.
9 P. Berl. 10621, 7.
10 Eg. Ostr. UCL. ⟨2⟩.
11 P. Len. 1116A rt. 79.
12 Sinai 310b; 317b.
13 Sinai 2 p. 196; p. 198.
14 Edfou II, 61.

krkr (Wb. V. 136. 5), occurring in the so-called 'Poem on the King's chariot'[1], is translated 'Steinhaufen' by the Wörterbuch, 'Geröll' by Erman[2], and 'splinters' by Peet and Dawson[3]. The word clearly refers to the small pieces[4] into which the hills disintegrate at the stroke of the king's right arm, and should perhaps be rendered 'rubble'. In form *krkr* is akin to other reduplicated words such as *sksk*, *ḳnḳn*, *tštš* and *gmgm*, but may be a loan-word as Burchardt[5] and the Wörterbuch suggest.

gmw (Wb. V. 169.13) occurs only once, in the phrase *gmw n ḫꜣst*, which Gardiner[6] translates 'precious stones (?) of the desert'. This seems altogether likely, since the *gmw* are in a bag, and precious stones are frequently shown in this way. Moreover, *gmw n ḫꜣst* would be a suitable description of stones found at random on the surface rather than mined, and may be compared with the expressions *mfkꜣt*, *wꜣḏ*, and *shrt n ꜥr*[7]. A similarly written word found elsewhere[8] is not in fact connected, and is in all probability a writing of *gmỉ* (Wb. V. 170.4)[9].

dbyt (Wb. V. 434.13) is clearly a word meaning 'base' or the like, as is recognised by the Wörterbuch, though, unlike *spt*, it may refer to either stone[10] or wood[11], and cannot therefore be taken as the designation of a type of stone block.

dp (Wb. V. 447.13) is known from only two examples, both of which suggest a species of large stone block[12]. In one context[13], three *dpw ꜥꜣw* are destined for the temple of Mut, and in the other[14] ⟨*ỉnr*⟩ *dp* is listed parallel to ⟨*ỉnr*⟩ *ḥṯ* and *ỉnr ḏbt* as the designation of a type of block, the dimensions of which are given[15]. Whether *dp* was distinguished by its shape or destination as well as size it is impossible to tell, and the Wörterbuch's 'Art großer Block aus Stein' cannot be improved upon.

ḏḥwt (Wb. V. 484.10) is confined to two similar passages in Anastasi I[16], where Gardiner translates 'boulders', as opposed to *ꜥr*, 'pebbles', or, more probably, 'small rocks'. This is almost certainly correct, and is appropriate to the determinative (□ııı), and it is interesting to note that elsewhere[17] two boulders are referred to as *ỉnr*.

ḏbt (Wb. V. 553.7f.), apart from being the regular word for mud-brick, is also employed as a term for brick, block, or bar of other materials, including metals, semi-precious stones, building stone, and other substances, including salt and natron. In the case of base metals,

[1] Ed. Soc. Ant. 916; cf. ZÄS. 18, 94; JEA. 19, 167.
[2] ZÄS. 18 p. 95.
[3] JEA. 19 p. 169.
[4] Cf. also Edfou II, 206, 15–16, where *krkr* must refer to lumps of incense material.
[5] Fremdworte II p. 52 no. 1020.
[6] Hier. Texts I p. 41.
[7] Cf. above p. 26.
[8] P. Northumberland II rt. 8.
[9] Cf. Barns, JEA. 34 p. 42; Caminos, LEM. p. 441.
[10] Anast. I, 15, 3.
[11] Ostr. Gard. 3, 6 = Hier. Ostr. 22, 2 rt. 6.
[12] But. cf. *dp* (Wb. V. 443. 6), which may be the same word.
[13] P. Tur. PR. 4, 7.
[14] Ostr. UCL. ⟨1⟩.
[15] One measurement 6 cubits, the others lost.
[16] Anast. I, 23, 3; 24, 2.
[17] Urk. IV, 894.

such as copper[1], and lead[2], *ḏbt* probably refers to the bars or ingots into which the metals were cast for purposes of transport and trade, and this is confirmed by an instance where bars shaped ⊂⊃ are described as *ḏbt*[3]. When used of gold and precious stones, however[4], *ḏbt* usually refers to the small blocks placed in foundation deposits, and *ḏbt* of *biꜣ*[5] may be something similar. Moreover, in an ostracon dealing with types of blocks[6], *ỉnr ḏbt* is parallel to ⟨*ỉnr*⟩ *dp* and ⟨*ỉnr*⟩ *ḥt*, and evidently specifies a block of a particular size[7]. Friable materials like faience[8], natron[9] and salt[10] were also compressed into brick form, and so too was fat[11]. The distinction between *ḏbt* and *ỉnr* as words for block is not altogether clear, but may be largely one of size, and with certain exceptions already noted, the blocks referred to as *ḏbt* may have been of the size and shape of normal bricks, whereas *ỉnr* could be of more or less any dimensions. Furthermore, while several substances were cast, moulded, or compressed into *ḏbt*, only very rarely does *ỉnr* refer to anything other than stones which could be extracted in blocks.

---

[1] Urk. IV, 708; 1321; 1322; Anast. IV, 17, 7; Auswahl 12, 33.
[2] Urk. IV, 708; 718; Auswahl 12, 33.
[3] Urk. IV, 637.
[4] Edfou II, 32; Ombos I, 141, 181; cf. also Düm. Baug. 52.
[5] Totb. Nav. 148, 16; 183, 14 etc.
[6] Ostr. UCL. ⟨2⟩.
[7] 2 cubits × 2 cubits × 1 cubit; cf. P. Kah. 23, 38; 23, 39.
[8] Louvre E 3239, 10–1; Gr. Oase 18.
[9] Harr. 16b, 2; 65b, 1; Abyd. III, 54, 10; Ann. Serv. 16, 196.
[10] Harr. 16b, 3; 19b, 5; 71b, 13 etc.
[11] Anast. V, 21, 8; P. Ch. B. V, 8, 4.

CHAPTER 2

METALS

The first group of materials to be considered in the present work consists of the metals and alloys, which, as has already been noted, were not strictly distinguished from stones and other minerals by the Egyptians themselves, since originally they were found native and worked cold in the same manner as the latter. Thus it was not felt inconsistent to interpolate lapis lazuli and turquoise between the noble and base metals in tribute lists and the like, and both metals and semi-precious stones were similarly referred to as *ꜥꜣt*.

In discussing the metals it will be convenient to divide them into two main groups, the noble metals, gold, silver, and *dꜥm*, which were evidently regarded as varieties of the same basic substance, distinguished only by colour, and the base, copper, bronze, iron, lead etc., which were more clearly differentiated. It will, moreover, be necessary to deal with copper and iron apart from the other base metals, if only because previous treatment of the problem has been so confused by the reading of [hieroglyphs] as *biꜣ*, that it is more or less impossible to separate the two, and to consider them apart would only result in the unnecessary repetition of many arguments.

Gold

Ancient Egyptian gold varied considerably both in purity and colour[1], generally containing some proportion of silver, sometimes a little copper, and occasionally traces of other metals. Not infrequently, the amount of silver present was sufficient to make the alloy pale yellow or even predominantly white in colour, and although such distinctions were recognised at least as early as the beginning of the 4th dynasty, it seems likely that before that time all gold of whatever colour was referred to in the same way.

The common term for gold, which until the latter part of the New Kingdom was the only name for all varieties other than electrum or silver, was *nbw* (Wb. II. 237.6f.), Coptic **ⲚⲞⲨⲂ**, normally written simply with the ideogram 𓋞, representing a collar of gold beads[2]. The nature of the sign was perhaps realised by Brugsch[3], who, although translating *nbw* as 'das schmelzbare', 'das worauf man (Schmucksachen) formt', took it as the word both for gold and gold-collar. Lepsius[4] on the other hand, accepted Rosellini's explanation of it as a sack for washing gold, but noted that it sometimes had the form of a collar. The value *nbw* is certain, not only from the Coptic, but also from examples of related words where *b* is written, and from a single instance of a wholly phonetic spelling[5].

Not until sometime during the Middle Kingdom does gold appear to have been considered more valuable than silver, for it is at this period that *nbw* first precedes *ḥḏ* in the enumera-

---

[1] It also varied in carat, some being much higher than the modern average of 14; cf. Ransom-Williams, Jewelry p. 25.

[2] Gardiner, Grammar³ p. 505 S. 12.

[3] Wb. p. 748.

[4] Métaux p. 3.

[5] ZÄS. 8 p. 20. Apparently [hieroglyphs]. For the *w*, cf. Pyr. 602c, [hieroglyphs], and Koptos pl. 19 lower, [hieroglyphs].

tion of minerals[1]. This order is henceforth the normal one, although the archaistic *ḥḏ nbw* still occurs with some frequency in the New Kingdom[2] and even later[3]. The evidence for the gold-silver ratio is treated more fully in the discussion of silver[4], but it may here be observed that in the Harris papyrus, and other documents of similar date, there is a clear appreciation of the relative value not only of gold and silver, but also of the various different kinds of gold[5].

nbw is mentioned so often and in such a variety of connections, that a complete treatment of the textual evidence is neither possible nor altogether profitable. Gold is regularly listed among articles of tribute, and is shown as the raw material, either in heaps or in sacks[6]. It is said to come from many different sources, including Coptos[7], Kush[8], ʿAmu[9], Ombos[10], Edfu[11], *ḏw wʿb*[12], Pwenet[13], and other localities[14], and an extant map shows in detail the gold-bearing region of the eastern desert[15]. Blocks of *nbw* are occasionally mentioned[16], and from gold are made vessels of various kinds[17], statues and parts of statues[18], jewellery[19], and a wide range of other objects[20], while it was also employed for plating and inlay[21]. Gold was beaten into thin plates known as *pꜣḳ*[22], and hammered gold is also referred to as *nbw m ḳmꜣ*[23] and *nbw m ḳḥḳḥ*[24]; *nbw m ḳnḳn*[25] is apparently gold in lumps or nuggets, and *nbw ḳwḳw*[26] may possibly be granular gold, though this cannot be certain. References to the 'washing' of gold are frequent[27], and although this may refer to the recovery of alluvial gold, it is more likely to denote 'the tedious, but essential, process of washing the powdered ore on sloping tables'[28], particularly since there are parallel references to 'washed' silver. But alluvial gold is certainly meant by *nbw n mw*[29], and *nbw ..... m ḳmꜣ n mw*[30], the former being distinguished from gold obtained (by mining) from other sources. Whether *nbw wʿb*[31] is washed gold, refined gold, or simply gold of ritual purity it is impossible to tell, though the last seems probable. Scenes depicting the various operations connected with gold-working are relatively common, and are discussed in detail by Klebs[32].

---

1 Berl. 1204; Cairo 28023; 28027; other early exx. Admon. 3, 2; Kamose 13–14; Urk. IV, 423; cf. also Urk. IV, 45; 96.

2 Eg. Urk. IV, 98; 1007; 1213; Wen. 2, 11; 2, 29; Anast. III, 8, 6; III, A 3; IV, 16, 2 etc.

3 Eg. Dend. Mar. IV, 36, 49–50.

4 Cf. below p. 41–2.

5 Cf. eg. Harr. 13b, 5; P. B. M. 10068 rt. 2, 2; 2, 5; 5, 20 etc.

6 Huy 23; Düm. H. I. I, 32; cf. Thes. 1274 etc.

7 Urk. IV, 931; Harr. 12a, 8; Düm. H. I. I, 32; Not. Descr. I, 365.

8 Urk. IV, 931; Harr. 12a, 8; Düm. H. I. I, 32.

9 Urk. IV, 436; Luxor ⟨560⟩.

10 Düm. H. I. I, 32.

11 Düm. H. I. I, 32.

12 Rec. Trav. 16, 51.

13 Urk. IV, 337.

14 Cf. Lucas, Materials p. 261; Schiaparelli, Geografia p. 36f.; p. 62f.

15 Probably in the Wadi Fawakhir, cf. Goyon, Ann. Serv. 49 p. 337f., where previous identifications are discussed.

16 Dend. Mar. I, 22.

17 Urk. IV, 637; Koller, 3, 7; Harmhab Kron. 25; Cairo 28024 etc.

18 Urk. IV, 666; Harr. 5, 12; 25, 10; P. Tur. PR. 32, 5; Piehl Inscr. II, 116 etc.

19 Cairo 28024; 28090.

20 Cf. Urk. IV, 457; 666; 669; Harr. 7, 13; Karn. Mar. 42, 7 etc.

21 Urk. IV, 168; Abyd. (Caul.) 19; Sin. B 308; Sh. S. 65 etc.

22 Mereruka pl. 30; Kawa III, 21; VI, 14; cf. Lepsius, Métaux p. 11–12.

23 Harr. 5, 11; Dend. Mar. III, 10; ZÄS. 45, 70.

24 Harr. 6, 5; cf. 13a, 9.

25 Harr. 62b, 5.

26 P. B. M. 10054 rt. 3, 7; cf. also *nḳr* (Wb. II. 344.12–16) which seems to refer to gold dust.

27 B. M. 569; LD. III, 140c; Kuban 10; Auswahl 22; Karn. ⟨741⟩; Red. C. 2; 6; 17; 18.

28 Ransom-Williams, Jewelry p. 18.

29 Düm. H. I. I, 32.

30 Luxor ⟨560⟩.

31 Urk. VIII, 134.

32 Reliefs AR. p. 84; MR. p. 110; NR. p. 115.

Several specific varieties of gold are mentioned in the texts, and these require particular treatment. *nbw ḥr ḫꜣst.f*, known from a number of examples[1], was regarded by Lepsius[2] as 'l'or avec sa gangue', and this is apparently also the view of the Wörterbuch, which translates 'Golderz'; Brugsch[3] and Gensler[4], however, preferred to translate *ḥr ḫꜣst.f* as 'aus seinem Lande'. Bissing[5] noted the parallel *nbw ḥr ḫꜣst tn*, and maintained that *nbw ḥr ḫꜣst.f* must mean 'Gold aus seinem Berg', and not 'Gold in der Scholle', and Macadam[6] has pointed out that, by analogy with *ꜥš ḥr ḫꜣst.f*, the expression should mean 'gold of the gold country'[7]. That the latter interpretation is correct is shown by such variations as *nbw ḥr ḫꜣst.f n ꜥmw*[8], and *nbw ḥr ḫꜣst.f n ꜥmw m ḳmꜣ n mw*[9], where a translation 'gold-ore' is not possible, and also by parallel expressions such as *nbw ḥr ḫꜣst tn*[10]. Moreover, a specific term for gold-ore does seem to have existed, and will be discussed below.

The somewhat similar term *nbw n ḫꜣst*[11], rendered 'or de la roche', 'or des montagnes', by Lepsius[12], must mean simply 'gold of the desert' or 'gold of the hills', perhaps with reference to mined gold[13], and *nbw n tp(w) (n) ḫꜣswt*[14] either 'gold of the best of the hill country' or, by comparison with *ꜥš n tp ḫtyw*, possibly 'gold from the mountain-top', though the significance of such a translation is not at once apparent.

nbw ḥr ḏꜥt.f (Wb. V. 535.1–2; V. 535.12) is confined to three examples[15], though *ḏꜥm ḥr ḏꜥt.f*[16] and possibly *ḫwdtf ḥr ḏꜥtt*[17] also occur, and *ḏꜥꜣt* is once found as the name of a kind of land[18]. The meaning depends entirely upon the interpretation of *ḏꜥt*, which Gardiner[19] thinks refers to a vein of rock from which gold was obtained. The expression would then be translated 'gold from the gold-bearing rock'[20] or the like, though it is also possible that it means 'gold with its vein', and is in fact the real term for 'Golderz', 'l'or avec sa gangue'. If this were so, a parallel *ḥmty ḥr ḏꜥt.f* might also be expected to occur, though its absence may perhaps be explained by assuming that *ḥmty* refers both to copper and copper-ore. No very satisfactory conclusion is possible, and the form of the word *ḏꜥt* gives no clue to the true meaning, but it is tolerably certain that the expression connects gold with the rock or ore from which it was obtained.

Other varieties of *nbw*, occuring principally in the New Kingdom, appear to represent different qualities of gold, chiefly distinguished by purity or colour. The most important of these is *nbw nfr* (Wb. II. 237.11), which Lepsius[21] realised was not merely a general term of praise, since it was found in contrast to other types of gold, suggesting that it meant pure gold as distinct from raw native gold. This also appears to be the view of Säve-Söderbergh[22], who takes *nbw nfr* as the designation of gold of particular refinement, and of Ransom-Williams[23], who points out that the term is not found before the 20th dynasty, which would fix a date for the separation of gold from silver, though it may only refer to richer grades of native gold.

---

[1] Not. Descr. I, 365; 366; 710; Düm. H. I. I, 30; I, 32; Luxor ⟨560⟩ etc.
[2] Métaux p. 6. [3] Wb. p. 1629. [4] ZÄS. 8 p. 142.
[5] Stat. Taf. p. 24. [6] Kawa I p. 39. [7] Cf. also *X ỉr n mwt. f.*
[8] Not. Descr. I, 710 – sic Wb., though the ref. is incorrect.
[9] Luxor ⟨560⟩. [10] Kuban 11; cf. ib. 9 *ỉw wn nbw ꜥšꜣ ḥr ḫꜣst.*
[11] Harr. 12a, 8; Mem. Inst. XXIV, 108; Düm. H. I. I, 32; cf. *nbw n ḫꜣst gbtyw* Harr. 12a, 8; Urk. IV, 931. [12] Métaux p. 6. [13] Cf. Kuban 6 *pr nbw m ḏw.*
[14] Rec. Trav. 22, 118; 29, 162; Piehl Inscr. I, 128e; Urk. IV, 933.
[15] Müller Res. I pl. 14; Edfou VI, 201, 7; Dend. Chass. IV, 159, 9.
[16] Urk. IV, 1422. [17] Edfou VI, 202, 7 – *ḏꜥ tt* is here probably a place-name.
[18] Onom. 12, 4. [19] Onom. 2 p. 218.
[20] Cf. also the similar *nbw ḥr mn. f*, Junker, Große Pylon p. 269, 13.
[21] Métaux p. 9. [22] Äg. und Nub. p. 211. [23] Jewelry p. 26.

There can be no doubt that *nbw nfr* was a specific kind of gold of particular purity, since it was contrasted with *nbw* and *nbw ḥḏ*[1], with *nbw ḥḏ* and *ḥḏ*[2], with *nbw n sp 2* and *nbw ḥḏ*[3], and with *nbw n ḫꜣst*[4]. Moreover, whereas the value of ordinary gold was approximately twice that of silver during the New Kingdom[5], *nbw nfr* was $3\frac{1}{3}$ times as valuable[6], i.e. more than $1\frac{1}{2}$ times as valuable as common gold. Finally, there occurs in Harris[7] the expression *nbw nfr n ktmt*, which almost certainly identifies *nbw nfr* as an especially pure variety of gold. That the purity was the result of refining seems unlikely, and references to *nbw nfr ḥr ḫꜣst.f*[8], and *nbw nfr n ḫꜣst*[9] would appear to indicate that *nbw nfr* was merely a naturally occurring native gold of more than normal purity and containing only a minimal amount of silver. The colour would therefore be somewhat different from that of ordinary gold, and this might account for the specific comparison of the appearance of *smꜣw nt 6* with that of *nbw nfr n ḫꜣst*[10].

Thus, even in those instances where there is nothing to suggest a specific meaning, or where it seems merely to take the place of *nbw*[11], *nbw nfr* is to be regarded as fine native gold. The only case in which it is likely to mean refined gold is when *nbw nfr* is stamped upon coins in the Graeco-Roman period[12], though this is scarcely relevant to the meaning in the New Kingdom, and may simply indicate that the gold is unadulterated.

Somewhat analogous to *nbw nfr* is *nbw ḥḏ* (Wb. II. 237.9–10), which is found in only two papyri[13], where it is evidently the name of a particular kind of gold distinct from *nbw nfr*[14], *nbw*[15], *nbw n sp 2*[16], and *ḥḏ*[17], to all of which, except *ḥḏ*, it was apparently inferior in quality. The literal meaning must be 'white gold', and since it cannot in these contexts refer to silver, it seems that Ransom-Williams[18] is right in regarding it as a term for electrum, no other word for which occurs in either papyrus[19]. But electrum is a somewhat imprecise designation, covering both natural and artificial alloys of gold and silver, varying in colour from pale yellow to greyish-white according to the relative proportions of the two metals which are present. In these particular instances, *nbw ḥḏ* probably refers to a native alloy, and presumably one which was white rather than pale yellow in colour, and it may therefore be preferable to translate *nbw ḥḏ* as 'white gold', while remembering that in modern metallurgy the term has a precise significance which it cannot have had in Egypt[20]. The use of the expression 'white gold' would perhaps indicate that at this period the Egyptians recognised electrum as a gold-silver alloy, which they may not have done before, and that *ḥḏ* referred only to silver which was almost pure.

nbw wꜣḏ, which seems to occur only once[21], is probably gold of another colour. The context, referring to heaps of *nbw wꜣḏ n ꜥmw*, is of no assistance in establishing the meaning,

---

[1] P. B. M. 10068 rt. 5, 20. [2] P. B. M. 10068 rt. 2, 2; 2, 5.
[3] Harr. 13b, 5. [4] Harr. 12a, 9. [5] Cf. below p. 42.
[6] P. Tur. 1881 rt. above cols. I–II [4]. [7] Harr. 47, 5; cf. 13a, 5.
[8] Harr. 33a, 5; Düm. G. I. I, 46. [9] Harr. 47, 4; 68b, 6; Düm. H. I. I, 34.
[10] Harr. 47, 4; cf. however ib. 6, 9, where it is compared simply with *nbw*.
[11] Düm. H. I. II, 42.
[12] Cf. Maspero, Rec. Trav. 22 p. 225; Chassinat, BIFAO. 1 p. 78; 7 p. 165; Rec. Trav. 40 p. 131; Naville, CRAI. 1925 p. 278.
[13] P. Harris; P. B. M. 10068. Three earlier writings of *nbw ḥḏ* are clearly the original form of the word *ḥḏ*, silver; cf. below p. 41.
[14] P. B. M. 10068 rt. 2, 2; 2, 5; 5, 20; Harr. 13b, 5. [15] P. B. M. 10068 rt. 5, 20.
[16] Harr. 13b, 5. [17] P. B. M. 10068 rt. 2, 2; 2, 5.
[18] Jewelry p. 30.
[19] The meaning of *ḏꜥm* in Harris is discussed below, p. 48–9.
[20] Modern 'white gold' is a gold-platinum or gold-nickel alloy, cf. Forbes, Metallurgy p. 153.
[21] D. el B. III, 74.

and either 'fresh gold' or 'green gold' is possible as a translation. But whereas *wꜣḏ* is best rendered 'fresh' when referring to *ʿntyw* or *ryt*, such an interpretation has no very obvious significance in relation to gold. 'Green gold', on the other hand, is quite intelligible, and would most naturally refer to gold with a fairly high admixture of silver, and traces of copper, which would render it green or greenish-grey in colour[1]. That gold of this nature was known to the Egyptians is quite possible, though the colouration must surely have been fortuitous, and not the result of deliberate alloying or any other method of artificial tinting.

Two other varieties of gold, confined to the Harris papyrus, are *nbw n sp 2*[2] and *nbw n sp 3*[3] (Wb. II. 237.12–13), regarded by the Wörterbuch as 'Gold zweiter (dritter) Qualität'. The contexts are of little significance, though it appears that *nbw n sp 2* was of less value than *nbw nfr* yet more than *nbw ḥḏ*[4], and that *nbw n sp 3* was of a distinctive colour[5]. The suggestion that the two terms refer to gold refined twice or three times[6], can hardly be maintained, since if so, a term *nbw n sp 1* might be expected to occur, and gold twice refined would surely be more valuable than *nbw nfr*, pure native gold. Moreover, there is no archaeological evidence to suggest that gold was in fact refined before the Persian period[7]. The Wörterbuch interpretation, on the other hand, is appropriate to the distinction between *nbw nfr*, *nbw n sp 2* and *nbw ḥḏ*, *nbw nfr* perhaps being regarded as gold of first quality. In default, therefore, of further evidence, *nbw n sp 2* and *nbw n sp 3* may by regarded as gold of second and third quality rather than as having any reference to refining.

nbw nʿʿ (Wb. II. 208.7) is a species of gold the nature of which is by no means clear. The term is of rare occurrence[8], though it is probably related to a similar word *nʿʿ* found elsewhere[9]. The Wörterbuch translates *nʿʿ* as 'bunt sein', 'mehrfarbig', which suggests that *nbw nʿʿ* may be coloured gold[10], though an instance of *nʿʿ* in parallelism with *nfr*[11] might point to its being simply another term for 'good gold' or the like. But *nʿʿ* also has the meaning 'smooth', so that *nbw nʿʿ* may be 'smooth gold', perhaps with reference to gold leaf, and this is to some extent supported by unpublished evidence communicated to me by Prof. Černý[12]. The inscription on a box (C. 330) from the tomb of Tutankhamūn states that it contains shawabtis of *nbw nʿʿ* and *mry* wood, the actual shawabtis being of wood (as yet unidentified) coated with gesso and gold leaf. From this it would appear that *nbw nʿʿ* is thin leaf gold, though how far this is applicable to *nʿʿ* alone is doubtful. In the absence of further evidence, no definite translation is possible, though identification with either gold leaf or tinted gold seems likely.

Mention must also be made of *nbw n ḳniw*, and *nbw n sṯni* (Wb. IV. 358.16–18), both of which are apparently confined to a single occurrence. The former, used for the kilt of a statue[13], may mean simply 'gold of yellow colour', though it is also possible that *ḳniw* is the word for orpiment, and that the expression alludes to the common ancient belief that gold

---

1 Cf. Forbes, Metallurgy p. 153.

2 Harr. 13a, 11; 13a, 13; 13b, 5; 33a, 6; 52a, 8; 68b, 6; 70a, 6.

3 Harr. 78, 5.

4 Harr. 13b, 5; 33a, 6; 52a, 8; 70a, 6.

5 Harr. 78, 5.

6 Cf. eg. Forbes, Metallurgy p. 156.

7 Cf. Lucas, Materials p. 262.

8 P. Tur. PR. 104, 3, 3–7.

9 Auswahl 12, 9; 12, 56; Urk. IV, 1235.

10 Possibly the peculiar rose-pink variety used for some objects of New Kingdom date, the tint being either natural or artificial; cf. Wood, JEA. 20 p. 62.

11 Anast. V, 10, 5 = Sall. I, 3, 7; cf. also *nʿʿ* 'der best von etwas' (Wb. II. 208.11).

12 I am grateful to Prof. Černý for permission to refer to this inscription, which he will shortly publish.

13 P. Tur. PR. 32, 5.

could be extracted from that material[1]. *nbw n sṯnỉ* is also the material of a statue[2], and since the term cannot be satisfactorily explained as it stands, it seems possible that it may be a mistake for *nbw n ḳmꜣ*, having been wrongly expanded by the sculptor into *sṯnỉ* instead of *ḳmꜣ*[3]. But the word may be correct in the form *nbw n sṯnỉ*, in which case the meaning cannot be deduced. Gold in some form may also be indicated by *nbyt* (Wb. II. 243.1), which occurs in a single medical context[4], though there are also other words similarly written with which it may be connected.

The expression *tỉ n nbw*, found in Amenemope[5] is altogether problematic, and the exact interpretation of the passage in which it occurs is obscure. The general meaning seems to be "if *tỉ n nbw* is 'passed off' as pure gold (*ktmt*), in the morning it is lead" the exact rendering of the verb *ꜥšgꜣiw* being uncertain[6]. Concerning *tỉ n nbw*, Lange[7] comments 'wohl ein Fehler'; Erman vermutet 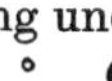, Grapow macht mich darauf aufmerksam, daß *ṯḥnt* 'Fayence' oft in griech. Zeit mit Gold zusammen genannt wird; eine Verschreibung dieses Wortes wäre eher denkbar', while Griffith[8] thinks it 'might perhaps be an expression for stucco, bronze or other material used as a basis for gilding'. It is perhaps possible that what is apparently *tỉ n nbw* is in fact *ḥsmn*, or *ḥsmn (n) nbw*, and refers to bronze, or gold debased with bronze, but the passage is too confused to admit of any worthwhile conclusion, and the most that can be said is that it indicates the counterfeiting of gold.

In addition to the different varieties of *nbw*, there are also other terms for gold occurring in the New Kingdom, and especially the Graeco-Roman period, which must here be discussed.

ktmt (Wb. V. 145.6–13), which first appears in the Harris papyrus, is evidently a term for (pure) gold derived from the Hebrew *ketem*[9]. This was recognised by both Brugsch[10] and Lepsius[11], though neither was altogether certain whether it actually represented gold or a similar metal. Dümichen[12], however, gave instances where the use of *ktmt* was parallel to that of *nbw*, *sꜣwy* and *rwy*, and Lepsius[13] was then convinced that gold was meant, though doubting whether the word was in fact borrowed from Hebrew, since *ketem* was not the regular Hebrew term for gold, and did not seem typically Semitic in form. But Brugsch[14] wholly accepted Dümichen's conclusions, recognising the *ktmt* of Harris as the earlier form of the Ptolemaic *ḳtm*, and Chabas[15] also supported the identification as gold, quoting examples of *ktmt* parallel to *nbw* and *ḏꜥm*.

There can be little doubt that *ktmt*, first occurring in the New Kingdom[16], and written in group-writing with slight orthographic variations, is in fact borrowed from Hebrew *ketem*. All the early examples have *k* (), and only in the Ptolemaic period is the initial consonant changed to *ḳ*. That *ktmt* is a kind of gold is clear not only from the parallelisms

---

[1] It may alternatively indicate that the Egyptians tried the experiment of fusing gold with orpiment to improve the colour or to increase the quantity, cf. Wood, JEA. 20 p. 62.

[2] Louvre C, 15, 8 = Pierret Inscr. II, 29.

[3] For a possible parallel cf. Hearst 9, 13, where *pꜣ yt* seems to be for *ḳmyt*.

[4] P. Med. Berl. 5, 2.

[5] Amenemope 18, 12, written 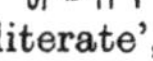.

[6] Probably lit. 'smear' or 'obliterate', perhaps referring to some process of staining a base metal to look like gold.

[7] Amenemope p. 90.

[8] JEA. 12 p. 215 n. 2.

[9] Cf. Burchardt, Fremdworte II p. 53 no. 1036; JAOS. 73. p. 151.

[10] Wb. p. 1484.

[11] Métaux p. 4.

[12] ZÄS. 10 p. 44.

[13] ZÄS. 10 p. 113.

[14] Wb. Erg. p. 1237; ZÄS. 12 p. 149.

[15] ZÄS. 12 p. 2.

[16] Apart from a number of references in Harris, *ktmt* occurs in P. Mag. Harr. 8, 4; Med. Hab. 107, 15, and then in Amenemope 18, 12.

cited by Dümichen and Chabas[1], but also from Harris, where *nbw nfr n ktmt* is mentioned[2]. This would seem to indicate that *ktmt* was in fact a particularly pure species of gold, which appears to be confirmed by the passage in Amenemope[3], already cited above, where the reference is evidently to the passing off of some base imitation as pure gold, *ktmt*. Moreover, in some few instances[4], *ktmt* is determined with 𓋞, which is only used for varieties of gold.

It seems likely, therefore, that *ktmt* was first introduced as a specific term for pure gold or gold of good quality, which would be consistent with the careful differentiation of separate kinds of gold in Harris. In the Graeco-Roman period, however, it is possible that this precise meaning had been abandoned, and that *ktmt* (*ḳtm*) was simply used as a term for gold in general.

sꜣwy (Wb. IV. 13.10–15) was recognised by Lepsius[5] as a word for gold occurring in the Ptolemaic texts, and although it is possible that the term is found earlier, the examples are so uncertain that they are best considered separately. Lepsius suggested a connection between *sꜣwy* and Coptic ˢⲤⲀ, ᴮⲤⲀⲒ[6], giving it the meaning 'le beau métal', and this view was supported by Dümichen[7], who translated 'das schöne Metall', and Kuhn[8], who compared Sanskrit *suvarṇa*, 'das schönfarbige', as a word for gold. Sethe[9], however, adopted a different interpretation, seeing *sꜣwy* as a development of the word for $^2/_3$, meaning gold with $^2/_3$ gold content, i.e. of $^2/_3$ purity. But this view depends on identification of *sꜣwy* and *rwy*, which Clère[10] has shown to be untenable, and there is in fact nothing to connect *sꜣwy* with the word for $^2/_3$, also read *rwy*.

Neither the meaning nor the origin of *sꜣwy* can be established with any degree of certainty, and the textual evidence is only sufficient to define it as a word for gold of some description. *sꜣwy* occurs as a variant of *nbw* in the tribute tables at Dendera and elsewhere[11], and is frequently parallel to *ʿrḳwr*, a Ptolemaic term for silver[12]. It also appears to be distinguished from *ktmt*[13], and is referred to as *sꜣwy m ỉrw.f*[14] which might suggest that it had a precise meaning. In the majority of instances, however, *sꜣwy* seems to be used simply as an alternative to *nbw*, and in this respect the word is similar to *ʿrḳwr*, which replaces *ḥḏ*, without any apparent difference of meaning. On the other hand, it is possible that *sꜣwy* has some precise meaning which cannot now be appreciated.

The origin of the word has been traced by Sethe[15] to the Pyramid texts, and this view is accepted by the Wörterbuch, which quotes references to *sꜣwy* from that source and from certain texts of New Kingdom date. The supposed examples of *sꜣwy* in the Pyramid texts are two in number[16], and to these may be added a somewhat similar occurrence in the Coffin texts[17]. The identification with *sꜣwy* is based on the first of these, where a word written *sꜣꜣ* is determined with 𓋞, but there is nothing to show that this word refers to gold or the golden colour of the sun[18], and indeed, on the analogy of the second passage, where 𓋞 seems to be the determinative of a *swḥ* vestment, it is more likely to be a garment

[1] Eg. Emb. Rit. 3, 16; 3, 19; 8, 4. [2] Harr. 47, 5; cf. 13a, 5. [3] Amenemope 18, 12.
[4] Eg. Philae ⟨1633⟩ Phot. 217; Dend. Mar. III, 44r; III, 21u; Edfou I, 283; I, 537; Mammisi 59, 19.
[5] Métaux p. 4. [6] Crum 315a–b; Peyron 187. [7] ZÄS. 10 p. 44.
[8] ZÄS. 11 p. 21. [9] Zahlen p. 95. [10] Arch. Or. 20 p. 631; p. 641.
[11] Dend. Mar. I, 70; I, 71; II, 12; Düm. G. I. II, 63; II, 66; IV, 170 etc.
[12] Dend. Mar. IV, 20; Edfou I, 46; II, 32 etc.
[13] Dend. Mar. III, 17h; cf. Edfou II, 260, 17.
[14] Dend. Chass. IV, 139; Urk. VIII, 56; Dend. Mar. I, 70.
[15] Zahlen p. 95. [16] Pyr. 706b; 1416a. [17] Coff. Texts V, 386e.
[18] Cf. Sethe, Kommentar III p. 297; Mercer, Pyr. Texts II p. 347.

of some description. The *s3wy* of the Coffin texts passage[1] may, however, be a material, since it is parallel to *mfk3t* in the phrase *nb mfk3t nb s3wy*, but this interpretation is by no means certain, and the word may well have some other meaning. Altogether, it seems most unlikely that any of these is to be taken as an early example of *s3wy*, 'gold', though it is not impossible that the Ptolemaic word is derived from a misinterpretation of one or the other.

Equally problematic are the alleged examples of *s3wy* in texts of New Kingdom date, where a word occurs written [2]. That this is to be read *s3wy* there is no evidence whatsoever, nor is there anything to suggest why it should not be read *nbwy*, the most obvious interpretation and one advanced long ago by Brugsch[3]. What *nbwy* then represents, it is not easy to see, since it is clearly something more than a variant writing of *nbw*[4], but it seems just possible that it is in fact a term for silver. Thus it occurs where *ḥḏ* would be expected in an enumeration of minerals, *nbw, nbwy, ḫsbḏ, mfk3t, ḥmty km*[5], and again following *ḏʿm*[6], and is also found in the phrase *s3tw.s nb m nbwy*[7], where similar texts usually have *ḥḏ*. However, it may well be some species of gold of inferior quality, akin to silver, and no certain conclusion is possible.

What does become clear is that the evidence for the existence of *s3wy* before the Graeco-Roman period is tenuous in the extreme, and that the word is almost certainly a Ptolemaic creation, though its origin is obscure[8].

rwy (Wb. II. 240.5–6) is undoubtedly confined to the Graeco-Roman period, when it occurs as a variant for *nbw*. Brugsch[9] regarded it simply as a variant writing of the latter, comparing  and , and Lepsius[10] also inclined towards *nbw*, though noting that the reading was uncertain, and that the normal meaning was $^2/_3$; such too appears to be the view of Junker[11]. Dümichen[12], on the other hand, transliterated *sr*, rendering 'das edle, ausgezeichnete Metall', which was then accepted by Lepsius[13]. Sethe[14] clearly identified the word with *s3wy*, which he also took to be the correct reading of the word for $^2/_3$, and regarded it as the name of a particular quality of gold or electrum having only $^2/_3$ gold content. But Clère[15] has pointed out that the word for $^2/_3$ is to be read *rwy*, and that therefore *r-2* or *rwy*, and not *s3wy* is the reading of this word also. *rwy* and *s3wy* are thus quite distinct, though they may be more or less synonymous. Clère does not, however, altogether reject the possibility that *rwy* may in fact be a graphic variant of *nbw*.

The reading is without doubt *rwy*, and the supposed evidence for regarding the word as a writing of *nbw* can be shown to be invalid[16]. There is in fact no case in which  cannot

---

[1] Written thus: .

[2] Urk. IV, 168; 637; 875; 1150; Karn. Mar. 15, 20; Karn. ⟨715⟩; ⟨717⟩; ⟨719⟩; = Karn. I, 33; I, 25; I, 26; = Urk. IV, 1669; 1668; 1668.

[3] Wb. p. 748.

[4] Cf. Urk. IV, 875; Karn. ⟨715⟩ = Karn. I, 33 = Urk. IV, 1669, where *nbw* and *nbwy* occur together. Karn. I, 26  may be for *nbw*, cf. Pyr. 602 c  and Koptos pl. 19 lower 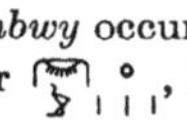, but Karn. ⟨719⟩ (Sethe) has *nbwy*, and so does Urk. IV, 1668.

[5] Urk. IV, 875; cf. Karn. ⟨715⟩ = Karn. I, 33 = Urk. IV, 1669.

[6] Urk. IV, 168; Karn. ⟨719⟩ = Karn. I, 26 = Urk. IV, 1668.

[7] Karn. ⟨717⟩ = Karn. I, 25 = Urk. IV, 1668; Urk. IV, 1150.

[8] An alleged instance of *s3wy* from Dyn. 22, LD. III, 258b, 8 = Bub. Port. pl. 22, may be dismissed. Caminos' interpretation of the passage, Chronicle p. 129, is impossible, and 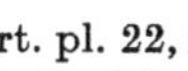 must be a substantive, probably to be read *mss* or *msms*.

[9] Wb. p. 748; p. 751.

[10] Métaux p. 4.

[11] Schriftsystem p. 9.

[12] ZÄS. 10 p. 44.

[13] ZÄS. 10 p. 113.

[14] Zahlen p. 94.

[15] Arch. Or. 20 p. 630; p. 641.

[16] I am grateful to Prof. Fairman for extensive notes on this question, which are incorporated here.

be given the value *r* or *rw*; thus [1] is *rwnwt*, and [2] is either *rnnwt*, *rnwt* or even *rwnt*[3]. Moreover, there is some slight evidence to show that the reading in the case of the word for gold was *rwy* also, since in one instance[4] it is preceded by *r*, which cannot easily be explained unless it is the initial radical written out, and in another[5], a writing *ry* ( ) would seem from the context to be *rwy* and not *ryt*, 'ink'. In view of these facts, it is impossible to maintain the reading *nbw*, and the term must be read *rwy*, and probably therefore connected with the word for $^2/_3$.

The meaning may well be gold of $^2/_3$ purity, as suggested by Sethe, though there is no positive evidence to support this interpretation. *rwy* occurs in the tribute tables[6] as a variant of *nbw*, and is also found parallel to both *nbw*[7] and *s3wy*[8] and in conjunction with *r-ḫmt*[9]. All that can be deduced from this is that it represents gold, though it may equally well be either a specific variety, or merely an artificial term used instead of *nbw*.

How *rwy* came to be used as a term for gold it is not easy to suggest, unless the assumption that it means '$^2/_3$ gold' is correct. However, two very tentative explanations may be offered, neither of which is more than the merest hypothesis. It is perhaps possible that both *rwy* and *r-ḫmt* originated in the same way from a misunderstanding of the terms *nbw n sp 2* and *nbw n sp 3*, or that *sp 2* and *sp 3*, carelessly written, gave rise to the confusion. On the other hand, *rwy* may conceivably have been the result of a different mistake, some such grouping as [10] having been taken as the name of a material, when in fact was merely part of the statement of quantity of a material which preceded *nbwy*. *rwy* having then been created in this way, *r-ḫmt* was simply invented to correspond to it. But neither of these explanations is very satisfactory, and they are here advanced with the utmost reserve.

Finally, mention must be made of two other terms, *wḥr* (Wb. I. 351.12) and *msny* (Wb. II. 146.1) which may possibly be connected with gold. The former is apparently ἅπαξ in the Dendera amulet list[11], where it occurs as the material for seven amulets[12]. The nature of these amulets is of no particular significance in establishing the meaning, but a heading above the table is suggestive. This states that the table is a list of the 104 amulets *nw nbw ꜥ3t nb(t)*, and while at least 16 precious stones are instanced below, there is no mention of gold, unless it be *wḥr*. It is therefore probable that *wḥr* is gold, and Petrie[13] suggested that the word was derived from Greek χρυσός. This is hardly likely, but *wḥr* may perhaps originate in Greek ὠχρός 'pale yellow', with the meaning 'the pale yellow stuff' in reference to gold[14]. On the other hand, it may arise from a misunderstanding of in such expressions as *nbw ḥḏ ḥnꜥ s3w n ꜥ3t nb n m3ꜥ*[15], or from some obscure word play. At all events, it seems reasonable to suppose that, whatever its origin, *wḥr* refers to gold.

---

[1] Edfou III, 124, 16; III, 184, 15; Mammisi 122, 14; 163, 10.

[2] Edfou I, 211, 12; III, 271, 17; IV, 378, 13–14; IV, 384, 17; VIII, 136, 3; Mammisi 165, 4; Dend. Chass. II, 33, 2; Dend. Mar. III, 47c–d. Cf. alphabetically written exx. Edfou IV, 102 6; 242, 13; 270, 9; 270, 17; 298, 3; 298, 13.

[3] Note especially , Edfou VIII, 116, 3, suggesting that is alphabetic *r* (BIFAO. 43 p. 74 n. 2) though this probably derives from *rw* or *rwy*.

[4] Dend. Chass. IV, 165, 2.

[5] Dend. Chass. IV, 154, 7.

[6] Dend. Mar. I, 70; I, 71; II, 12; Düm. G. I. II, 66; II, 69.

[7] Dend. Chass. IV, 181, 19.

[8] Dend. Mar. IV, 20.

[9] Dend. Mar. IV, 15; IV, 20; Dend. Chass. II, 113, 15; IV, 41, 4; IV, 181, 19; Edfou II, 283, 4 etc.

[10] Cf. Karn. ⟨719⟩ = Karn. I, 26 = Urk. IV, 1668:

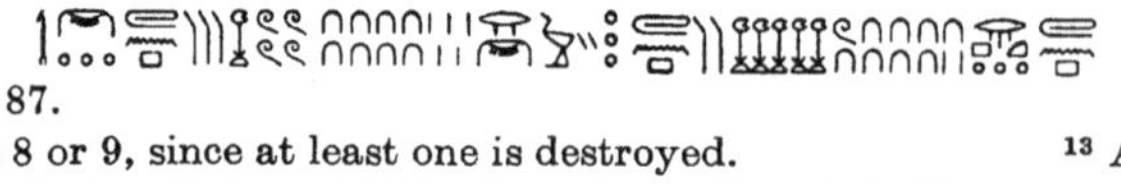 etc.

[11] Dend. Mar. IV, 87.

[12] Possibly in fact 8 or 9, since at least one is destroyed.

[13] Amulets p. 6.

[14] Cf. Greek ὤχρα, Syriac *ûkrâ*, and Akkadian *ukuru*, Campbell Thompson, Dictionary p. 265–6.

[15] Thes. 929: .

msny is also confined to one context[1], where it occurs three times. The passage concerns a statue of Amūn *nty msny n nbw*, and other gods *n nbw nty msny* and *nty msny*. The form of these expressions is confusing, but the only possible interpretation seems to be that *msny* is the name of some kind of gold or of gold treated in a particular way. A connection with *msn* 'spin' (Wb. II. 144.12–15) is feasible, and *msny* might then refer to gold filigree, though the point cannot be pressed, and, in fact, no very clear interpretation is possible.

The existence of so many separate terms representing different varieties of gold is hardly surprising in view of the great frequency with which the metal was employed in Egypt from the earliest times. But the deveploment of the nomenclature of gold is of little significance apart from that of the other noble metals, which must first be considered before attempting to summarise the whole.

Silver

Before the importation of the metal from abroad, or its much later production from its ores, all silver used in Egypt, save perhaps for a small quantity found native, was merely an alloy of gold, akin to the pale electrum, but containing so large a proportion of silver that it was predominantly white in colour[2]. The Egyptians themselves seem to have recognised this fact, and evidently regarded gold and silver as two forms of the same mineral, distinguished only by colour, referring to the latter as 'white gold'.

The earliest form of the name, already noted by Bissing[3], was *nbw ḥḏ*, which dates from the beginning of the 4th dynasty, and is known from only three examples[4]. That this term designates silver and not merely electrum is clear, since *ḏʿm* also occurs in the Medum inscription, and there is therefore no connection between it and the later *nbw ḥḏ* (Wb. II. 237.9–10) which appears in the 20th dynasty[5]. By the 5th dynasty[6], however, *nbw ḥḏ* had become transformed into a specific term *ḥḏ* (Wb. III. 209.9f.), Coptic ϩⲁⲧ, with the hieroglyph for gold as the determinative, and this continued to be the regular and only word for silver down to the Graeco-Roman period[7], when two other names, *ʿrḳwr* and *r-ẖmt* were also used.

Silver was comparatively rare until the New Kingdom, such items as the anklets of Hetephras, and the cups and other objects in the Tôd treasure being exceptional, and even in the tomb of Tutankhamūn there were few articles of silver. Thus, for a long period, silver was considered more valuable than gold, and this is reflected in the order in which the two metals are listed in earlier texts[8]. *nbw* first precedes *ḥḏ* in one or two examples of Middle Kingdom date[9], and this gradually becomes the normal order, though in the New Kingdom[10], and even later[11], the archaistic *ḥḏ nbw* is still fairly frequent[12].

---

[1] Urk. III, 122.

[2] Cf. Lucas, Materials p. 278f.; Forbes, Metallurgy p. 175.

[3] Act. Or. IV p. 138. Meyer, Gesch. I[3] § 225 took  as *ḥḏ*, which it is not.

[4] Medum 13 (twice); Abydos II, 12, 279.

[5] Cf. above p. 35.

[6] Sahurēʿ II pl. 18 and Text II p. 93; pl. 61 and Text II p. 126; Urk. I, 235; Ann. Serv. 38, 96. For an example which may be earlier cf. Goyon, Nouvelles Inscriptions p. 50, Inscr. 12, *ḫrp smntyw ḥḏ* (). If this graffito is in fact of the archaic period as Goyon suggests, the theory proposed for the origin of the term *ḥḏ* may require modification.

[7] Cf. above p. 39 for the possibility that *nbwy* may mean silver.

[8] Cairo 28024.

[9] Berl. 1204; Cairo 28023; 28027; other early exx. Admon. 3, 2; Kamose 13–14; Urk. IV, 423; cf. Urk. IV, 45; 96.

[10] Eg. Urk. IV, 98; 1007; 1213; LD. III, 127b; Wen. 2, 11; 2, 29; Orb. 14, 8; 15, 5 etc.

[11] Eg. Dend. Mar. IV, 36, 49–50.

[12] Forbes, Metallurgy p. 195, states that in the earlier Assyrian tribute lists silver precedes gold, though this does not indicate that the former was more valuable, since the gold-silver ratio was 10 to 1.

There is, however, more exact evidence of the relative values of gold and silver in the New Kingdom, which has been conveniently tabulated by Černý as follows[1].

	nbw nṯr	*nbw*	*ḥḏ*	*ḥmty*	*ḏḥty*	*Date*
a)	—	2	1	—	$^1/_2$	Dyn. 12 and/or Hyksos.
b)	—	$1^2/_3$	1	—	—	Dyn. 18; Thut. III–Amen. II.
c)	—	—	1	$^1/_{96}$	—	Dyn. 18.
d)	—	—	1	$^1/_{100}$	—	Dyn. 19; Ramses II.
e)	—	—	1	$^1/_{104}$	—	Dyn. 19; Ramses II.
f)	$3^1/_3$	—	1	$^1/_{60}$	—	Dyn. 20; Ramses IX Yr. 7.
g)	—	2	1	—	—	Dyn. 20; Ramses IX Yr. 17.

a) P. Rhind. 62. b) P. Boul. XI col. 1 l. 5. c) P. Boul. XII. d) JEA. 21, 140f. e) P. Wien 34, von Berg. pl. 1. f) P. Tur. 1881 rt. cols. I–II. g) P. B. M. 10068 rt. 4, 14.

From these more or less precise indications, it is clear that for a considerable time the gold-silver ratio remained practically constant at 2 to 1, and Forbes[2] states that this was still the ratio in the Persian period, when 13 to 1 was normal in other parts of the empire. Alexander attempted to enforce a ratio of 10 to 1, but not until the reign of Ptolemy II did the standard 13 to 1 ratio exist in Egypt[3].

The textual evidence for *ḥḏ* is considerable, though there were not apparently as many distinct varieties of silver as there were of gold. *ḥḏ ḥr ḫꜣst.f*[4], parallel to *nbw ḥr ḫꜣst.f*, must mean 'silver from the silver-country' rather than the Wörterbuch's 'Silbererz'[5], while *ḥḏ wꜥb*[6] is self-explanatory, and can only refer to silver either naturally pure, or refined. *ḥḏ iꜥ*[7], literally 'washed silver', seems to imply some process of washing the powdered ore[8], though whether such an interpretation is possible as early as the 5th dynasty may perhaps be doubted. *ḥḏ m ḳmꜣ*[9], and also *ḥḏ m ḳhḳh*[10] both indicate hammered silver, whereas *ḥḏ m̄ ḳnḳn*[11] is evidently silver in lumps. Apart from these, no other kind of silver appears to be mentioned.

ḥḏ is shown in heaps and as a raw material[12], and is frequently brought as tribute, from Retenu[13], Alasia[14], Sengar[14], Cyprus[14] and *tꜣ-nṯr*[15], which corresponds to Lucas' observation[16] that the greater part of New Kingdom silver was of foreign origin. On the other hand, it is also said to come from *wꜣwꜣt*[17], *istrn*[18] and the cerastes nome[19], although, as far as can be ascertained, neither silver nor silver ores are actually found in Egypt[20]. Blocks of *ḥḏ* are sometimes mentioned[21], and from *ḥḏ* were made vessels of various types[22], statues[23], offering

[1] Prices and Wages p. 906, slightly adapted. NB. It is possible that the Rhind passage (a) is merely a mathematical exercise with hypothetical values.

[2] Metallurgy p. 185–6; p. 218; cf. Milne, JEA. 24 p. 245.

[3] Forbes. Metallurgy p. 185–6; cf. Milne. JEA. 15 p. 150.

[4] Harr. 33a, 8; 46, 7; Müller Res. I pl. 14.

[5] Cf. above p. 34.

[6] Harr. 6, 11; 48, 4; Karn. Mar. 42, 8; Theb. Grab. 65 ⟨618⟩.

[7] Ann. Serv. 38, 223, 13; cf. Sahurēꜥ II pl. 18; Text II p. 93.

[8] Cf. Ransom-Williams, Jewelry p. 18.

[9] Harr. 5, 12; 47, 5.

[10] Harr. 13b, 14.

[11] Harr. 13b, 15.

[12] Urk. IV, 686; Düm. H. I. I, 32; I, 34; Karn. ⟨826⟩ etc.

[13] LD. III, 127b.

[14] Müller Res. II, 91–2.

[15] Urk. IV, 1007.

[16] Materials p. 280.

[17] Dend. Mar. I, 70; I, 71; II, 12.

[18] Düm. G. I. II, 63; IV, 102; IV, 170.

[19] Theb. Temp. Gr. Zeit ⟨49 n.⟩.

[20] Lucas, Materials p. 279.

[21] Flor. Sch. 1617; Emb. Rit. 10, 18; Rec. Trav. 27, 225.

[22] Urk. IV, 22; 634, 635, 637; 669; 688; 733; 766; 952; Harr. 4, 7; 13b, 9; Cairo 28024 etc.

[23] Urk. IV, 23; 634; 666; Champ. Mon. 288, 2; Dend. Mar. IV, 38.

tables[1], mirrors[2], rings[3], amulets[4], jewellery[5], and other objects[6], as well as inlay and plating[7]. *ḥḏ* was also the standard material of the unit of value *šʿty*, and, like Greek ἄργυρος and Coptic ϩⲁⲧ, *ḥḏ* was used to refer to money in a quite general sense[8].

It has been suggested that there was some connection between Egyptian *ḥḏ* and Hittite *ḫat*, which has a similar meaning, and this seems not unlikely. However, there can be no question as to which was the loan-word[9], since the Egyptian term is without doubt of much greater antiquity, and its etymology is clear[10].

In the Graeco-Roman period, as has already been mentioned, silver was known by two other names, *ʿrḳwr* (Wb. I. 213.10), and *r-ḫmt* (no Wb. ref.), both noticed by LEPSIUS[11], and DÜMICHEN[12], who were, however, uncertain as to the correct reading of the former.

ʿrḳwr was first recognised by BRUGSCH[13], who read it correctly, and derived it from Greek ἄργυρος, in which he was followed by PLEYTE[14], who conceded that  was *ʿrḳ* and not *sḥ* in this instance. The reading is confirmed by a passage quoted by BRUGSCH[15], in which there is a pun on the sound *ʿrḳ*, and since the word is not easily interpreted as a purely Egyptian formation, it seems reasonable to suppose that it was derived from Greek ἄργυρος, just as *ktmt* came from the Hebrew *ketem*, though it must be admitted that few other Greek words were adopted in this way. That *ʿrḳwr* is a term for silver is quite evident from the tribute tables at Dendera and elsewhere[16], where it occurs as a variant of *ḥḏ*, and from the Rosetta stone[17], where hieroglyphic *ʿrḳwr* is parallel to Demotic *ḥt*. Moreover, *ʿrḳwr* also replaces *ḥḏ* in lists of minerals[18], and is found balancing *nbw*[19] and *sꜣwy*[20]. What is not altogether clear is whether *ʿrḳwr* is simply an alternative to *ḥḏ*, or whether it has a specific meaning such as 'pure silver', since either would be equally appropriate to the contexts in which it occurs. It seems, however, possible that in the Greek period the Egyptians became aware that *ḥḏ*, embracing all kinds of silver, was a somewhat unsatisfactory term for the purest varieties, and adopted ἄργυρος to refer to these. Strangely enough, there do not appear to be any examples of *ʿrḳwr* with the meaning 'money', though both *ḥḏ* and ἄργυρος are thus used.

r-ḫmt, read *ru* or *rua* by LEPSIUS[21], and *ruu* by DÜMICHEN[22], was identified by BRUGSCH[23] with *ryt*. SETHE[24], however, pointed out that the reading was *r-3*, and regarded the word as a term for silver with $^1/_3$ gold content. The same reading, *r-3* or *r-ḫmt*, is adopted by CLÈRE[25], who takes it to be a near synonym for *ʿrḳwr*, comparing the connection between *sꜣwy* and *rwy*; he thinks it possible that *r-ḫmt* was either a technical term, or a 'poetical' variant for *ḥḏ*, or even an obscure writing of the latter. In fact, no more satisfactory conclusion seems possible. *r-ḫmt* occurs in the tribute tables[26] as a variant of *ḥḏ*, and is also

1 Urk. IV, 636; 870; Harr. 5, 12.
2 Cairo 28024; 28087.
3 Urk. IV, 701; 892.
4 Harr. 68b, 10.
5 Cairo 28027; Urk. IV, 39.
6 Harr. 13b, 11; Urk. III, 105; Edfou I, 569; Kenam. 17–18; Leid. V, 38; Totb. Nav. 30B etc.
7 Urk. IV, 658; 659; 692; Harr. 7, 1; Kenam. 13; Six. Temp. 12, 3; 12, 25.
8 Cf. Peet, Griff. Stud. p. 124–5; Mél. Masp. I p. 185; Černý, op. cit. p. 914.
9 Cf. Schroeder, OLZ. 18 p. 5; p. 79.
10 Cf. Möller, OLZ. 18 p. 78; Bissing, Act. Or. IV p. 138.
11 Métaux p. 16.
12 ZÄS. 10 p. 43.
13 Wb. p. 1018; cf. p. 1273; Erg. p. 271.
14 ZÄS. 7 p. 82.
15 Dend. Mar. I, 71; Wb. Erg. p. 271.
16 Dend. Mar. I, 70; I, 71; II, 12; Düm. G. I. II, 63; II, 66; II, 69; IV, 170.
17 Rosetta 14.
18 Dend. Mar. III, 32; III, 37i.
19 Edfou I, 572; Dend. Mar. III, 64 etc.
20 Edfou I, 439; Dend. Chass. IV, 159; Dend. Mar. IV, 20; Thes. 1375; Edfou IV, 391, 12.
21 Métaux p. 17.
22 ZÄS. 10 p. 43.
23 Wb. p. 849.
24 Zahlen p. 83.
25 Arch. Or. 20 p. 631; p. 641.
26 Dend. Mar. I, 70; I, 71; Düm. G. I. II, 66; II, 69.

found parallel to both *ḥḏ*[1] and *ˁrḳwr*[2], and balancing *rwy*[3]. Thus it is clearly a term for silver of some description, bearing the same relation to *ḥḏ* as *rwy* does to *nbw*, and may have some such meaning as that suggested by Sethe. That it is merely a sportive writing of *ḥḏ* is unlikely, since the analogous *rwy* can be shown not to be an orthographic variant of *nbw*, and the two are evidently on a similar footing. A possible theory as to the process by which *r-ḫmt* came to mean silver has already been advanced above (p. 40), though it is to be treated with the utmost circumspection.

It has been shown above that, with the exception of the Ptolemaic loan-word *ˁrḳwr*, the terms for silver are closely related to, and derived from, those for gold, of which silver was thought to be a sub-species. It may be suggested, further, that one or two of the varieties of gold already discussed were perhaps nearer to electrum or even silver, the fact that they were classed as *nbw* merely serving to underline the unity of the noble metals in the eyes of the Egyptians, at least until the end of the New Kingdom.

ḏˁm

ḏˁm (Wb. V. 537.13f.) has been the centre of much controversy, the main point at issue being whether the word represents a specific material, a form of gold, probably electrum, or whether it is merely an artificial, 'affected' or 'poetical' name for gold, with no precise meaning. The problem has been unnecessarily complicated by the mistaken identification of *ḏˁm*, at first read *wꜣsm*, with Greek ἄσημος, and the subsequent assumption that because this supposed connection is invalid, the translation 'electrum' is therefore to be abandoned. In fact, there is no logical nexus between the two, and the meaning of *ḏˁm* can be ascertained simply from a consideration of the textual evidence.

The discussion of *ḏˁm* has a long and unfortunate history, which must briefly be summarised here. Champollion[4] first read **ⲚⲞⲨⲂ ⲢⲰϪⲒ**, and translated 'or pur' or 'poudre d'or', but Birch[5] read *ṫam*, and this was followed by other early Egyptologists, including Dévéria[6], de Rougé[7] and Chabas[8], though the latter also[9] read *num*, and took the word as a variant writing of *nbw*. Brugsch[10] originally read the group as *sm* or *ỉsm*, and took it to be gold, and this was adopted by Le Page Renouf[11], who noted that 𓌂 could be *wꜣs*, but preferred to read *sm(w)* as Brugsch, rather than *wꜣsm(w)*. In his Wörterbuch[12], Brugsch read *wꜣsm*, and translated copper or bronze. *wꜣsm* was clearly a metal, akin to gold either in value or appearance, but not an alloy like orichalc, because the expression *wꜣsm n ḫꜣst* occurs; since doors etc. were overlaid with it, it was probably copper.

Lepsius[13] discussed *ḏˁm* at some length, noting previous opinions and realising that 𓌂 had two values, though he accepted Brugsch's reading *wꜣsm* in this case, on the grounds of an apparent writing 𓇋𓋴𓌂𓅓𓋞. Lepsius recognised that the metal was connected with gold, but rejected the 'pure gold' of Champollion and Birch, because *wꜣsm* was always listed after gold. On the other hand, it preceded *ḥsbḏ* and *mfkꜣt*, and was not, therefore, copper, and was moreover found where silver would be expected. It was not an alloy, because it was taken straight from the rock, *wꜣsm n ḫꜣst*. Lepsius quoted Pliny and Strabo to the effect that all gold contains some silver, and proposed to identify *wꜣsm* with electrum, which would be less valuable than gold, and might well replace silver. It would be the

[1] Dend. Chass. IV, 181, 19.
[2] Dend. Mar. IV, 20.
[3] Dend. Mar. IV, 15; IV, 20; Dend. Chass. II, 113, 15; IV, 41, 4; IV, 181, 19; Edfou II, 283, 4 etc.
[4] Grammaire p. 90.
[5] In Bunsen, Egypt² I p. 576.
[6] Mém. I p. 291; 295; 317.
[7] Inscr. d'Ahmes p. 132.
[8] Oeuvres Div. II p. 273.
[9] Pap. Harris p. 226.
[10] ZÄS. 2 p. 66; p. 68.
[11] ZÄS. 6 p. 47.
[12] Wb. p. 351; p. 768; p. 906; cf. also p. 1021, where he reads *us*.
[13] Métaux p. 12.

colour of brass, and more brilliant than gold, and hence appropriate as the covering for the pyramidia of obelisks. The word *wꜣsm* occurs only in texts of the 'classical' period, or in archaistic survivals, and this could be explained by the gradual advance in the technique of separating gold from silver.

Lepsius' conclusions were accepted by Dümichen[1], who noted the parallelisms of *wꜣsm* and *nbw* at Dendera, and maintained that since by the Ptolemaic period electrum had declined in use, the word properly designating it could then be used of gold in general, both electrum and silver being nothing more than forms of gold to the Egyptians. Kuhn[2] also favoured electrum, and quoted Sanskrit parallels for the use of separate words to denote gold and electrum. In the supplement to his Wörterbuch, Brugsch[3] reassessed the position, and deduced that since *wꜣsm* nowhere occurs in Harris in a list of metals, whereas *nbw*, *nbw nfr*, *nbw ḥḏ* and *ḥḏ* all do, it was probably a compound. From a comparison of certain passages[4] he argued that *wꜣsm* was an alloy of *nbw nfr*, *ḥḏ* and *ḥmty km*, and suggested 'Kupfergold' as the most suitable translation, pointing to the use of *wꜣsm* for overlay, and the parallelism of *wꜣsm* with *nbw*.

Le Page Renouf[5] contested Lepsius' findings, proposing to translate *wꜣsm* as copper, because it was a true metal and not a compound, and because it was neither gold nor silver, and its frequent occurrence suggested something less valuable. Referring to the use of *wꜣsm*, he quoted St. Ephrem concerning the two obelisks of Heliopolis said to be capped with 'white copper', a statement supported by Dionysius of Antioch, and also by the Arab writers, who mention the capping 'of copper like to gold', and refer to the staining of the obelisks by the green oxide of the metal. Renouf thought it most unlikely that large obelisks were entirely covered with gold or electrum, and preferred to interpret the expression *bꜣk m ḏꜥm* (*wꜣsm*) as referring to the metal, copper, with which the shafts were sculptured. He found the phrase *wꜣsm ndb m ḥmty* more intelligible if *wꜣsm* were copper or bronze, and regarded *wꜣsm ḏꜥb* as χάλκος κεκαυμένος. But he was unable to explain the occurrence of *wꜣsm* in rings, probably as 'money', and dismissed too lightly the difficulty inherent in the existence of *ḥmty* as a more usual term for copper. Against the translation 'electrum' he further observed that it was doubtful whether there would have been any distinction between gold and electrum in Egypt.

Chabas[6] also opposed electrum, since he was unable to find any example in which *wꜣsm* differed from *nbw*, and regarded instances where the two were juxtaposed as parallelism rather than opposition. Thus *nbw*, *ktmt*, and *wꜣsm* are found together, and since the first two are in all probability identical, the third must be also. Elsewhere[7], Chabas modified his view to the extent of admitting that a technical distinction between *wꜣsm* and *nbw* was not impossible, though contending that in general it was not felt, since the Egyptians affected repetitions and poetical devices, and varied words without varying the meaning. The uses of *wꜣsm* and *nbw* are exactly similar, and on the Rosetta stone, *wꜣsm* is rendered by χρυσός, whereas *nbw* is rendered χρυσός on the Canopus decree.

But in general, the reading *wꜣsm* and the translation 'electrum' were accepted, and Pleyte[8] went so far as to attempt to show that *wꜣsm* was the origin of Greek ἄσημος, which he thought, wrongly, meant electrum, a view accepted by Wiedemann[9], and upheld by him even after the reading *ḏꜥm* had been proved.

---

[1] ZÄS. 10. p. 45. [2] ZÄS. 11 p. 21. [3] Wb. Erg. p. 336.

[4] Harr. 26, 11 with 33a, 5; 33a, 8; 33a, 14.

[5] ZÄS. 11 p. 119. Renouf's arguments are reiterated by Naville, Rev. Arch. 20 (1924) p. 1f.

[6] ZÄS. 12 p. 2. [7] Études[2] p. 17.

[8] Asemos, in Mededeelingen Amsterd. Akad. 3rd. Ser. III p. 211. [9] Alt. Aeg. p. 344.

The reinstatement of the reading *ḏꜥm*, previously recognised by Birch and others was due to Gardiner[1], who outlined Brugsch's arguments in favour of *wꜣsm*, and pointed out that the confusion arose from regarding the enclitic 𓇋𓋴 as part of the word. 𓌀 certainly has the value *wꜣs*, but if *ḏꜥm* were *wꜣsm*, it would be a quadriliteral substantive, which is improbable – hence the alternative suggestion *wsm* – and some variants with fuller spelling might be expected to occur, which they do not. Moreover, the writing 𓌀𓋞 could only be *wꜣs*, since no omission of final *m* is known, and the existence of this *m* elsewhere shows that 𓌀 must have a syllabic value, of which it is the last letter. The 𓌁 sceptre is called both *ḏꜥm* and *wꜣs*, the former having a spiral handle, while the latter is straight, and there can be no doubt that the name of the metal is to be read *ḏꜥm*, which explains the presence or absence of the final *m*. Gardiner then pointed out that once the word was read *ḏꜥm* and not *wꜣsm*, the fancied connection with ἄσημος vanished altogether.

Gardiner's establishment of the reading *ḏꜥm* was further confirmed by Sethe[2] who quoted examples to show that *ḏꜥm* was written with the twisted sceptre, and *wꜣs* with the straight, and has since gained universal acceptance. Of the meaning of the word Gardiner said nothing, but in his commentary on Sinuhe[3] expressly attacked Lepsius' interpretation as electrum, claiming that it rested on no evidence, and maintaining that in 'poetical' texts *ḏꜥm* is nothing more than a synonym for gold. A similar view was held by Ransom-Williams[4], who noted that the connection between *wꜣsm* and ἄσημος was invalid, and thought that the translation 'electrum' was therefore also incorrect, although it might at times seem appropriate. The distinction between *ḏꜥm* and *nbw* was not, she claimed, one of composition or colour, and *ḏꜥm* was merely a 'poetical' or 'affected' word for gold, both *ḏꜥm* and *nbw* being equally applicable to a gold-silver alloy not pale enough to be called silver.

This interpretation of *ḏꜥm* has since found considerable favour, though the Wörterbuch seems to regard it as a specific kind of gold 'von hellerer Farbe', and Balcz[5] still translates 'Elektron' without question. Moreover, Bissing[6], in the most recent article devoted to the subject, upholds the meaning electrum. He points out that most Egyptian gold had a 20 or 25 per. cent. silver content, and thinks that such gold was called *ḏꜥm* as distinct from the darker gold called *nbw*. The identification with ἄσημος is, of course, a myth, because in any case the Greek word means simply 'ohne Zeichen', and is a quite normal formation with alpha-privative. The apparent identity of *ḏꜥm* and *nbw*, stressed by Chabas, is to be explained, as Dümichen realised, by the fact that in the Greek period *ḏꜥm* had ceased to have any real meaning, and was nothing more than a synonym for *nbw*. A somewhat similar development can be observed in the case of Greek ἤλεκτρος, which originally meant electrum, but by the time of Herodotus had changed its gender and referred to amber.

Previous discussion of *ḏꜥm* has confused rather than clarified the issue, and the problem must be approached, as far as possible, from the beginning. The reading *ḏꜥm* is now established beyond any doubt, and the alleged connection with Greek ἄσημος, which does not in any case mean electrum, is thereby destroyed. But it is a non sequitur to assume that the identification of *ḏꜥm* as electrum is therefore necessarily incorrect, and a careful reassessment of the evidence is required.

That *ḏꜥm* is in fact a species of gold, and not copper or the like, cannot seriously be questioned for a number of reasons. First, the word is regularly determined with 𓋞, or written

[1] ZÄS. 41 p. 73.
[2] ZÄS. 44 p. 132.
[3] Sinuhe p. 117.
[4] Jewelry p. 30.
[5] Mitt. Kairo 3 p. 75.
[6] Act. Or. IV p. 138.

as [1], the sign being used only for the noble metals, all of which were regarded as varieties of gold. Second, *ḏʿm* occurs together with *nbw* in the captions accompanying scenes of gold-working[2], where it can only refer to gold or gold ore of some description. Third, *ḏʿm* is found replacing *nbw*, and in parallelism with it and other terms for gold, and fourth, the uses of *ḏʿm* are strikingly similar to those of both *nbw* and *ḥḏ*.

Instances where *ḏʿm* occurs in place of *nbw* are fairly frequent, particularly in enumerations of minerals[3], where, however, it also seems to replace *ḥḏ* in one or two cases[4]. More common are passages in which *ḏʿm* is parallel to *nbw*, for example in religious texts[5], and especially in inscriptions of the Graeco-Roman period[6]. The exact similarity of the two words in many Ptolemaic texts led CHABAS and others to assume that *ḏʿm* and *nbw* were identical in meaning, a view supported by a passage in the Rosetta stone[7] where *ḏʿm* is translated by χρυσός. In fact, there can be little doubt that in these late texts, *ḏʿm* had lost its original meaning, and had become an alternative name for gold, and this may also be the case in religious and quasi-poetical contexts. However, there are other and earlier passages where *ḏʿm* is found together with *nbw* in which this certainly does not apply, and where *ḏʿm* must have a precise significance. In the Koller papyrus[8] *ḏʿm* is clearly distinguished from *nbw* and *nbw nfr*, while elsewhere it is mentioned together with *nbw* and *ḥḏ*[9], and *nbw* alone[10]. In one passage *ḏʿm n ḫȝst* and *nbw ḥr ḫȝst.f* are apparently contrasted[11], and in another the hill country produces *nbw* and the mountains *ḏʿm*[12], a similar distinction being also drawn in two other contexts[13]. Moreover, there are references to statues made partly of *nbw* and partly of *ḏʿm*[14], where the two metals must be regarded as different in some way.

But the most convincing proof that *ḏʿm* is not merely a 'poetical' or 'affected' term for gold comes from early examples of the word in offering lists and elsewhere[15], where literary artifice is out of the question, and from the captions to the scenes of gold-working already mentioned. Furthermore, it is reasonable to suppose that *ḏʿm* has a precise meaning in the many instances when it is cited as the material for some specific object, though the nature of these objects is not in itself significant. Thus *ḏʿm* is used for vessels[16], for statues and parts thereof[17], for offering tables[18], furniture[19], mirrors[20], jewellery[21], and a variety of other articles[22]

---

[1] NB. An exception, ZÄS. 37, 95, written (Wb. V. 539.7), occurring in a single medical context, Eb. 88, 20, may be the metal *ḏʿm*, as Stern and Joachim assume, though the identification cannot be certain.

[2] Eg. Beni Has. II, 4; D. el Geb. I, 14; Urk. IV, 367, 8; Mereruka pl. 30 etc., cf. Klebs, Reliefs AR. p. 84; MR. p. 110; NR. p. 115. Montet, Scènes p. 285 suggests that should also in these instances be read *ḏʿ m*, but without justification.

[3] Eg. Urk. IV, 168; 820; Karn. R. III 90, 25; Karn. I, 26 = Urk. IV, 1668; Med. Hab. 11, 3; Ombos I, 64, 70.

[4] Eg. Rec. Trav. 29, 165.

[5] Eg. Totb. Nav. 172, 19; Mythe d'Horus IV; Emb. Rit. 3, 17; 3, 18; 8, 4; Düm. Baug. 29.

[6] Eg. in the phrase *nbw nṯrw ḏʿm nṯrwt* (Wb. V. 539.1); cf. the similar parallelism between *nbw* and *ḥḏ* (Wb. II. 239.7).

[7] Rosetta 8.

[8] Koller 3, 8.

[9] Urk. IV, 1260; cf. Berl. 1204.

[10] Urk. IV, 337.

[11] Düm. H. I. I, 31.

[12] LD. III, 141b; cf. LD. III, 140b, 2.

[13] Luxor ⟨560⟩; Sethe 17, 36 [5] *ḏw n nbw ḏʿm*.

[14] Sin. B 308; Theb. Grab. 65 ⟨619⟩; cf. Harr. 7, 13.

[15] Eg. Medum 13; Lesest. 71, 9; 75, 15.

[16] Medum 13; Urk. IV, 173; 1160.

[17] Rec. Trav. 29, 165; Sin. B 308; Theb. Grab. 65 ⟨619⟩.

[18] Urk. IV, 173; 869.

[19] Kuban 8; Dend. Mar. IV, 71; Urk. IV, 427.

[20] Amamu 21–2; El Arabah 26.

[21] Amamu 21–2; Cairo 28027; 28088; 28091.

[22] Urk. IV, 657; Rec. Trav. 16, 51; 22, 118–9; LD. III, 210; Harr. 25, 11 etc.

as well as for plating and inlay[1], particularly of obelisks[2]. It also occurs in rings[3], and as dust or lumps in bags[4], presumably for use as currency.

As a species of gold, *ḏᶜm* is normally coloured yellow[5], though a mirror and a collar of *ḏᶜm* are also shown as white[6], which may be significant. On the other hand, some objects of *ḏᶜm* are painted red[7] and other colours[8], so that the value of these colour indications may be doubted.

Several different types of *ḏᶜm* are mentioned, distinguished either by quality or provenance. *ḏᶜm ḥr ḏᶜt.f*[9], parallel to *nbw ḥr ḏᶜt.f*, may be either *ḏᶜm* from the *ḏᶜm*-bearing rock, or *ḏᶜm* with its ore, *ḏᶜt* being evidently a term for the vein from which the metal or ore was extracted. *ḏᶜm n ḫꜣst*[10] is simply '*ḏᶜm* from the hills or desert', and *ḏᶜm n tp(w) ḫꜣswt*[11] '*ḏᶜm* of the best of the hill-country'[12]. *ḏᶜm n ᶜmw* is also specifically mentioned[13], and *ḏᶜm* comes from Kush[14] and Pwenet[15]. Fine quality *ḏᶜm* is referred to as *ḏᶜm n ḳn*[16], and possibly also *ḏᶜm wᶜb*[17], though *wᶜb* may have purely ritual significance, and an expression *ḏᶜm mꜣᶜ* also occurs[18]. The meaning of *mꜣᶜ* in this case is not altogether clear, and the epithet does not appear with any other word for gold. It seems probable, however, that *ḏᶜm mꜣᶜ* designates the native gold alloy *ḏᶜm* as opposed to an imitation produced artificially. This is particularly likely since in three passages[19] *ḏᶜm mꜣᶜ n tp ḫꜣswt* is mentioned, which shows *ḏᶜm mꜣᶜ* to have been a native product. Lastly, a single reference[20] to *ḏᶜm mꜣ* can be quoted, though it is possible that *mꜣ* is a mistake for *mꜣᶜ*, since unlike 'new turquoise', the meaning of 'new *ḏᶜm*' is not apparent, unless perhaps it refers to a freshly made alloy.

From these references it appears that the kind of gold known as *ḏᶜm* was of differing quality, and came from various sources, and that it was primarily a naturally occurring material, a fact confirmed by other passages[21]. But it was also counterfeited in some way, at least during the New Kingdom, when the expression *ḏᶜm mꜣᶜ* first occurs, and this can only refer to adulteration with other metals, or to the reproduction of a native alloy by artificial means.

Now there is, in fact, positive evidence for the nature of *ḏᶜm* in the Harris papyrus, as was realised by Brugsch[22] and Breasted[23], though the point seems to have been ignored by the majority of other scholars. Among the various donations to the temple of Heliopolis, mention is made[24] of a splendid scale or balance (*mḫꜣt*) of *ḏᶜm*[25], and nowhere else in the papyrus is a balance referred to, save in the subsequent enumeration of certain materials 'for the balance'. These include gold (*nbw nfr* ⟨*ḥr*⟩ *ḫꜣst.f*)[26], silver (*ḥḏ ḥr ḫꜣst.f*)[27], black copper (*ḥmty km*)[28], a block of hewn wood[29], and a post of *mrw*-wood[30], the obvious inference being that such were

---

[1] Urk. IV, 53; 339; 422; 425; 427; 521; Westc. 5, 9; Luxor ⟨410⟩; Six. Temp. 12, 6; LD. III, 152; V, 1b; B. M. 586 etc.

[2] Urk. IV, 366; 426; 590. [3] Urk. IV, 335; 338; 526. [4] Urk. IV, 1096; 1099.

[5] Medum 13; Cairo 28027; 28088; 28092; Berl. 45; Amamu 21–2; cf. Lepsius, Métaux pl. 1.

[6] El Arabah 26; cf. Sin. B 308 where the kilt of a statue might be white.

[7] Cairo 28091, 28. [8] Cairo 28091, 32; 39. [9] Urk. IV, 1442. [10] Düm. H. I. I, 31.

[11] Urk. IV, 362; 421; 429; 766; 1708; Rec. Trav. 22, 118–9; 29, 162f. l. 15; Nauri 8.

[12] *ḏᶜ m n tp(w) ḫꜣswt nbt*, Urk. IV, 362; 1708, must be '*ḏᶜm* of the best of every land'.

[13] Urk. IV, 384; 524. [14] Med. Hab. 11, 3. [15] Urk. IV, 335; 338.

[16] Urk. IV, 367; cf. Wb. V. 49.15. [17] Red. C. 3.

[18] Urk. IV, 167; 849; Karn. I, 24; Med. Hab. 103, 4; 182, 1; Abyd. Mar. II, 3; Rec. Trav. 29, 165.

[19] Nauri 8; Urk. IV, 1706; 2029. [20] Luxor ⟨405⟩ = Urk. IV, 1709.

[21] Eg. LD. III, 140b, 2 = Red. A. 2; 141b; Luxor ⟨560⟩ etc.

[22] Wb. Erg. p. 336. [23] AR. IV p. 144 n. b. [24] Harr. 26, 11.

[25] *ḏᶜm* is only found in three other passages in Harris, twice for overlaying doors (4, 1; 4, 12), and once for the figures on a door overlaid with *nbw* (7, 13), where a contrast must be implied.

[26] 33a, 5, *dbn* 1278, $9^{2}/_{3}$. [27] 33a, 8, *dbn* $1891^{1}/_{2}$. [28] 33a, 14, *dbn* 67, 3.

[29] 34a, 14; 71a, 6. [30] 34b, 2; 71a, 9.

the materials required for the balance of *ḏʿm*, and that *ḏʿm* was therefore an alloy of gold and silver with a small admixture of black copper. This was the conclusion reached by BRUGSCH, and there seems to be no reason for doubting its validity. That the Egyptians were able to alloy gold and silver in the 20th dynasty is altogether probable[1], and it is known that earlier in the New Kingdom gold was debased with copper[2].

But if man-made *ḏʿm* was an alloy of gold and silver with traces of copper, it follows that the naturally occurring metal was of similar nature, ie. essentially electrum, which is found native and was widely used in Egypt[3]. Electrum, generally pale yellow in colour, is sufficiently characteristic to be distinguished from the darker yellow and even reddish gold, and the pale greyish-white silver, and it is only natural that the Egyptians should have had a specific name for it.

Moreover, the translation of *ḏʿm* as electrum[4] agrees well with certain facts. Thus, references to *ḏʿm* are most common during the period when electrum was widely used, and become less frequent in the late period, when the use of electrum was apparently declining[5]. By the Graeco-Roman epoch, *ḏʿm* had ceased to have any real meaning, and was employed simply as a synonym for *nbw*. The colouring of objects of *ḏʿm* on Middle Kingdom coffins is also perhaps explained, since while yellow would be the most common colour of the alloy, it might also be white, and possibly, in some circumstances, of other hues. Furthermore, the extensive use of *ḏʿm* for covering doors, columns, obelisks etc. is shown to be apposite, since electrum is harder and therefore more durable than gold, and would be better suited for such purposes.

The use of electrum for covering obelisks is perhaps confirmed by the statement that Ashurbanipal carried off from Egypt two obelisks cast in *zaḫalî ibbi* of 2500 talents weight ie. about 70 tons. The nature of *zaḫalû* is uncertain, but CAMPBELL THOMPSON[6] thinks it may mean electrum, comparing *zaḫalî ibbi* with λευκὸς χρυσός. The obelisks were probably of stone overlaid with electrum, thought it is just possible that they were solid metal[7]. On the other hand, the statements of the Arab writers and others concerning the obelisks of Heliopolis might suggest that the metal there employed had a considerable copper content. Thus the mention of 'copper like to gold' and reference to the staining of the shafts with green copper oxide would point to a gold-copper alloy[8], while ST. EPHREM's 'white copper' might indicate an alloy of electrum and copper such as that implied in Harris.

In general, however, it seems practically certain that *ḏʿm* refers to electrum[9], and this is perhaps the only possible translation in texts of the Old and Middle Kingdoms, when a more complex alloy is out of the question, and also in those instances where *ḏʿm* is clearly a natural product. But during the New Kingdom, an artificial *ḏʿm* was also produced, which appears to have been either man-made electrum, or a gold-silver-copper alloy, and may even have included gold adulterated in other ways. Towards the end of the New Kingdom, possibly as a result of improved metallurgical techniques, new terms were introduced for the different grades of gold, and *ḏʿm* therefore became obsolete as an all-embracing word for electrum. Thus in the late period, and more particularly in Graeco-Roman times, *ḏʿm* had no precise

[1] Cf. Ransom-Williams, Jewelry p. 29–30.

[2] Lucas, Materials p. 263. [3] ib. p. 267.

[4] Brugsch's 'Kupfergold' is hardly acceptable, since the proportion of copper in the *ḏʿm* of Harris is so small. Furthermore, it is not likely that a natural alloy would contain sufficient copper to make it distinguishable from electrum.

[5] Cf. Lucas, Materials p. 268. It seems that in the 20th dynasty electrum was recognised as a species of gold and called *nbw ḥḏ*, and from this time *ḏʿm* ceases to have any real meaning.

[6] Dictionary p. 60. [7] Cf. Rev. d'Ég. 8 p. 47. [8] Cf. Lucas, Materials p. 263.

[9] Ie. gold with 20–25 per cent. silver content, though the proportion may be even higher.

meaning, and was simply another name for gold, a fact which has given rise to much unnecessary confusion regarding the earlier significance of the term.

Having considered the various terms for gold and silver, and also the meaning of the word *ḏʿm*, it is now necessary to glance at the terminology of the noble metals as a whole, and to observe as far as possible its gradual development. The earliest stages of this development cannot, unfortunately, be traced, but it may be inferred that at some remote period there was only one term for all gold of whatever purity, ie. including electrum and silver, and that this was *nbw*[1], the sign for which was subsequently used as the determinative of all the other terms with the exception of Ptolemaic *ʿrḳwr*, *rwy* and *r-ḫmt*. By the beginning of the 4th dynasty, however, and possibly even earlier[2], two other terms, *ḏʿm* and *nbw ḥḏ* were already in use, the one evidently referring to the pale yellow electrum, the other to alloys white enough to be classed as silver, and called simply *ḥḏ* from the 5th dynasty onwards[3].

This rudimentary classification of gold and its alloys as either *nbw*, *ḏʿm* or *ḥḏ* according to external characteristics was maintained until the New Kingdom, when greater contact with the outside world, the importation of minerals from abroad, and possibly also improvements in metallurgical techniques, led to the realisation that a simple division of this kind was altogether inadequate. Thus, although *nbw* remained the most general word for gold of all kinds, with *nbw nfr* and *ktmt* as specific terms for the purer varieties, and *ḥḏ* continued to refer to silver, *ḏʿm*, which by this time covered not only all grades of electrum, but also, apparently, certain other artificial alloys of gold and eg. copper, gradually ceased to have any precise meaning. Instead, the naturally occurring electrum was referred to as *nbw ḥḏ*, and *nbw n sp 2* and *nbw n sp 3* were applied to other inferior types of gold or gold-siver alloys.

How long this particular system prevailed it is impossible to tell, and it may even be confined to the 20th dynasty alone. In the Greek period, at all events, the terminology was again different, and of the older words only *nbw*, *ḥḏ* and *ktmt* survived with their meaning more or less unchanged, while *ḏʿm* became a mere 'literary' synonym for *nbw*. But two pairs of closely parallel terms, *sȝwy* and *ʿrḳwr*, and *rwy* and *r-ḫmt*, were introduced into the language to refer either to gold and silver in general, or to specific varieties of each. The exact significance of these words is as yet obscure, and while they may well have had some more or less precise meaning, it is equally likely that they were merely used to diversify the inscriptions in which they occur.

[hieroglyphs] and *biȝ* (Wb. I. 436.1f.)

Although it is abundantly clear that the two words here discussed are quite distinct, the one referring to copper, the other to iron and meteoric material in general[4], it has nevertheless seemed preferable to treat them together, in as much as they have been persistently confused in the past, and are not separated in the Wörterbuch. The reading of [hieroglyphs], and its relationship with *biȝ*, constitute one of the most involved problems of Egyptian lexicography, the inherent complexity of which has been increased by its connection with the perennial controversy as to the antiquity of the use of iron in Egypt. The whole question must be approached, as far as possible, with a fresh mind, but the volume of previous discussion is such that a somewhat extended summary must first be given.

---

[1] For early examples of of [hieroglyph] cf. H. Petrie, Eg. Hier. nos. 713—20.

[2] For a possible example of *ḏʿm* on an ivory tablet of Dyn. 1 cf. Breasted, History of Egypt p. 43 fig. 27.

[3] But cf. above p. 41 n. 6.

[4] Cf. below p. 166.

The earliest comprehensive discussion was that of Lepsius[1], who quite clearly distinguished [hieroglyph] from *bỉꜣ*, reading the former *χomt*, and regarding it as a word for copper, while *bỉꜣ* (*bȧa*) was 'minéral dur, pierre'. Lepsius noted the various forms of the ideogram [hieroglyph], which Champollion and Birch took to be a crucible[2], and contended that whereas the earlier [hieroglyph] might thus be interpreted, [hieroglyph] and the later forms could not. The value of the sign was difficult to determine, since it was always used ideographically. Champollion proposed no reading, but compared Coptic ⲂⲈⲚⲒⲠⲈ and translated 'iron'; de Rougé and Birch both read *ba*, the former rendering iron or steel, the latter wood, iron or brass; Chabas gave no value, but interpreted as bronze or iron, and Dümichen translated simply 'metal', again without proposing any reading. Brugsch in his Wörterbuch wavered between reading *ȧpot* and translating 'iron'[3], and reading *ṯhst*[4] and translating 'bronze'[5]. Elsewhere[6] he proposed to read *bȧ*, and reverted to iron as a translation, though in his Wörterbuch he regarded *bȧ* as a species of hard stone[7].

Lepsius cited passages from Edfu where the ideogram [hieroglyph] or [hieroglyph] appears to stand for the name of Horus' lance, written phonetically *ḥmt* and *ḥms*, and therefore proposed that the word be read *ḥmt* also. Its position in the lists, following gold and silver, suggested copper, which was confirmed by Coptic ϨⲞⲘⲦ, and by the translation of [hieroglyph] by χαλκός in the Canopus decree. The other Coptic word for copper, ⲂⲀⲢⲰⲦ, did not occur in hieroglyphs, but its first syllable must have been *bȧa*; ϨⲞⲘⲚⲦ ⲚⲂⲀⲢⲰⲦ, χαλκολίβανος, was evidently copper, ⲂⲀⲢⲰⲦ being a more precise term than ϨⲞⲘ(Ⲛ)Ⲧ 'Erz'. Lepsius then noted *χomt ḥr ḥꜣst.f*, 'cuivre dans sa gangue', which he contrasted with *χomt stfw*, 'le cuivre fondu ou purifié', and *χomt km*, 'cuivre noir', which was also copper smelted or purified ready for use, comparable to Greek χαλκὸς μέλας and modern 'black copper', which has undergone one fusion. *χomt* probably included not only copper, but also various alloys, *χomt set* (*sṯt*), 'bronze Asiatique' being a variety of particular value. [hieroglyph] was used as the determinative for a number of words representing objects of copper.

Following Lepsius' treatment of the evidence, Dümichen[8] read *χmt*, and translated 'Erz, Kupfer, Bronze', noting *ṯhst* and *ḥwdtf* as Ptolemaic variants. Brugsch[9] also read *χmt*, comparing Coptic ϨⲞⲘⲚⲦ, and carefully distinguishing *bỉꜣ*. The latter he took to be iron[10], the material of which the vault of heaven was thought to be, and noted the two species *bỉꜣ n pt* and *bỉꜣ n tꜣ*. Lepsius himself[11] added a further note on *χomt set* (*sṯt*), pointing to its use for door hinges, and concluding that it was not the finest copper, but a particular variety of light colour in contrast to *χomt km*. It may in fact have been iron, though the Greek χαλκὸς λευκός was brass.

The translation of *bỉꜣ* as iron was proved by Déveria[12], who observed that it was the material of the implements used in 'opening the mouth', and that the Louvre possessed a *nwꜣ* of wood with an iron blade[13], and a *msḫtyw* entirely of iron. Iron was connected with the vault of heaven, *bỉꜣ*, a similar relationship being apparent in σίδηρος, siderites, and sidus-sideris as well as Coptic ⲂⲈⲚⲒⲠⲈ. *bỉꜣ ḳsy*, discussed below, he also regarded as a form of iron.

---

[1] Métaux p. 45.

[2] Gensler, ZÄS. 8 p. 137, also regarded [hieroglyph] as a crucible, and read *ba*. He thought *ba* was a general word for metal, particularly copper or bronze, and that it was the origin of ⲂⲀ- in Coptic ⲂⲈⲚⲒⲠⲈ, ⲂⲀⲢⲰⲦ ⲂⲀⲤⲞⲨⲢ, ⲂⲀⲤⲚϬ, and ⲀⲂⲀϬⲈⲒⲚ.

[3] Wb. pp. 51; 571; 618.

[4] Wb. p. 1592.

[5] Wb. p. 751.

[6] Sonnenscheibe, eg. p. 44.

[7] Wb. pp. 373; 375.

[8] ZÄS. 10 p. 101.

[9] Wb. Erg. p. 925.

[10] ib. p. 413.

[11] ZÄS. 10 p. 113.

[12] Mél. d'Arch. 1 p. 2.

[13] Cf. Chabas, Oeuv. Div. 4 p. 389 for a text expressly stating that five *nwꜣ* implements are to have blades of *bỉꜣ* (written [hieroglyph]).

Chabas[1] distinguished [hieroglyphs] and *bi͗ꜣ*, though unable to suggest the phonetic value of the sign [hieroglyph], which he regarded as a crucible. [hieroglyphs] was bronze, as was shown by the reference to its golden colour in Harris, and its use for weapons; [hieroglyphs] *stt* was the most valuable species. *bi͗ꜣ*, on the other hand, was a quite separate word meaning primarily 'mineral', but also referring to a very solid and very hard metal. This could only have been copper or iron, and since few objects of *bi͗ꜣ* are mentioned, the former is ruled out. Chabas quoted Dévéria's conclusions at length, noting the connection with the celestial vault and the interchange of *bi͗ꜣ* and *bi͗ꜣ n pt*. *bi͗ꜣ* meant iron or steel, *bi͗ꜣ n pt* steel, and *bi͗ꜣ n tꜣ* 'quelque métalloïde qui n'appartenait pas nécessairement a l'espèce fer'.

Thus the majority of scholars during the last century differentiated between [hieroglyphs] and *bi͗ꜣ*, translating the one copper, the other iron, and many accepting Lepsius' reading of the former as *χomt*, ie. *ḥmt*. Bissing[2], however, discussing the [hieroglyph] determinative of *bi͗ꜣ*, pointed out that the shape of the mould for fitting spouts onto vessels was ◇, and claimed therefore that the determinative referred to copper soldering, and that the meaning iron could not be retained for *bi͗ꜣ*. This view was also supported by Möller[3], who held that the variant determinatives [hieroglyph] and [hieroglyph] might show in [hieroglyph] a form of spout.

Spiegelberg[4], on the other hand, still regarded *bi͗ꜣ* as the word for iron, comparing the phrase *bi͗ꜣ pr m stẖ* with Manetho's statement that the Egyptians called iron 'the bones of Seth[5]. He opposed translating *bi͗ꜣ n pt* as meteoric iron, and thought that it and *bi͗ꜣ n tꜣ* were two obscure species of iron contrasted with one another. Remarking upon the relative rarity of *bi͗ꜣ* compared with [hieroglyphs], Spiegelberg suggested that [hieroglyphs] *km* must also have been a word for iron, citing a Mexican parallel in support of this idea[6].

Gsell[7] also maintained a clear distinction between [hieroglyphs] and *bi͗ꜣ*, being inclined to identify the latter as iron from its reference to things in or from heaven, and the evident connection with meteorites. He noted Pliny's statement that basanites was so called because of its iron colour and hardness, and observed that *bi͗ꜣ* and *bi͗ꜣ n pt* were apparently interchangeable. The word [hieroglyphs], copper, he read *ḥmt*, following Erman[8], and took *ḥmt km* to be impure copper in contrast to pure red copper, *ḥmt stfw*. *ḥmt m smꜣw nt 6* was a form of bronze, probably brass since it was of a golden colour. Thomas[9] also regarded *bi͗ꜣ* as iron, and evidently separated the two words.

Sethe[10] assumed that the word for copper, written [hieroglyphs] in the Old Kingdom, was to be read *bi͗ꜣ*, and that this was the old word for ordinary metals in contrast to the noble metals. *bi͗ꜣ* occurs in ⲂⲈⲚⲒⲠⲈ, 'metal of heaven', ⲂⲀⲢⲰⲦ, 'grown metal', so called because it was found in the mountains, and ⲂⲀⲤⲚϬ, 'metal of . . .'. Coptic ϨⲞⲘⲚⲦ must be derived from *ḥmwty*, 'smith', and probably signifies bronze as opposed to ⲂⲀⲢⲰⲦ, pure copper. *bi͗ꜣ* is also the word for mine, and [hieroglyph] represents a lump of metal on a sledge. Such was essentially the view of Gardiner[11], who, commenting on the sign [hieroglyph] which he took to be an ingot of metal, stated that it was 'early perhaps read *bi͗ꜣ* and later *ḥmt* (?)'. 'The reading *bi͗ꜣ*', he observed, 'would be confirmed for early times if the compound word written with this sign *JEA*. 4 pl. 9 1.10 proved to be really *bi͗ꜣ rwḏ*, the Coptic *barōt* 'copper'. The supposed later reading *ḥmt* (?) is based solely on Coptic *homent*, *homt*'.

---

[1] Études[2] p. 36; p. 44.
[2] ZÄS. 38 p. 150.
[3] ZÄS. 38 p. 151.
[4] Rec. Trav. 26 p. 166.
[5] PSBA. 13 p. 38.
[6] Cf. Forbes, Metallurgy p. 427, where it is suggested that the Egyptians thought of iron as a form of black unrefined copper, only later recognising its true origin. Nahuatl *tliltic teputzli*, 'black copper', meant iron.
[7] Eisen, Kupfer und Bronze p. 45.
[8] Glossar p. 83.
[9] Cairo Sc. Journ. 7 p. 138.
[10] ZÄS. 53 p. 51 JEA. 1 p. 234.
[11] Grammar[(1)] p. 478.

Paton[1] clearly distinguished *bi3* from copper, *ḥmt*, and regarded it as a generic term for what is found in the ground by deep mining, ie. 'ore'. Most examples seem to refer to metal, but some, eg. *bi3 ḳsy*, are evidently stone. Petrie thought *bi3 n pt* embraced both metal and iron ore, haematite, which would explain the statuettes of *bi3 n pt* mentioned in Harris. Paton concluded that the proof for or against the rendering of *bi3* as iron was not decisive, though appearing to favour such a translation himself. Daressy[2] likewise took *bi3* to mean metal in general, and in particular iron, the metal p a r e x c e l l e n c e, observing, further, that *bi3 n pt* is mentioned in such quantity that it cannot be meteoric and must refer to iron of any kind.

The reading *bi3* for 𓈞𓈒𓏥, proposed by Sethe, was adopted by the Wörterbuch, and by Grapow[3] in discussing the figurative uses of the word. He noted the various references to mountains or walls of *bi3*[4], hearts of *bi3*[5], and treading heavily like *bi3*[6], the description of the vizierate as *bi3*[7], and the closing formula of late Egyptian letters which runs 'apply yourself . . . with the firmness of *bi3*'[8], the implication in each case being firmness or durability.

The problem of iron in Egypt was discussed at great length by Wainwright[9], with particular reference to *bi3*. Noting the *nw3*, *msḫtyw* and *mḏtft* chisels of *bi3*, he connected the 'opening of the mouth' with the force of the thunderbolt or lightning-stroke; *bi3* was therefore the material that comes from thunderbolts, ie. iron. Wainwright quoted Dévéria's examples of implements of iron, adding the small chisels, probably *mḏtft*-chisels, from the tomb of Tutankhamūn, which had blades of iron. He regarded the mineral *bi3* as identical with the metal, and remarked upon the variant determinatives 𓈞 and ⋆, which express explosive power and sidereal origin respectively. The explosive power of *bi3* explains why it was considered sufficient merely to present blocks of the material to the mouth of the dead man. The various meanings of the root *bi3* can be shown to be connected with the vault of heaven from which the meteoric iron fell; thus [hieroglyph] depicts a meteorite on a sledge, and the use of *bi3t* to refer to quartzite is appropriate to its appearance, which is much like that of a rusty meteorite. *bi3* is not to be confused with the word for copper, which is always written with an ideogram, 0 or Δ, and later 𓈞. Thutmose III received considerable quantities of 𓈞𓈒𓏥 as tribute, but only once did he acquire objects of *bi3*. The weapons of Horus and Seth, said to be of *bi3*, were clearly meteorites, shooting stars or thunderbolts, and the ropes of *bi3* mentioned in the Pyramid texts may be similarly explained. Plutarch's statement that the Egyptians called iron 'bone of Typhon' is curiously illustrated by the finding at Qau of hippopotamus bones 'mineralised, heavy and black, and presenting a metallic lustre and appearance'; these were probably considered to be some kind of iron. Iron did not in fact become at all common until the New Kingdom, when the term *bi3 n pt* first occurs. This replaces *bi3* in the ritual of the 'opening of the mouth', and is applied to both meteoric and man-made iron, its reference to a sidereal origin being only indirect. Wainwright concluded that *bi3* meant 'iron, or rather meteoric material in general, whether iron, or stone, or a conglomerate of the two'.

Weill[10] maintained that the root *bi3* represented 'copper' or more generally 'metal', a 'precious' or 'marvellous' thing, a 'quarry' or 'mineral mass' or 'block of stone', in particular quartzite sandstone, and lastly a 'hard' or 'solid' thing; from these basic meanings developed 'transport', 'excellence' etc. The phonetic complement was often defective, and

---

[1] Records III, 2 pp. 93–6; IV pp. 14–16. [2] Ann. Serv. 22 p. 165.
[3] Bild. Ausdr. p. 59. [4] Cf. below p. 55. [5] Paheri 3, 10.
[6] P. Tur. PR. 136, 3. [7] Urk. IV, 1087. [8] Cf. below p. 55 n. 18.
[9] JEA. 18 p. 3; cf. Antiquity 10 p. 5; cf. also Hertz, Anthropologie 35 p. 75f.
[10] Rev. d'Eg. 3 p. 69.

the word was written [hieroglyph] or [hieroglyph] + determinative, or [hieroglyph] + determinative in the New Kingdom. The determinatives vary, the most common being [hieroglyph] and [hieroglyph], two forms of the same object, [hieroglyph], exactly similar to the elephant tusk, and [hieroglyph]. In the Pyramid texts the determinative was [hieroglyph], [hieroglyph] being constant, and [hieroglyph] varying in form to [hieroglyph], 'une lame métallique, peut-être le tranchant de hache en bronze', occasionally reversed [hieroglyph], and sometimes degenerating into [hieroglyph] [1]. This gives a clue to the reading of the word for 'copper', 'bronze' or 'metal', written with an ideogram [hieroglyph] or [hieroglyph], without phonetic indications. *bi͗ȝ* has long been considered the most likely reading, though the Wörterbuch does not accept it without reservation, and Gardiner is doubtful. Weill regarded it as certain that [hieroglyph] and [hieroglyph] were simply equivalents of *bi͗ȝ*, and pointed to the use of [hieroglyph] both as a determinative for *bi͗ȝ*, and as an ideogram. Comparing [hieroglyph] with [hieroglyph] he concluded that [hieroglyph] and [hieroglyph] represented the same object, namely a crucible, *bḏ*, and observed that [hieroglyph] was gradually replaced by [hieroglyph], an earthenware crucible. The determinative [hieroglyph] was a simplification of [hieroglyph], though it is, in fact, found before the latter. However, [hieroglyph] was a complicated hieroglyph, and some abbreviation was inevitable, leading to confusion with the tusk. In hieratic, [hieroglyph] became [hieroglyph], and in the New Kingdom the [hieroglyph] determinative was revived.

The ideogram for copper has also been discussed by Strachmans[2] and Junker[3], the former pointing out that although the [hieroglyph] sign was commonly held to be an ingot of metal, eg. by Gardiner, it was in fact a crucible, as could be shown by comparison with similar crucibles depicted in scenes of metalworking listed by Klebs. Junker examined the ideograms for' Erz' and 'Erzarbeiter' in greater detail, showing that the two were originally distinct, the one [hieroglyph], the other [hieroglyph]. The former was always [hieroglyph] or [hieroglyph] in early examples[4], only becoming like the latter towards the end of the Old Kingdom. [hieroglyph] represented a furnace (Schmelzofen), and [hieroglyph] a crucible (Tiegel) in such a furnace; [hieroglyph] was a furnace with one crucible, and the signs were closely related. But Junker was unable to suggest a plausible reading for the word for copper.

Now the previous discussion of [hieroglyph] and *bi͗ȝ*, though lengthy, has not, in fact, produced any very satisfactory conclusions. A distinction between the two has generally been accepted, though Weill's article tends to obscure this, and [hieroglyph] is usually taken to be copper. The meaning of *bi͗ȝ* is still, however, the subject of some doubt, and the question of the reading of [hieroglyph] is no nearer solution than it was a century ago. In treating the problem anew, it will first be shown that whatever the reading of [hieroglyph], the two terms have quite distinct meanings, and that there is nothing in common between the determinatives of *bi͗ȝ* and the ideograms for copper. With these facts established, the value of [hieroglyph] may then be examined without prejudice.

Before dealing with the meaning of each in detail, it may be as well to point out a number of ways in which [hieroglyph] and *bi͗ȝ* are distinguished, showing that they were regarded as two separate words, and suggesting that there was no reason (such as the reading) why they should be confused. Thus [hieroglyph] and *bi͗ȝ* are found together in lists of minerals[5], and [hieroglyph] is also found with *bi͗ȝ n pt*[6], and in the annals of Thutmose III, whereas [hieroglyph] occurs many times,

---

[1] These forms, represented in hieroglyphic fount, are most inaccurate, and comparison with the hand copies in Sethe, Pyr. reveals that they are quite distinct from the copper ideograms; cf. Wainwright, JEA. 18 p. 12.

[2] Mél. Cumont 2 p. 967. [3] Mitt. Kairo 14 p. 89.

[4] Eg. 1. Medum 13. 2. Giza H. 2 fig. 219. 3. LD. II, 49. 4. Giza J. 4 taf. IX and p. 72. 5. Urk. I, 229, 2. 6. Mem. Inst. I (3) p. 328. 7. Atlas I, 402. 8. Giza J. 9 abb. 24 and taf. 126.

[5] Sehel 16; cf. also *ṯḥst* and *bi͗ȝ*, Düm. G. I. II, 63–4; IV, 102f.; IV, 170f.

[6] Harr. 40b, 11; LD. III, 194, 10; P. Ch. B. IV vs. 2, 7; Geog. Pap. fr. 16.

objects of *bἰꜣ* are only once mentioned[1]. From such examples it would appear that the two were quite distinct, and this is borne out by other considerations. The typical use of *bἰꜣ* is for the various implements used in the 'opening of the mouth'[2], in which connection [ḥmt] is never mentioned[3], though it is applied to knives[4], and appears as the determinative of weapons and numerous other objects[5] with which *bἰꜣ* is not associated. Moreover, [ḥmt] is frequently employed for overlaying doors and the like[6], for which *bἰꜣ* is not used, and is adopted as a standard of value[7], while *bἰꜣ* never is. Scenes of metalworking involving [ḥmt] are quite common[8], whereas no evidence for the metallurgy of *bἰꜣ* is forthcoming from any source, and, lastly, the different varieties of [ḥmt] in no way coincide with those of *bἰꜣ*. Thus [ḥmt] *ḥr ḫꜣst.f*, [ḥmt] *stt*, [ḥmt] *km*, [ḥmt] *rwḏ*, [ḥmt] *m smꜣw nt 6* and *ḫꜣw nw* [ḥmt] are all well known[9], while no similar expression involving *bἰꜣ* can be quoted, and, on the other hand, *bἰꜣ n pt*[10] is not confused with any species of [ḥmt].

Indeed, there is only one point at which [ḥmt] and *bἰꜣ* seem to coincide at all, namely in the figurative expressions discussed by Grapow, though not even in all of these. *ḏw n bἰꜣ*[11] is parallelled by *ḏw n* [ḥmt][12], and *sbty n bἰꜣ*[13] or *bἰꜣ n pt*[14] by *sbty n* [ḥmt][15], Demotic *sbt* [ḥmt][16] and Coptic ⲤⲞⲂⲦ Ⲛ̄ϨⲞⲘⲚⲦ[17], but the common phrase found in the closing formula of many Late Egyptian letters seems invariably to contain [ḥmt] and never *bἰꜣ*[18].

Mention must here be made also of two possible instances of *bἰꜣ* and [ḥmt] in parallel texts, though neither is sufficiently clear to be of any real importance. In one Ptolemaic text[19] hieroglyphic *bἰꜣ* is allegedly parallel to Demotic *ḥmt* (sic.), though the latter is in fact in lacuna and the reading is restored, and in an earlier inscription[20] *bἰꜣ* is replaced by [hieroglyph], which may well have been [hieroglyph], though [hieroglyph] or [hieroglyph] are both possible. In view of the extreme uncertainty of these two examples, no conclusion can be based on them, though their existence cannot be ignored completely.

That [ḥmt] refers to copper and copper ore there can be no doubt. The group regularly follows the noble metals, *ḥsbḏ*, and *mfkꜣt* in the lists of minerals, only giving place to *ḥsmn* or [ḥmt] *km*, and is clearly the most common of the base metals. It is the material for tools and weapons[21], vessels of various kinds[22], statues[23] and numerous other objects[24], and is widely used for covering doors and the like[25]. [hieroglyph] is used as a determinative for objects of various types, and for the word *ḥsmn*, 'bronze', and [ḥmt] is transferred as a term for

---

[1] Urk. IV, 638; 733 (the same).
[2] Cf. below p. 58.
[3] Cf. however Theb. Grab. 121 ⟨785⟩.
[4] Cf. below n. 21.
[5] Cf. below n. 21–4.
[6] Cf. below n. 25.
[7] Cf. below p. 56 n. 3.
[8] Cf. below p. 56 n. 4.
[9] Cf. below p. 56f.
[10] Cf. below p. 59.
[11] Harr. 77, 7; 58, 10; LD. III, 211; Med. Hab. 44; Champ. Mon. IV, 332.
[12] LD. III, 130b, 17; Kadesh 10; Israel. 3; Abyd. Mar. II, 4, 9; P. Tur. PR. 137, 6.
[13] Urk. II, 36; Abyd. Mar. I, 52, 17; Thes. 630; cf. ib. 628.
[14] LD. III, 187e, 7; Urk. IV, 1233; Kadesh 125.
[15] Not. Descr. II, 76; cf. Urk. IV, 1087; Naukratis 2.
[16] Petubastis 12, 13; P. Insinger 11, 15.
[17] Jer. 1, 18; 15, 20; Mem. Inst. IV p. 821.
[18] P. Bol. 1094, 1, 5; 1, 6; 1, 10; 4, 3; 4, 8; Anast. VI, 53; VI, 64; Lansing vs. 10; Sall. I, 4, 6; cf. P. Tur. PR. 134, 13; 135, 1 etc.
[19] Philensis II, 11.
[20] Marriage Stela: Abu Simbel 33 = Amara x + 12 [6].
[21] Cairo 28083; 28092; Karn. Mar. 54, 58; P. B. M. 10054 rt. 2, 10; Inscr. Hier. Char. 18, 5631, 4 etc.
[22] Harr. 33b, 1; Medum 13; Urk. IV, 707; 686; Nastesen 35; 38; LD. III, 258ab, 14 etc.
[23] Harr. 73, 13; Goshen 5, 4; Dend. Mar. III, 9a.
[24] Eg. Urk. III, 123–4; IV, 640; Harr. 77, 8; P. Berl. 10629, 4; P. B. M. 10068 rt. 2, 6 etc.
[25] Urk. IV, 53; 387; 437; Abyd. Mar. I, 50a, 16; II, 11a; Rec. Trav. 22, 118–9; Med. Hab. ⟨482⟩ etc.

(copper) objects[1], and particularly (copper) weapons[2], recalling the duality of the word *ds*. Moreover 𓈓𓈒𓏥 is also used as a standard of value, and in some instances has the meaning of money or valuables[3].

But more positive evidence for the meaning copper are the cases where 𓈓𓈒𓏥 occurs in the captions to scenes of metalworking[4], and the occasional references to metallurgical processes, since only copper and the noble metals were worked to any great extent. Thus, the melting of copper is referred to by the expressions *wdḥ* 𓈓𓈒𓏥[5], and 𓈓𓈒𓏥 *m gꜣš*[6], the latter having a unique determinative showing a crucible pouring molten metal, and 𓈓𓈒𓏥 *stfw*[7] is probably also molten copper or copper that has been purified in some way. There are, too, references to copper metallurgy in the Smith papyrus, where the term 𓈓𓈒𓏥 *n ꜥḏnt* occurs[8], which Breasted[9] thought might refer to molten copper, and where the convolutions of the brain are compared to the slag which forms on the surface of this copper[10]. Furthermore, the ideograms 𓈓, 𓈔 and 𓈕 are themselves suggestive of copper smelting, the first representing a crucible[11], and not an ingot of metal as Gardiner maintains, the second either a crucible or a furnace with crucibles, and the third also a crucible of a different form[12].

Certain expressions involving 𓈓𓈒𓏥 which occur in the medical papyri are also suggestive of copper, in particular *ẖꜣw nw* 𓈓𓈒𓏥 (Wb. III. 361.9–10), which is almost certainly the Greek λεπίς χαλκοῦ, Latin aeris squama[13], a common medicament consisting of the small flakes which are hammered off in forging ingots of copper. Whether the Egyptian term also covers the classical ἄνθος χαλκοῦ, flos aeris, which forms on the surface of rapidly cooling copper, it is impossible to say, though it seems unlikely. Other copper compounds mentioned in medical contexts include *wꜣḏ* 𓈓𓇅𓇅𓈒𓏥[14] and *ḏꜥbt nt* 𓈓𓈒𓏥[15], both probably forms of verdigris, and *ḥs* 𓈓𓈒𓏥[16], the exact meaning of which is uncertain.

Various different species of 𓈓𓈒𓏥 are specifically referred to, though not all can be identified with certainly. 𓈓𓈒𓏥 *ḥr ḫꜣst.f*[17] is clearly 'copper from the copper-country' as realised by Brugsch[18], Gensler[19] and others[20], and 𓈓𓈒𓏥 *n ḫꜣst*[21] simply 'copper of the hills or desert'. 𓈓𓈒𓏥 *m smꜣw nt 6*, discussed in detail below (p. 65) is evidently an alloy of a golden colour, and can only be interpreted as a species of bronze or the like. 𓈒𓏥 *mꜣ n bꜣt*, and 𓈒𓏥 *psḏ n ỉhwiw*, both ἅπαξ in the same text[22], cannot satisfactorily be identified, since the literal translations 'new copper' and 'shining copper' are of little significance. Both were evidently particular varieties of copper or copper ore obtained from different sources, and it is just possible that the latter may have been a kind of bronze or a copper ore containing tin.

---

[1] Harr. A 1, 4; P. Wien 6, 2, 10; Inscr. Hier. Char. 18, 5631, 9.

[2] Hor. & Seth 9, 2; 9, 6; 9, 7; 11, 6; 13, 10; Orb. 12, 5; Apoph. 31, 1; Sall. IV, 3, 3; P. Tur. PR. 120, 11 etc.

[3] Urk. IV, 1064; P. B. M. 10403, 2, 2; 3, 16; 10053 rt. 5, 7; 10068 vs. 1, 4; 1, 9.

[4] Cf. LD. II, 49–54; Beni Has. II, 6–7; D. el Geb. I, 13–14; Giza Wb. ⟨1⟩ etc.; cf. Mitt. Kairo 14 p. 90–1.

[5] P. Ch. B. III, 8, 19a.

[6] Abydos (Calv.) IV, 62d; cf. Wb. V. 156.5–6.

[7] Urk. IV, 708; cf. Wb. IV. 342.5–6 & 9.

[8] Smith 2, 20; 2, 25; 4, 10.

[9] Pap. Smith p. 167.

[10] Smith 2, 20f.

[11] Correctly recognised by Champollion, Birch, Gensler, Chabas, Strachmans, Weill, and Junker.

[12] For crucibles cf. Forbes, Metallurgy p. 132; p. 339–40.

[13] Cf. Chassinat, P. Méd. Copte p. 128; Ebbell, Pap. Ebers p. 133; Lüring, Med. Kent. p. 96.

[14] Eb. 59, 12; 59, 16; 79, 3.

[15] Eb. 106, 2.

[16] Eb. 85, 20, literally 'copper excrement', cf. Grapow, Bild. Ausdr. p. 129.

[17] Urk. IV, 718; 719; 733; 1322; 1571; Hier. Ostr. 75, 3; Totb. Leps. 145, 73; Anast. IV, 17, 7.

[18] Wb. p. 1629.

[19] ZÄS. 8 p. 142.

[20] Cf. Bissing, Stat. Taf. p. 24, who points out that *ḥr ḫꜣst.f* means 'aus seinem Berg' and not 'in der Scholle'.

[21] Piehl II Inscr., 95.

[22] JEA. 4 pl. 9.

stt, literally 'Asiatic copper', described by LEPSIUS[1] as 'une variété à laquelle on attachait un grand prix', and regarded by BRUGSCH[2] as synonymous with *ṯḥst* as a term for pure copper, was apparently a species of copper or copper alloy of a characteristically light colour[3]. It was chiefly employed for covering doors etc.[4] and also for door hinges[5] the latter perhaps suggesting a particularly hard copper or even bronze. On the other hand, *sṯt* follows *ḥsmn* in lists of minerals[6] so that it cannot be ordinary bronze, though WAINWRIGHT[7] may be right in suggesting that it was bronze smelted from a natural mixture of copper ore and tin ore as opposed to that made by melting together metallic copper and tin. Copper and tin ores are found together near Byblos, which would account for the name 'Asiatic' copper, though *sṯt* might also refer to Sinai, and in later texts is probably an indication of quality rather than provenance. In default of further evidence, no certain identification is possible, but it seems likely that *sṯt* was a copper alloy akin to bronze, and WAINWRIGHT's interpretation may well be correct.

km, literally 'black' or 'dark' copper, was taken by LEPSIUS[8] to be copper that had been smelted and purified, the equivalent of Greek χαλκὸς μέλας, and modern 'black copper' that has undergone a first fusion. BRUGSCH[9], however, thought it was a term for pure copper, like *sṯt*, and SPIEGELBERG[10] regarded it as a name for iron, citing a Mexican parallel. GSELL[11] rightly opposed this view, and preferred to interpret *km* as impure copper. Although it is true that among certain primitive peoples iron is called black copper, this cannot be the case with *km*, which is still found in texts of Graeco-Roman date[12], when there can be no question of iron being mistaken for copper, and which moreover, is used for purposes such as the covering of doors etc.[13] to which iron is quite unsuited.

Black copper is copper that has been smelted from its ore, but which has not yet been purified and still therefore contains traces of gold, silver and other metals. It was probably in this form that much of the copper was traded in the ancient world, and it is significant that *km* was brought as tribute from Syria[14], Keftiu[15] and elsewhere. The term first occurs in the New Kingdom[16], when the different stages in the production of copper from its ores were evidently recognised, and in the enumeration of minerals is usually found before [17], *ḥr ḫ3st.f*[18], *rwḏ*[19], and even *ḥsmn*[20], though the last frequently precedes it[21]. This would seem to indicate that *km* was quite highly esteemed[22], perhaps because it was more durable than pure copper, which would also account for its use in overlaying doors and the like. Although the identification of *km* as black copper cannot be proved absolutely, it seems reasonable to suppose that such was in fact the meaning of the term, the only alternative being to assume a particular species of dark copper or copper alloy of which nothing is now known.

---

1 Métaux p. 51. 2 Wb. Erg. p. 417.

3 Urk. IV, 1150; cf. Birch, ZÄS. 15 p. 33; Lepsius, ZÄS. 10 p. 113.

4 Urk. IV, 56; 766; Karn. Mar. 42, 11; Düm. Baug. 16; LD. III, 152c; Thes. 1375; Petos. 61, 18; 62, 4; 81, 50 etc. For a dubious early example cf. Urk. I, 294, 11.

5 Koptos 20a, 19; Ann. Serv. 18, 145, 1.70. Lepsius, ZÄS. 10 p. 113, thought this might indicate iron.

6 Urk. IV, 817; 818; 821. 7 Antiquity 17 p. 96; cf. also Schaeffer, JEA. 31 p. 92f.

8 Métaux p. 48. 9 Wb. Erg. p. 418.

10 Rec. Trav. 26 p. 166. 11 Eisen, Kupfer und Bronze p. 50.

12 Eg. Dend. Mar. III, 42b; IV, 36; Düm. Res. 39; Baug. 29.

13 Urk. IV, 168; 174; 422; cf. B. M. 523; 524, where obelisks are capped with it.

14 Urk. IV, 744. 15 Urk. IV, 1098. 16 Earliest ref. apparently Urk. IV, 98.

17 Urk. IV, 744; Harr. 21b, 14; 62a, 15; the reverse Harr. 52a, 4. 18 Urk. IV, 1571.

19 Karn. R. III pl. 90. 20 Urk. IV, 744; Karn. I, 27 = Urk. IV, 1668; Urk. IV, 1729.

21 Urk. IV, 1571; cf. Rec. Trav. 29, 162.

22 Cf. the expression *km m3ʿ*, Urk. IV, 168, which might suggest that it was counterfeited.

rwḏ seems to be known from only four examples, one dating back to the Old Kingdom[1], the others occurring in texts of the New Kingdom[2]. The reading of the term, and its possible relationship with Coptic ⲂⲀⲢⲰⲦ are of considerable importance in determining the value of , and will be discussed more fully below. The meaning alone is here in question, and this must be literally 'hard copper', *rwḏ* being undoubtedly the adjective 'hard', and not derived from the verb *rd* 'grow' as Sethe[3] suggested in the case of Coptic -ⲢⲰⲦ. To what precise species of copper or copper alloy the term refers it is difficult to determine, though since it was hard it was evidently not pure copper. It may therefore have been a variety produced from ores containing tin or other impurities, and thus somewhat similar to bronze, or, like *km*, a species of copper which was only partially refined. On the other hand. it is just possible that *rwḏ* was copper artificially hardened by hammering[4], though this is perhaps unlikely, and can scarcely be applied to the one early example. In fact, no very satisfactory conclusion is possible, and the non-committal 'hard copper' is to be preferred as a translation.

The very fact that so many different varieties of are mentioned is in itself conclusive proof of the meaning copper, since only the metallurgy of the noble metals and copper was sufficiently advanced, even in the New Kingdom, for many separate kinds to be recognised.

In contrast to , *biꜣ* has a very limited range of application, nor does its meaning appear to be the same in all instances. For the sake of convenience, references to *biꜣ* as the material of the two blocks used in 'opening the mouth' and of the *ḥnt* vessels which occur in offering lists are treated separately in another chapter[5], together with certain other cases in which the word seems to have a fairly wide significance. The present discussion will deal only with those cases in which *biꜣ* must have a more precise meaning. Thus *biꜣ* is the material for the *nwꜣ*[6], *msḫtyw*[7] and *mḏtft*[8] used in 'opening the mouth', and for the spear of Horus and other weapons mentioned in religious contexts[9]; it also occurs in lists of minerals[10], where a specific material must be indicated, and reliefs and the like are said to be *ḫt m biꜣ*[11]. Of interest, too, are the unique vessels of *biꜣ* acquired by Thutmose III[12]. Other references to *biꜣ*, including fingers of *biꜣ*[13], *biꜣ* which opens the mouth[14], *biꜣ* which comes from Seth[15], ropes of *biꜣ* by which the ascent to heaven was made[16], and *biꜣ* as the material of walls[17], the heavenly throne[18] and the like[19] are of doubtful interpretation and probably refer to meteoric material in general[20].

There can be little doubt that in the case of the *nwꜣ*, *msḫtyw*, and *mḏtft* implements, *biꜣ* represents iron, since, as was pointed out by Dévéria, Chabas and Wainwright, similar

---

[1] JEA. 4 pl. 9, written .
[2] Karn. Mar. 34, 6 = Urk. IV, 1732; LD. III, 200d, 14; Karn. R. III pl. 90.
[3] ZÄS. 53 p. 51 = JEA. 1 p. 234.
[4] Cf. Lucas, Materials p. 246.
[5] Cf. below p. 166.
[6] Pyr. 13c; Leid. K. 15; Lib. Fun. taf. 57b, 6; 63.
[7] Pyr. 13c; 14a; Düm. G. I. I, 81, 20; Lib. Fun. taf. 57b, 6; 58; 63.
[8] Totb. Nav. 23, 5; Amenemhet pl. 17; p. 59; Lib. Fun. taf. 60; 70; I p. 127; 128; 129; JEA. 17, 31 etc.
[9] Totb. Nav. 108, 7; Apoph. 22, 9; 22, 10; Urk. VI, 59, 2; Dend. Mar. IV, 85; ZÄS. 59 pl. 47; 48 etc.
[10] Sehel 16; Düm. G. I. II, 63–4; IV, 102f.; IV, 170f.; cf. Urk. VIII, 56, *biꜣ m ḏt. f.*
[11] Edfou VI, 203, 4–5; I, 553; Dend. Chass. IV, 159; Dend. Mar. III, 30; Thes. 266, 14.
[12] Urk. IV, 638; 733.
[13] Pyr. 1983e.
[14] Pyr. 14; Petos. 82, 71.
[15] Pyr. 14a; Lib. Fun. taf. 58.
[16] Pyr. 138c; Coff. Texts I, 271c; Cairo 28092; Rec. Trav. 29, 149; Text. Relig. 22.
[17] Totb. Nav. 109, 7; 149, 7.
[18] Pyr. 1562b–c; 1992c; cf. Totb. 153A, Nu 20, 13.
[19] Dend. Mar. II, 48; 49; Düm. Baug. 7 etc.
[20] Cf. below p. 166 and JEA. 18 p. 3f.

implements are known which are either wholly of iron or have blades of that material. Moreover, references to reliefs carved with *bỉꜣ* are consistent with the use of iron tools in the Graeco-Roman period, and it is probable that the spear of Horus and other similar weapons were thought to be of iron, particularly in later times when copper had been superseded by iron for implements of all kinds[1]. The vessels of *bỉꜣ* brought from *Tinꜣy* by Thutmose III are less certainly iron, and the yellow colour is strange[2], though they were evidently metal, since the handles were of silver, presumably welded on[3]. It may perhaps be significant, as Forbes observes[4], that the temple treasure of Mishrifé-Qatna included objects of iron, for although the vessels are unlikely to have come from that very source, they were in all probability looted from the treasure of some town on the confines of Asia, where iron objects were more common than in Egypt.

But the most conclusive proof that *bỉꜣ*, in some instances at least, refers to iron, is the replacement of *bỉꜣ* by *bỉꜣ n pt*, Coptic ⲂⲈⲚⲒⲠⲈ, iron, in certain texts relating to the 'opening of the mouth'[5]. That *bỉꜣ n pt*, which first occurs in the 19th dynasty, designates iron has never seriously been doubted, except by Lepsius[6] who thought it was not a metal at all. Brugsch[7] noted the connection with Coptic ⲂⲈⲚⲒⲠⲈ and took it to be meteoric iron and later iron in general, contrasting it with *bỉꜣ n tꜣ*, while Dévéria[8] translated 'substance du ciel' and thought it might be either meteoric iron or magnetic iron or both. Daressy[9] rendered 'métal du ciel' and favoured iron, and Spiegelberg[10] curiously regarded *bỉꜣ n pt* not as meteoric iron, but as 'heaven-iron', an unspecified variety contrasted with 'earth-iron'. Gsell[11] realised that the name arose from the relationship of iron to meteorites, and remarked upon the identity of *bỉꜣ* and *bỉꜣ n pt*, and Ball[12] also took the expression as proving that metallic iron was first obtained from meteorites, though the name was later transferred to iron from terrestrial sources; a similar view was held by Sayce[13]. Wainwright[14] noted the replacement of *bỉꜣ* by *bỉꜣ n pt*, and observed that although the latter meant 'iron from heaven', its late appearance and its application to terrestrial iron showed that it had only indirect reference to the meteoric origin of early iron.

It is quite clear that *bỉꜣ n pt* is the origin of Coptic ⲂⲈⲚⲒⲠⲈ and that the meaning is iron, and the term may be compared to the Sumerian AN.BAR[15] which has a similar significance. *bỉꜣ n pt*, as already mentioned, first appears in the 19th dynasty, at a time when the Egyptians were beginning to import manufactured iron, and is used exactly as *bỉꜣ*, not only in reference to the *mḏtft* chisels[16], but also in the metaphorical *sbty n bỉꜣ*[17]. Other mentions of *bỉꜣ n pt* are not numerous, but include a few cases in which it is contrasted with 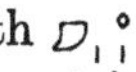[18], being once described as *bỉꜣ n pt n Rṯnw*[19], which indicates the Asiatic origin of most iron in the New Kingdom. In Harris[20] 6784 statuettes of the Nile god are said to be of *bỉꜣ n pt*, and in the Berlin medical papyrus[21] *bỉꜣ n pt* crushed in water is prescribed, which Gsell took to refer

[1] Cf. however, Wainwright's suggestion that the weapons of Horus and Seth were thunderbolts.
[2] Urk. IV, 638.
[3] Urk. IV, 733.
[4] Metallurgy. p. 453.
[5] Eg. LD. Text III, 301; P. Tur. PR. 118, 3; Pap. Ani Budge pl. 15 1.4.
[6] Métaux p. 55, on the grounds of the writing of LD. III, 194, 10.
[7] Wb. p. 1722–3; Erg. p. 413.
[8] Mél. d'Arch. 1 p. 1.
[9] Ann. Serv. 22 p. 165.
[10] Rec. Trav. 26 p. 166.
[11] Eisen, Kupfer und Bronze p. 47.
[12] Cairo Sc. Journ. 8 p. 204.
[13] Antiquity 2 p. 226.
[14] JEA. 18 p. 3; Antiquity 10 p. 5.
[15] Campbell Thompson, Dictionary p. 80; Wainwright, Cairo Sc. Journ. 8 p. 177.
[16] Eg. LD. Text III, 301; P. Tur. PR. 118, 3; Pap. Ani Budge pl. 15 1.4.
[17] Urk. II, 36; Abyd. Mar. I, 52, 17; *sbty n bỉꜣ n pt* LD. III, 187e, 7; Kadesh 125.
[18] P. Ch. B. IV vs. 2, 7; LD. III, 194, 10; Geog. Pap. fr. 16.
[19] P. Ch. B. IV vs. 2, 7.
[20] Harr. 40b, 11.
[21] P. Med. Berl. 7, 12.

to 'eisenrosthaltiges Wasser'. Lastly it must be noted that in two inscriptions at Hammamat[1] there occurs a title *iry n bỉꜣ n pt*[2], which the Wörterbuch translates 'Verfertiger eiserner Werkzeuge', in all probability correctly.

Although *bỉꜣ n pt* means literally 'iron from heaven', and should logically refer to meteoric iron alone, it is clear that it has a wider significance and includes all iron. This is shown by, for example, the reference to *bỉꜣ n pt n Rṯnw*, which can only be imported man-made iron, and the problem then arises why this term alluding to the meteoric origin of iron was introduced at a time when man-made iron was first being used to any great extent. The most plausible explanation seems to be that with the greater knowledge and use of iron, the older term *bỉꜣ*, referring not only to iron but also to 'meteoric' material in general[3] was felt to be inadequate, and a specific word for iron as opposed to the other materials embraced by *bỉꜣ* was required. *bỉꜣ n pt* was therefore created, probably on the model of Sumerian AN.BAR, or because all iron, whether meteoric or not, was traditionally connected with the vault of heaven.

The only other term which requires comment is *bỉꜣ n tꜣ*, known from only one, or at most two, examples[4], the significance of which is by no means clear. It may be no more than an artificial designation balancing *bỉꜣ n pt*, though it is possible that it has a specific meaning and refers to telluric iron which occurs in basalt and other rocks as grains or nodules[5]. Dümichen's ingenious suggestion[6] that [hieroglyph] stood for *rd* and that *bỉꜣ n rd* was the origin of Coptic ⲂⲀⲢⲰⲦ need hardly be considered.

From an examination of the different meanings of [hieroglyph] and *bỉꜣ* and their compounds, it becomes quite clear that, save in a very few figurative uses, the two are in no way parallel, and this is borne out by the evidence of the writings. The determinatives of *bỉꜣ* are quite distinct from the ideograms for copper, as was pointed out by Wainwright, and Weill's attempt to identify certain of them is specious[7]. Efforts have been made to explain [hieroglyph], [hieroglyph], [hieroglyph], [hieroglyph], [hieroglyph], [hieroglyph], [hieroglyph] and the other less common forms on the assumption that *bỉꜣ* was copper, but Wainwright's view that the majority are indicative of the meteoric origin of *bỉꜣ* is more convincing[8]. The ideograms [hieroglyph], [hieroglyph], [hieroglyph], [hieroglyph], etc. are, on the other hand, quite clearly derived from copper metallurgy, the first being perhaps a lump of ore, the second and third crucibles, and the fourth a furnace.

In no instance is any of the common determinatives of *bỉꜣ* used to write the word for copper, nor is any of the earlier copper ideograms applied as a determinative to *bỉꜣ*. The only apparent confusion arises in the Graeco-Roman period, when the sign [hieroglyph] is employed both as an ideogram for copper and as a determinative for *bỉꜣ*. But it is clearly a purely coventional sign, perhaps in the former case representing a variant of [hieroglyph] such as [hieroglyph], [hieroglyph], or [hieroglyph][9], and in the latter standing for [hieroglyph] or [hieroglyph] (= [hieroglyph]), and it is doubtful how much real importance can be attached to this apparent inconsistency.

Having thus established that there was no confusion in either the meaning or the orthography of [hieroglyph] and *bỉꜣ*, the problem of the reading of the former may now be considered.

---

[1] Hamm. 54; 57. [2] *bỉꜣ n pt* written [hieroglyph].

[3] Cf. below p. 166 for the probable range of *bỉꜣ*.

[4] Thes. 1375; Lib. Fun. I, 127 [hieroglyph] probably for [hieroglyph].

[5] Cf. Forbes, Metallurgy p. 380; Lucas, Materials p. 268.

[6] ZÄS. 10 p. 105.

[7] Thus, for example, Weill claims that [hieroglyph] and [hieroglyph] are identical, though the first, a det. of *bỉꜣ*, is in fact [hieroglyph], while the second, an ideogram for copper, is [hieroglyph]. His alleged [hieroglyph] as det. of *bỉꜣ* is, moreover, a deformed [hieroglyph], ie. [hieroglyph]. These forms should have been drawn out and not represented in hieroglyphic fount!

[8] JEA. 18 p. 12. [9] Note also an early example of [hieroglyph], Dec. Roy. pl. III, 2 and p. 29.

The word is invariably written with the ideogram in one form or another[1], even in Demotic, when the sign for copper is a transcription of [hieroglyphs], and there is therefore no direct evidence for the reading, which is deduced almost entirely from Coptic. [hieroglyphs] is undoubtedly masculine, as is shown by Late Egyptian examples with the article *pꜣ*[2], and by the expression [hieroglyphs] *ḥr ḫꜣst.f*, but this is not in itself significant, since whether the word is *biꜣ* (Coptic ⲂⲀ-) or *ḥmt* (sic.) (Coptic ϨⲞⲘⲚⲦ), the gender is the same.

The reading *biꜣ* is based entirely upon the supposed identity of [hieroglyphs] *rwḏ* and Coptic ⲂⲀⲢⲰⲦ, copper, the meaning of which has long been recognised, though its derivation has been variously explained. Gensler[3] took it to be ⲂⲀ-ⲢⲰⲞⲨⲦ, literally 'das Grünmetall' in reference to the patina of copper, and Dümichen too evidently thought it was *biꜣ n r(w)d*[4]. Lepsius, who distinguished ⲂⲀⲢⲰⲦ, copper, from ϨⲞⲘⲚⲦ 'Erz', and noted the term ϨⲞⲘⲚⲦ ⲚⲂⲀⲢⲰⲦ, χαλκολίβανος, presumed that the ⲂⲀ- was *biꜣ*[5], but later suggested[6] that ⲂⲀⲢⲰⲦ was a place-name, ie. Beirut. Sethe, who proposed the identification of ⲂⲀ- with [hieroglyphs][7], held that ⲂⲀⲢⲰⲦ meant 'grown metal' and referred to copper which came from the hills as opposed to iron which fell from the sky. The identification of ⲂⲀⲢⲰⲦ with [hieroglyphs] *rwḏ*[8], however, is only possible if ⲢⲰⲦ- means 'hard', since there can be no doubt as to the significance of *rwḏ* in the few hieroglyphic examples. If this is so, then the Coptic word for copper in general is derived from an Egyptian term meaning hard copper and referring to a particular, and rare, species, which seems prima facie unlikely. Nor is it any more satisfactory to assume that [hieroglyphs] means 'ore' in this case and that ⲂⲀⲢⲰⲦ is therefore 'hard ore', since this is scarcely a suitable name for copper. In fact, Sethe's interpretation of ⲂⲀⲢⲰⲦ as 'grown metal' seems the most likely derivation of the term, and unless, therefore, a hieroglyphic example which is certainly [hieroglyphs] *rd* can be found, identification of the two is improbable. Moreover, it may be observed that neither ⲂⲀⲤⲚϬ nor ⲂⲀⲤⲞⲨⲢ can plausibly be explained on the assumption that ⲂⲀ- is [hieroglyphs], and some other origin must be sought for all three terms.

Against the reading of [hieroglyphs] as *biꜣ* is also the fact that, apart from a single late example of [hieroglyphs] for *bit*, 'character'[9], none of the words of the *bi-* or *biꜣ-* roots is ever written with the [hieroglyphs] ideogram, and it is impossible to believe that some such confusion would not have arisen had [hieroglyphs] had a similar value. Thus, although *nbw*, like [hieroglyphs], is regularly written with an ideogram, this ideogram is not confined to *nbw* alone, but occurs in the writings of other words of which *nb-* is the first syllable. The only logical conclusion is, therefore, that [hieroglyphs] was not transferred to the writings of other words of the *biꜣ-* root simply because its value was not *biꜣ*[10].

Coptic ϨⲞⲘⲚⲦ, ϨⲞⲘⲦ̄, was derived by Sethe[11] from *ḥmt*, 'Handwerk', and connected with [hieroglyphs], though such a derivation cannot satisfactorily be explained, nor is there any justification for reading [hieroglyphs] as *biꜣ* in [hieroglyphs], and as *ḥm(w)* in [hieroglyphs] as it implies. Moreover, ϨⲞⲘⲚⲦ cannot have originated in an Egyptian *ḥmt*, since it has a double closed vowel at the end, and retains its Ⲧ, which presupposes an original having two syllables, and

---

[1] Including: [hieroglyphs], etc.

[2] Eg. P. B. M. 10403, 2, 2; 3, 16; 10053 rt. 5, 7; 10068 vs. 1, 4; 1, 9; Hor. & Seth 9, 6; 9, 7; 11, 6; P. Tur. unpublished etc. An apparent example of *tꜣ*, P. Berl. 10655, 4, is in fact *dbn*, which is found in a number of abbreviated forms.

[3] ZÄS. 8 p. 137.

[4] ZÄS. 10 p. 105.

[5] Métaux p. 47.

[6] ZÄS. 10 p. 116.

[7] ZÄS. 53 p. 51 = JEA. 1 p. 234.

[8] Cf. Gardiner, JEA. 4 p. 36; Grammar p. 478.

[9] Ann. Serv. 18, 144, 1.59.

[10] It must be admitted that [hieroglyphs] does not in fact occur in the writing of any other word except [hieroglyphs], but if, as will be suggested, the reading is *ḥmty* and not *ḥmt*, there are no other words which could have been written with it.

[11] ZÄS. 53 p. 51.

with its 'final' *t* protected. The most likely form of the Egyptian word from which ϩⲟⲙⲛ̄ⲧ was derived is therefore *ḥmty* (*ḥmti*) or *ḥmtw*[1], and if ϩⲟⲙⲛ̄ⲧ is to be connected with , the latter must be read *ḥmty* and not *ḥmt*. That *ḥmty* is in fact the reading of cannot be proved directly, though it is strongly suggested by three passages from Edfu, where alliteration is clearly intended[2], and the presence of the radical *t* may conceivably be inferred from a few writings [3]. Such a reading is not, moreover, inconsistent with anything that is known about , and only conflicts with the alleged identity of *rwḏ* and ⲃⲁⲣⲱⲧ.

If then is read *ḥmty* or the like, and not *bἰ3*[4], the distinction between and *bἰ3* already observed in meaning and orthography is borne out, and no possible reason remains for supposing that the two words were in any way connected. *ḥmty* it is clear refers only to copper and copper ores, and *bἰ3* to iron, meteoric material in general, and probably some other substances thought to be meteoric[5]. That *bἰ3* has any connection with 'mines' or 'things mined' is unlikely, since the similar *bἰ3* (Wb. I. 438.12–13) is more easily explained as a name for Sinai than as a generic term for 'mining district' or the like[6]. Finally it must be noted that the translation of *bἰ3* as iron in some instances is justified by a few early examples of the use of the metal[7], but should not be taken to indicate that iron was employed to any great extent before a comparatively late date.

There now remain to be discussed some few terms representing other base metals, bronze, lead and the like, which do not form a related group and are therefore treated separately in alphabetical order.

ἰn3wḳ (no Wb. ref.) is apparently confined to one example in an inscription from Kawa[8], where Macadam[9] is no doubt right in regarding it as a mineral from its position between *ḥsbḏ* and *mfk3t*. The strange orthography might suggest a vox peregrina, and an identification with Akkadian *anâku*, as proposed by Macadam, is tempting. The latter is interpreted by Campbell Thompson[10] as 'tin, sometimes lead', the mixing of *erû*, copper, and *anâku* to make bronze suggesting tin, while cognate languages, Hebrew *ᵃnak*, Syriac *ankâ*, Persian *anâka*, and Arabic *ankᵘⁿ* point to lead[11]. Tin is perhaps more appropriate in the present context, since the term for lead, *ḏḥty*, was common enough, and lead would hardly be likely to separate *ḥsbḏ* and *mfk3t* in an enumeration of materials. On the other hand, tin would perhaps be sufficiently rare to do this, and its silver colour might explain its usurping the

---

[1] There are other possibilities, since ϩ could originate in *h* as well as *ḥ*, and ⲧ in either *d* or *ṯ*.

[2] *ḥms.n.ἰ Ḥmty m* (Edfou IV, 343, 8–9).

n msnty (*ἰ3wty*) *ḥms.n.f Ḥmty* (Edfou VI, 238, 15–239, 1).

msnty (*ἰ3wty*) *ḥw Ḥmty m* (Edfou VI, 239, 7).

I am grateful to Prof. Fairman for these references.

[3] Totb. Leps. 145, 73; P. Bol. 1094, 1, 5; 1, 6; 1, 10; the *t* in each case small and in ligature. Cf. also two writings of *km* with *kmt*, Harr. 62b, 16; P. Ch. B. V rt. 8, 13) which might suggest a *t* in .

[4] To assume early *bἰ3* and later *ḥmty* is an unhappy compromise, which begs the question of how, when, and why the value of the ideogram altered.

[5] Cf. below p. 166.

[6] Cf. Černý, Sinai 2 p. 1f.

[7] Cf. Lucas, Materials p. 270.

[8] Kawa III, 9.

[9] Kawa I p. 11.

[10] Dictionary p. 121.

[11] Cf. also Müller, OLZ. 2 p. 294.

position of *ḥḏ*[1]. Moreover, it seems likely that  represents *mꜣꜥ* as Macadam suggests[2], and if so, 'genuine tin' would make better sense than 'genuine lead'. However, no certain conclusion is possible, though the equation with *anâku* must surely be accepted.

ꜥgt (Wb. I. 235.6) seems to occur only twice, in the same inscription, among tribute from Assur. The first context[3] refers to 'a chariot worked in gold, its *ḏbw* of gold', and 'a chariot worked in gold, its *ḏbw* of *ꜥgt*', which would suggest that *ꜥgt* was a precious material, perhaps a metal, and possibly akin to gold. In the second passage[4], *ꜥgt* and *ḫt ꜥgt* are simply included in a list of materials, and the meaning is not apparent, though *ḫt ꜥgt* might point to something with which wood was covered[5]. Various translations have been proposed, Chabas[6] and later Jéquier[7] regarding *ꜥgt* as a kind of metal, Brugsch[8] translating 'Nagel', and Maspero[9] favouring 'émail', while the Wörterbuch has 'ein kostbares Holz'. Any connection with *ꜥgyt* (Wb. I. 235.7), *Mentha piperita*, or *ꜥgt* (Wb. I. 235.5) is altogether unlikely[10], and it seems probable that *ꜥgt* is in fact a foreign product, and perhaps even a loan-word. On the slight evidence, a species of metal or alloy appears suitable, and a specific alloy or gold imitation of Asiatic origin may perhaps be assumed, though some form of inlay (?niello) is also possible.

ḥsmn (Wb. III. 163.14–24), read *men* by most early Egyptologists, was first discussed by Lepsius[11] who noted that the word occurred only in 'classical' Egyptian, and not in the Graeco-Roman period, when it was apparently replaced by *ṯḥst*. The determinative suggested a translation copper or bronze, but Lepsius preferred to render iron, regarding the widespread use of iron as certain, and rejecting bronze because no people in antiquity separated bronze and copper. Dümichen[12] also favoured iron as a translation, though denying any connection between *ḥsmn* and *ṯḥst*, and quoted a late text where *ḥsmn ḥr nbw* was apparently parallel to *biꜣ pt ḥr nbw*. Chabas[13], on the other hand, pointed to the close similarity between the uses of *ḥsmn* and *ḥmty*, and concluded that the former meant bronze rather than iron, and Gsell[14] also took *ḥsmn* to be bronze, while regarding *ḥmty* as brass. The reading *ḥsmn* was established by Erman[15], who showed that the value of the sign  is certain from the various writings of the word *ḥsmn*, natron, in Ebers. He distinguished three words *ḥsmn*, quoting Brugsch's[16] comparison of *ḥsmn*, amethyst, with the Hebrew for ἤλεκτρον, and concluding that *ḥsmn* was evidently a bright shining metal, and hence not iron, Dümichen's supposed parallel between *ḥsmn* and *biꜣ pt* being invalid, because *ḥsmn* was similarly parallel to *ḥmty* in the same text. Moreover, Piehl's observation[17] of the white and red colour of objects of *ḥsmn* was against the equation with iron, and the nature of the objects themselves suggested bronze rather than iron. Since Erman's article, *ḥsmn* has generally been translated bronze, eg. by the Wörterbuch, Gardiner[18] and Bissing[19], and there can scarcely be any doubt that this is correct.

---

[1] Cf. Admon. 3, 2; Kamose 13–14; Hier. Ostr. 95, 1 rt. 3.

[2] Wb. has no example of the writing, though it does occur once, cf. below p. 176 n. 3.

[3] Urk. IV, 669.

[4] Urk. IV, 670.

[5] Cf. Wb. III. 339.13.

[6] Voyage p. 274; p. 366.

[7] BIFAO. 19 p. 64.

[8] Wb. p. 225.

[9] Rec. Trav. 2 p. 150; cf. also Paton, Records III, 2 p. 76 n. 180.

[10] Cf. Keimer, Gartenpflanzen p. 138; Jéquier, loc. cit.

[11] Métaux p. 52.

[12] ZÄS. 11 p. 46.

[13] Études[2] p. 40.

[14] Eisen, Kupfer und Bronze p. 44; p. 51.

[15] ZÄS. 30 p. 31.

[16] Wb. Erg. p. 853. Brugsch read  as *men*, Wb. Erg. p. 598.

[17] ZÄS. 28 p. 19.

[18] Grammar p. 582.

[19] Stat. Taf. p. 11.

In the majority of instances where *ḥsmn* occurs in a list of metals and semi-precious stones, it precedes *ḥmty*[1], though sometimes following it[2], and also being mentioned between *ḥmty km* and simple *ḥmty*[3]. *ḥsmn* is frequently determined with [hieroglyph][4], and once with [hieroglyph][5], which suggests a connection with copper, and in two medical contexts[6], *ḥsmn ḫ3w* seems to replace the well-known *ḫ3w nw ḥmty* (Wb. III. 361.9–10), which is perhaps significant. The colour of *ḥsmn* was apparently chestnut-brown, 'marron'[7], though the metal written [hieroglyphs] on Middle Kingdom coffins, and taken as *ḥsmn* by the Wörterbuch, may well be *ḥmty*[8].

Of the various objects said to be made of *ḥsmn*, the most significant are the weapons of different kinds[9], and the numerous vessels[10], for both of which bronze is the most appropriate material[11]. Other uses of *ḥsmn* are many and varied, and include overlay for doors etc.[12], statues[13], offering tables[14], mirrors[15], and miscellaneous objects[16]. The majority of references to *ḥsmn* are of New Kingdom date, when bronze was first used to any great extent, and the variety of the objects stated to be made of *ḥsmn* precludes any other metal except copper, which is, however, *ḥmty*. As bronze, *ḥsmn* would naturally take precedence over *ḥmty*, copper, and would be the principal material for weapons, tools etc.

Before the introduction of the term *ḥsmn*, bronze was in all probability confused with copper and called *ḥmty*, since until the Middle Kingdom it was almost unknown[17], and even then was by no means common, so that only with the great increase in the use of bronze from the 18th dynasty onwards did a specific name become necessary[18]. Although, however, the relatively late appearance of the word *ḥsmn* can thus be accounted for, its virtual absence from texts of the Graeco-Roman period[19] requires some explanation. That *ṯḥst* is a late equivalent of *ḥmty* or *ḥmty sṯt* rather than *ḥsmn* seems certain[20], though in some instances both *ṯḥst* and *ḥmty sṯt* appear to refer to objects which would naturally be of bronze[21]. It seems, therefore, possible that in later texts *ḥmty* includes both copper and bronze, and that *ḥmty sṯt* and *ṯḥst* may refer particularly to the latter, though why the more precise word *ḥsmn* was abandoned it is difficult to understand.

ḫwḏtf (no Wb. ref.) seems to be known from only two examples, both of Graeco-Roman date[22]. Both Dümichen[23] and Brugsch[24] regarded it as a term for copper, translating literally 'glänzend ist sein Leib oder seine Gestalt', and 'dessen Körper glänzt'. The former noted

---

[1] Eg. Urk. IV, 173; 817; 821; Cairo 34183, 18 etc., before *ḥmty km* Urk. IV, 1213; 1571; 1729.
[2] Urk. IV, 1046; Karn. Mar. 42, 16.
[3] Urk. IV, 744.
[4] Urk. IV, 817; 821; 1046; Karn. Mar. 42, 16; Kenam. 17; Pianchi 57; P. Mayer B 11; 12 etc.
[5] Cairo 20539 IIb 8.
[6] P. Ram. V lower 22; lower 71.
[7] Cairo 28024; 28027.
[8] Cf. Medum 13; Ransom-Williams, Per-Nēb p. 53.
[9] Urk. IV, 664; 718; 719; 726; 727; 891; 1344; Kenam. 17; Kamose 13–14 etc.
[10] Cairo 28024; 28027; Urk. IV, 667; 686; 1046; Amada 15; Amada Wien 141 etc.
[11] Cf. in particular Urk. IV, 686, where vessels of *ḥmty* and *ḥsmn* are mentioned together.
[12] Urk. IV, 170; 423; 705.
[13] Urk. IV, 1047; Harr. 41a, 11.
[14] Urk. IV, 817; Sonnenheiligtum III, 28.
[15] Cairo 28024; P. B. M. 10053 rt. 4, 13.
[16] Urk. IV, 639; P. B. M. 10068 rt. 2, 19; 2, 23; 2, 39; Kenam. 17; P. Mayer B 11; 12; Harr. A. I, 15; 17; Sall. 3, 2; P. Tur. PR. 102, 2, 5–8.
[17] Cf. Lucas, Materials p. 252–3.
[18] Apparent examples on Middle Kingdom sarcophagi, written [hieroglyphs] are dubious, since [hieroglyphs] may well be *ḥmty*.
[19] One or two exceptions, eg. Thes. 18–23.
[20] Cf. below p. 66.
[21] Cf. above p. 57; below p. 66.
[22] Düm. H. I. II, 50, 13 = Edfou VI, 202, 7; Düm. G. I. II, 77, 4 = Dend. Chass. IV, 159, 14.
[23] ZÄS. 10 p. 101.
[24] Wb. Erg. p. 904.

that in the Dendera text there is no mention of either copper or bronze among the minerals listed, and so thought that *ḫwḏtf* must refer to one or the other, the  determinative being also in favour of such an interpretation. In support of this he further cited a passage in another late text[1] where *ṯḫst* is described as *ḫwt* (sic.) *m ḏt.f*.

Now this comparison is based on the incorrect assumption that *ꜣḫ, ỉꜣḫ*, (Wb. I. 33.6) is to be read *ḫw*[2], and is therefore invalid. Moreover, the literal translations of Brugsch and Dümichen also presuppose the existence of a verb *ḫw*, 'glänzen', which does not in fact exist. Indeed, the element *ḫw* in the word *ḫwḏtf* can only be derived from *ḫwỉ* (Wb. III. 244. 10f.), 'schützen', and the name must therefore have some such meaning as 'protective of body'.

However, although the translation 'copper' proposed by Dümichen was based on a number of false premises, it is not necessarily incorrect. The use of *ḫwḏtf* for overlaying *dbḥt* shows that it was a metal, and the  determinative in the Dendera text strongly suggests copper or bronze. On the other hand, the Edfu text refers to it as *ḫwḏtf ḥr ḏꜥtt*, and if *ḏꜥtt* is the same word which occurs in connection with *nbw* and *ḏꜥm*[3], a species of gold might seem likely. No certain conclusion is, however, possible, and it may be that *ḫwḏtf* represents some obscure alloy, though the form of the name would perhaps imply a sportive term with no specific meaning.

smꜣw nt 6 (Wb. III. 450.16–17) is clearly a copper compound of some description, as is shown by two instances where it is called *ḥmty m smꜣw nt 6*[4]. Gardiner[5] takes *smꜣw nt 6* to be an alloy of six ingredients, a species of bronze of the colour of gold, probably somewhat like brass, while Caminos[6] simply translates 'six-fold alloy'. In fact, precise identification of this compound is scarcely possible, though it is quite clearly an alloy with a copper base, and probably a particular species of bronze. That it is common bronze is altogether unlikely from the relative rarity of the term[7], and because Egyptian bronze was simply copper and tin with accidental impurities and did not have six ingredients. Moreover, the golden colour of *smꜣw nt 6*[8] suggests a very characteristic metal perhaps akin to brass.

Brass itself may almost be ruled out for texts of New Kingdom date, since there is no evidence for it in Egypt before classical times[9]. However, it is not impossible that ores containing both copper and zinc may have been employed in making a kind of brassy bronze, to which other constituents may have been added. Alternatively, *smꜣw nt 6* could represent a copper-gold alloy[10], or something of the same nature as the classical orichalc[11], though if 'six-fold alloy' is correct, it was probably considerably more complicated than either. In default of more exact knowledge of the details of ancient metallurgy, it is not possible to advance a definite interpretation of *smꜣw nt 6*, and the most that can be said is that it was a specific compound of copper or bronze.

ṯḫst (Wb. V. 396.8–10) is virtually confined to texts of the Graeco-Roman period[12], where it is clearly a term for either copper or bronze. Brugsch[13] translated *ṯḫst* 'Erz', and regarded

[1] Düm. G. I. II, 74, 10.
[2] Cf. Brugsch, Wb. Erg. p. 903.
[3] Cf. below p. 207.
[4] Koller 1, 7; Harr. 45, 5; cf. also Harr. 52b, 10; 6, 9.
[5] Hier. Texts I p. 37 n. 24.
[6] LEM. p. 215.
[7] Harr. 1, 7; 6, 9; 45, 5; 47, 4; 52b, 10; P. Tur. PR. 32, 6; 7; 9; Koller 1, 7; Anast. IV, 16, 12.
[8] Harr. 6, 9 *m ỉwn nbw*; 47, 4 *m ỉwn n nbw nfr n ḫꜣst*.
[9] Lucas, Materials p. 256.
[10] Cf. ib. p. 263.
[11] Cf. C. R. 69, V no. 1, March 1955 p. 21; Forbes, Metallurgy p. 286; Isidorus 16, 20, 3.
[12] An earlier example is Amenemope 18, 8, 22nd or 26th dyn.; cf. also *ṯḫty* below.
[13] Wb. p. 1591.

in Ptolemaic texts as a variant writing, while LEPSIUS[1] took *ṯḥst* to be a late substitute for *ḥsmn*, which might be copper or bronze, but which he preferred to interpret as iron. DÜMICHEN[2] noted LEPSIUS' conclusions, but maintained that *ṯḥst* was merely a late variant for *ḥmty*, since it never occurs together with *ḥmty*, and cannot, on the other hand, be found used in the same way as *ḥsmn*. But *ṯḥst* does replace *ḥmty* in some contexts, and an expression *ṯḥst nfr n sṯt* parallel to *ḥmty sṯt* is known. Furthermore, the absence of any mention of copper from Ptolemaic lists would be strange. LEPSIUS[3] was convinced by DÜMICHEN's arguments, and rejected his original theory in favour of regarding *ṯḥst* as an alternative word for copper or bronze. CHABAS[4] also took *ṯḥst* as a variant of *ḥmty*, but translated bronze, and BRUGSCH[5] modified his first view, accepting *ṯḥst* as the equivalent of *ḥmty sṯt*. GSELL[6] likewise saw *ṯḥst* as a type of copper or bronze.

Now it is fairly clear that *ṯḥst* does in fact replace *ḥmty* or *ḥmty sṯt* in a number of instances, for example in a passage quoted by DÜMICHEN, where *ṯḥst nfr n sṯt* is used for covering wooden doors[7], and in one of the tribute tables[8], where *ṯḥst* and *ḥmty sṯt* seem to be variants. Moreover, the frequent use of *ṯḥst* for overlay[9], and its occurrence in the various tribute tables at Dendera and Philae[10], where no other word for copper is found, would suggest a term for copper and bronze, and the latter is more particularly implied by the use of *ṯḥst* for bolts[11], and its description as *ꜣḫ m ḏt.f*[12]. It seems likely, therefore, that in the Ptolemaic period *ḥmty* had a dual meaning, copper and bronze, and that *ḥmty sṯt* and *ṯḥst* while perhaps referring more specifically to the latter, could be applied equally to either.

That is to be read *ṯḥst* in texts of the Graeco-Roman period as BRUGSCH suggested, is out of the question, since there is evidence of alliteration to prove the reading *ḥmty* at this time[13]. It is therefore clear that the older term *ḥmty* and the compound *ḥmty sṯt* continued in use after the introduction of *ṯḥst*, and this can be parallelled by the occurrence of *nbw*, *sꜣwy* and *rwy*, and *ḥḏ*, *ʿrḳwr* and *r-ḥmt*. But the origin of *ṯḥst*, unlike that of some other late words, is not easy to discover.

ṯḥty (Wb. V. 396.11) is apparently ἅπαξ as the material for the *ḳrty* of a door[14]. It is evidently a metal, and although at first sight an identification with *ḏḥty* might seem likely, *ḳrty* of lead would be quite useless, and the WÖRTERBUCH suggestion that *ṯḥty* in fact stands for *ṯḥst* is altogether more satisfactory. On the other hand, *ṯḥty* may be a distinct and otherwise unknown word, and no definite conclusion is possible.

ḏḥ (Wb. V. 605.6) is a somewhat rare metal, occurring in only a few instances, and clearly to be distinguished from *ḏḥty*, lead. The word was not noted by LEPSIUS, but was recognised by BIRCH[15], who translated 'tin', and this was quoted by GLADSTONE[16], who, however, added that there seemed to be some doubt as to the metal meant. BRUGSCH[17] discussed *ḏḥ* in some

---

[1] Métaux p. 52. [2] ZÄS. 10 p. 101. [3] ZÄS. 10 p. 114.
[4] Etudes² p. 42. [5] Wb. Erg. p. 417. [6] Eisen, Kupfer und Bronze p. 44; p. 52.
[7] Düm. T. I. I, 111, 1–2. [8] Düm. G. I. II, 66–7.
[9] Düm. T. I. I, 111, 1–2; 6; Thes. 1375 etc.
[10] Dend. Mar. I, 70; I, 71; II, 12; Düm. G. I. II, 63–4; 66–7; IV, 170f.
[11] Düm. Baug. 16; Edfou VII, 19, 10.
[12] Düm. G. I. II, 74, 10. [13] Cf. above p. 62. [14] Urk. III, 68.
[15] In Wilkinson, Manners and Customs 2 p. 232; cf. also Hier. Pap. p. 12.
[16] PSBA. 14 p. 226.
[17] ZÄS. 30 p. 110; cf. Wb. Erg. p. 1385, where he has 'vielleicht Zinn'.

detail, referring to the bars of *ḏḥ* shown in the tomb of Rekhmirēʿ[1], and coloured 'bleu ardoisé' according to Champollion[2], and to the contrast between *ḏḥ* and *ḏḥty* in Harris, where Birch rendered 'tin' and 'lead'. Brugsch then quoted a small figure of a ram, seen with a dealer, which was made of magnetic iron, and bore an inscription, suggesting that it was to this dark slate-coloured mineral that *ḏḥ* referred. This identification was rejected by Müller[3] on the grounds that a distinction between magnetic iron and iron would not be significant, and latterly, Caminos[4] has stressed the distinction between *ḏḥ* and *ḏḥty*, without, however, offering any translation of the former.

The textual evidence is not very helpful, particularly since it is hardly possible in some instances to determine whether a writing such as *ḏḥw*[5] is for *ḏḥ* or *ḏḥty*. The only certain examples of *ḏḥ* are three in Harris[6] where it follows *ḏḥty*, and the label *ḏḥ* attached to the two slate-coloured bars in the tomb of Rekhmirēʿ[7], to which may perhaps be added the one quoted by Brugsch. Apart from these, what may be *ḏḥ* occurs in the Chester Beatty mineral list[8], and in two other accounts of materials[9], and an object of *ḏḥꜣw* (sic.) is mentioned with other articles of metal[10]. From these few references, all that can be deduced is that *ḏḥ* was a metal, obtained from Retenu[11], transported in bars or ingots[12], of less value than lead[13], and apparently blue-grey in colour[14].

Birch's translation 'tin', accepted by Wainwright[15], though attractive, can hardly be supported. Despite the fact that objects of tin are known from the New Kingdom[16], it is unlikely that the metal would be imported in bars at that date, or that it would be regarded as less valuable than lead, and the blue-grey colour given to *ḏḥ* is inappropriate to tin. Moreover, it is probable that the Egyptians, like the Assyrians[17], confused lead and tin, and it is significant that Coptic ⲐⲢⲀⲚ, Demotic *trn*[18], seems to be a new word. In addition, *ỉnꜣwḳ*[19], and possibly *ḏḥty ḥḏ*[20] are apparently used to refer to tin, which also suggests that no more specific word was known.

Brugsch's 'Magneteisen' is equally difficult, even assuming that his interpretation of the text on the ram is correct[21]. Magnetic iron and magnetite are rare in Egypt, and the importation of either in bars is most unlikely. Furthermore, magnetite is in all probability referred to as *bỉꜣ ḳsy*, and this term may also include magnetic iron, distinction between the two being improbable before a relatively late date.

It therefore seems that *ḏḥ* is to be regarded either as an unknown metal, or as some kind of ore, perhaps an oxide of iron, lead, manganese or tin, though the evidence is insufficient to warrant any more precise identification.

ḏḥty (Wb. V. 606.4–8). Demotic *thth*[22], Coptic ⲦⲀϨⲦ, was recognised as lead by Brugsch[23] and Lepsius[24], and the translation was accepted by Gensler[25], Gladstone[26] and the majority of later scholars without question. There can in fact be no possible doubt that *ḏḥty* refers

---

[1] Urk. IV, 1101. [2] Not. Descr. I p. 509. [3] OLZ. 2 p. 295.
[4] LEM. p. 218. [5] Anast. IV, 17, 7; P. Ch. B. IV vs. 7, 6; P. Tur. 1881 vs. col. I, 6. [7]
[6] Harr. 21b, 14; 40b, 14; 68a, 11. [7] Urk. IV, 1101. [8] P. Ch. B. IV vs. 7, 6.
[9] Anast. IV, 17, 7; P. Tur. 1881 vs. col. I, 6 [7]. Cf. however, ib. rt. col. III, 9, which is *ḏḥty* [8].
[10] Ostr. Nash 19, 3, 2 [9]. [11] Urk. IV, 1101. [12] Urk. IV, 1101.
[13] Harr. 21b, 14; 40b, 14; 68a, 11. [14] Urk. IV, 1101; cf. ZÄS. 30 p. 110.
[15] Antiquity 18 p. 62 n. 33. [16] Lucas, Materials p. 286.
[17] Cf. Campbell Thompson, Dictionary p. 116; p. 121; *anâku*.
[18] P. Dem. Mag. vs. 10, 2–3. [19] Kawa III, 9. [20] P. Med. Lond. 6, 7.
[21] may well be *ḏḥty*, the colour of the object having resulted in its being mistaken for lead.
[22] P. Insinger 23, 16. [23] Wb. p. 1656; cf. ZÄS. 30 p. 110. [24] Métaux p. 58.
[25] ZÄS. 8 p. 148. [26] PSBA. 14 p. 226.

to lead, the only question being whether or not it also embraces tin. MÜLLER[1] thought that such was the case, and interpreted *ḏḥty* as 'das dem *ḏḥt* gleiche' ie. like tin, assuming that the word referred to both lead and tin. Similarly, WAINWRIGHT[2], noting the importance of tin in the bronze age, and the absence of any Egyptian name for it, suggested that *ḏḥty*, primarily lead, may also have included tin.

The problem is incapable of definite solution, but comparative evidence would strongly suggest that lead and tin were confused. CAMPBELL THOMPSON[3] notes that lead, tin and antimony were often mixed in antiquity, and that Akkadian *anâku* refers to both lead and tin, and although it is clear that classical authors generally distinguish between the two, PLINY[4] and ISIDORUS[5] both refer to lead as plumbum nigrum, and tin as plumbum candidum. The assumption that *ḏḥ* is tin has been shown above to be false, and in fact the only terms which may refer specifically to that metal are *ḏḥty ḥḏ*[6] and *ỉnꜣwḳ*[7], though the former may be ceruse.

The textual evidence for *ḏḥty* is equally applicable to lead or tin, though the majority of instances in which it is found in enumerations of metals[8] must surely refer to lead, and the occurrence of *ḏḥty* in the name of a lead-mine[9] is self-explanatory. Moreover, ingots of *ḏḥty*[10] are more likely to be lead, and the context in Amenemope[11] where *ḏḥty* is used to express worthlessness can hardly refer to tin. On the other hand, the passage in the Rhind mathematical papyrus[12] where *ḏḥty* is half as valuable as silver, and a quarter as valuable as gold, if it is to be taken literally, may suggest that *ḏḥty* also means tin, since lead would scarcely have this value.

On the whole, it seems likely that *ḏḥty* originally meant lead, but that when tin was first introduced into Egypt, it was also loosely referred to as *ḏḥty*, or more specifically *ḏḥty ḥḏ*, only in later times being known as *trn*, a word which does not in fact occur in any hieroglyphic text. However, in most cases *ḏḥty* is still to be translated lead, since tin can only have been relatively rare.

Two possible examples of *ḏḥty* incorrectly written must finally be noted, namely the *dḥwy* of Ebers[13] (Wb. V. 480.10), and the *ṯḥ* of the Sehel inscription[14]. The former is taken as *ḏḥty* by WRESZINSKI and EBBELL[15], though STERN and LÜRING[16] identify it with the plant *ṯḥwy* (Wb. V. 323.1–4), which may be correct. The latter, which follows *nbw*, *ḥḏ* and *ḥmty*, is read *dḥt* by BRUGSCH[17] and translated accordingly, and although the reading is certainly *ṯḥ*, not *dḥt*, the context seems to point to an identification with *ḏḥty* and the translation, accepted by BARGUET[18], may therefore be accurate. The additional possibility that certain apparent writings of *ḏḥ* are in fact *ḏḥty* has already been mentioned, though in no case is there anything to suggest which is the correct interpretation.

The various metals discussed above are so diverse in nature that no summary of the evidence is possible, but it may be observed that with the exception of one or two rarely used metals, such as antimony and platinum, which may not have been recognised as such, all the metals employed by the ancient Egyptians can be identified lexicographically. On the other hand, there does appear to be textual evidence for certain alloys which have not as yet been confirmed archaeologically.

[1] OLZ. 2 p. 295. [2] JEA. 20 p. 32; cf. however, Antiquity 18 p. 57f., particularly p. 62 n. 33.
[3] Dictionary p. 116; p. 121. [4] HN. 34, 47, 156. [5] 16, 22, 1. [6] P. Med. Lond. 6, 7.
[7] Kawa III, 9. [8] Urk. IV, 706; 708; 718; 724; 744; Harr. 21b, 14; Cairo Wb. 375 etc.
[9] Ann. Serv. 54, 179–80. [10] Urk. IV, 706; 708; 718; 724; Düm. H. I. I, 34.
[11] Amenemope 18, 3. [12] Rhind 62 – though the values may be hypothetical.
[13] Eb. 63, 10. [14] Sehel 29. [15] Pap. Ebers p. 76.
[16] Med. Kent. p. 115. [17] Hungersnoth pl. 28. [18] Stèle de la Famine p. 31.

CHAPTER 3

BUILDING AND MONUMENTAL STONES

Limestone, sandstone, granite, quartzite, alabaster and *bḫn*, the six principal building and monumental stones used by the ancient Egyptians, have already been dealt with in considerable detail by SETHE[1], to whose conclusions, at least as regards the first three, little of consequence can be added. But no discussion of Egyptian minerals would be complete without some mention of these stones, which are therefore treated in the present chapter.

Limestone

The simplest and most general name for limestone, probably, as SETHE suggests[2], in distinction to granite, is *inr ḥḏ* (Wb. I. 97.12), 'white stone', which is used at an early date with reference to the limestone of Ṭurah[3], the classic source of the material in the Old Kingdom. But the term is only rarely found in this elementary form[4], and is more frequently qualified in some way, particularly in the New Kingdom, when *inr ḥḏ* also forms the basis of the name for sandstone, *inr ḥḏ (nfr) n rwḏt*. Thus *inr ḥḏ nfr* occurs[5], without any distinction of meaning[6], and *inr ḥḏ n r-ꜣw*[7], *inr nfr n r-ꜣw*[8] and *inr ḥḏ nfr n r-ꜣw*[9] are used of the limestone from Ṭurah, though without always that precise significance. Furthermore, a small limestone temple at Semneh is said to be of *inr ḥḏ nfr n tꜣ-sty*[10], which in all probability refers to limestone from the neighbourhood of Aswan.

But the most common designation of limestone, specifically that from the Ṭurah-Maʿṣarah quarries, though also used more generally, is *inr ḥḏ nfr n ʿin* (Wb. I. 97.12; I. 191.2–3). This expression is of very frequent occurrence both fully written[11], and with either *nfr*[12] or *ḥḏ*[13] or even both[14] omitted. It is the material of temples[15], tombs[16], offering tables[17], door-posts[18] etc., and is quite often applied by mistake to buildings of sandstone[19]. That the presence or absence of either of the adjectives has any real significance is altogether unlikely, and such variations can be parallelled exactly in the case of *inr ḥḏ nfr n rwḏt*.

---

[1] Bau- und Denkmalsteine. [2] Bau- und Denkmalsteine p. 7. [3] Urk. I, 99.
[4] Sin. B 196; Dend. Mar. III, 38; LD. III, 3a, b; 71a, b; P. Ch. B. IV vs. 7, 6; Hier. Ostr. 22, 1 rt. 6.
[5] Wien 150; Gr. Oase 10; Karn ⟨1034⟩.
[6] Daressy, Rec. Trav. 10 p. 143, translates *inr ḥḏ* as 'calcaire', and *inr ḥḏ nfr* as 'calcaire compact', though without justification.
[7] Louvre C 3. [8] Dend. Mar. II, 60d; I, 64b. [9] Dend. Mar. I, 59b.
[10] Urk. IV, 197; 212; 213; cf. 819 (Wadi Halfa).
[11] Eg. Urk. IV, 25; 855; 858; 863; D. el B. 134; Petos. 61, 18; 61, 25; 62, 4; 81, 49; Ann. Serv. 18, 145, 70; 22, 147, 18 etc.
[12] Urk. IV, 388; Memphis V, 79–80. [13] Dend. Mar. I, 64a; II, 60; Beni Has. I, 26.
[14] Cairo 20539; Lansing 12, 2; Harr. 8, 8; 57, 11; 57, 13; 58, 6.
[15] Urk. IV, 25; 388; D. el B. 134; Abyd. Mar. II, 3; Cairo 20539; Petos. 61, 18 etc.
[16] Louvre Apisstela 193. [17] D. el B. 8. [18] Lansing 12, 2.
[19] Urk. IV, 607; 855; 858; 863; Dend. Mar. I, 64a; cf. the similarly mistaken use of *inr ḥḏ nfr n r-ꜣw*, Dend. Mar. I, 64b etc.

Concerning the two chief sources of limestone, *r-ꜣw* and *ꜥin*, something must be said. The first is clearly the name of the Ṭurah quarries, and probably has some such literal meaning as 'die langgestreckte Mündung' suggested by SETHE[1], referring to the halls and galleries from which the stone was extracted. The form *tꜣ r-ꜣw*, which occurs once[2], was evidently the origin of the Greek Τροία, and the Arabic Ṭurah, and the identification can scarcely be doubted[3].

ꜥin, read *ꜥnw* by the WÖRTERBUCH (Wb. I. 191.1–3), presents a somewhat more difficult problem, and is discussed at length by SETHE[4], who prefers to read *ꜥin* by comparison with the related verb *ꜥin*[5] 'mit Kalkstein bekleiden', and the *ꜥyn* of Harris (Wb. I. 191.4–5), which must be the same word. He points out that *ꜥin* was regularly written with an eye, and not infrequently with the eye alone[6], and concludes that the Egyptians originally had a word for eye identical with the Semitic *ꜥain*, but that this no longer existed in historical times. Moreover, this ancient word, like its Semitic counterpart, also had a secondary meaning 'spring', represented by the oval ⬭ which encloses the eye, which is a pointer to the location of the region *ꜥin*. Springs are not very common in Egypt, but there are springs at Helwan, at the northern end of the Mokaṭṭam range, and at ʿAin es Ṣîra between Mokaṭṭam and Old Cairo. *ꜥin* was, furthermore, a district within the region 𓁹𓏏𓊖, though not identical with it as BRUGSCH[7] and ERMAN[8] supposed, and therefore lay to the south of Gebel Aḥmar, in the southern part of the ancient region of Heliopolis. It was also connected with *r-ꜣw*[9] the Ṭurah quarries, and was in all probability the name for the whole area of sulphur springs south of Heliopolis.

GARDINER[10] also deals with *r-ꜣw* and *ꜥin* in some detail, referring to SETHE's article, and pointing out that the latter is only rarely[11] determined as a place-name, unless this is implicit in the writing 𓁹 as SETHE supposes. He notes that the connection between 𓁹𓏏𓊖 and 𓁹𓏏𓊖 is based on a faulty copy[12], though admitting that a new reading of another passage[13] might suggest a relationship. Concerning the location of *ꜥin* he concludes that is must be 'either an earlier synonym of *r-ꜣw* or else a rather more extended term for *r-ꜣw* and its neighbourhood'.

Without more definite evidence it is impossible to place *ꜥin* with any greater accuracy, but since limestone came principally from the Ṭurah-Maʿṣarah quarries, more specifically referred to as *r-ꜣw*, it would appear likely that *ꜥin* was either a more ancient name for the same quarries, or a term for the area as a whole, perhaps including the region of the sulphur springs as SETHE suggests. That the reading was originally *ꜥin* or *ꜥinw*, and not *ꜥnw* can scarcely be doubted in view of the arguments advanced by SETHE.

In the majority of examples, *ꜥin* occurs in the phrase *inr ḥḏ nfr n ꜥin*, where SETHE takes it to be a place name, though as GARDINER points out, the Egyptians may have ceased to attach any very precise meaning to it. The oldest writing is 𓁹[14], later examples having

[1] Bau- und Denkmalsteine p. 6. [2] Harr. 37b, 3; cf. also Onom. 2 p. 127.

[3] Cf. Gardiner, Onom. 2 p. 126, for occurrences of the name *r-ꜣw* at Ṭurah and Maʿṣarah.

[4] Bau- und Denkmalsteine p. 7; cf. also Vycichl, OLZ. 52 p. 393f.

[5] Urk. I, 20.

[6] LD. III, 71a, b; Urk. IV, 43; Avignon 30 (Rec. Trav. 35, 204); Louvre Apisstela 5 (Rec. Trav. 22, 20); Louvre Apisstela 193; Urk. I, 272; Memphis V, 79–80; Not. Descr. II, 732 etc.

[7] Dict. géog. p. 974; p. 1390.

[8] Sitz. Berl. Ak. 1911 p. 931. The two cannot be identical; the reading of the former is *šns*.

[9] Cf. Mar. Mon. Div. 59; LD. III, 257a = Hamm. 93.

[10] Onom. 2 p. 128; p. 278. [11] Mar. Mon. Div. 59; Berl. 1416; Hamm. 93.

[12] Brugsch, Dict. géog. p. 1390 = Edfou IV, 39.

[13] Urk. I, 20 = Giza H. IV, 168, 𓁹 for ⬭. [14] Urk. I, 272; cf. Sahurēʿ I, 88–9.

[1], [2], and [3], and only a few cases being without the oval[4]. Alphabetic signs are frequently absent[5], or are written as [6], [7], [8], [9], or [10]. But although it is normally used in this way, *ʿin* does occasionally occur on its own as a word for limestone[11], which might suggest that it had lost its primary significance as a place-name and had become, like *rwḏt*, a designation of the stone itself.

This is particularly likely in Harris, where a term *ʿyn, inr n ʿyn* (Wb. I. 191.4–5) occurs[12], which can only be a variant for the more usual *ʿin*, and is thus interpreted by Wendel[13], Sethe[14], Burchardt[15] and others. In no case is there any suggestion that *ʿyn* is other than the actual name of the stone, and it therefore seems reasonable to suppose that by the New Kingdom, when *inr n ʿin* and *inr n rwḏt* were parallel terms for limestone and sandstone, the original meaning of *ʿin* had been forgotten.

That all the terms mentioned above refer to limestone has never been doubted, and the direct evidence from inscriptions in the Ṭurah-Maʿṣarah quarries and from buildings of limestone is considerable. It has, however, been noted that in some few cases, sandstone is incorrectly described as *inr ḥḏ nfr n ʿin*, and it is moreover possible that at an early date other white stones such as dolomite, marble or gypsum were confused with limestone and mistakenly referred to in this way. However, no definite proof of this is forthcoming, and the point cannot be pressed.

Sandstone

The regular name for sandstone, formed, as Sethe points out[16], on the same pattern as that for limestone, was *inr ḥḏ nfr n rwḏt* (Wb. I. 97.13; II. 413.2), though either *nfr*[17] or *ḥḏ*[18] or both[19] may be omitted without affecting the meaning[20], and *mnḫ* occasionally replaces *nfr*[21]. Moreover, in course of time, the genitival *n rwḏt* became shortened to *rwḏt*[22], and the latter then lost its feminine ending, being interpreted as an adjective qualifying *inr*, so that forms such as *inr nfr rwḏ*[23] and *inr ḥḏ nfr rwḏ*[24] occur. Finally, *rwḏt* alone was in some instances used to refer to sandstone[25], though it is more frequently a quite general term for hard stone, as pointed out above p. 23).

---

[1] Urk. IV, 25; Abyd. Mar. II, 3; Dend. Mar. I, 64a; LD. Text I, 20; LD. III, 71a, b.
[2] Urk. IV, 43; 214; 294; 295; 607; Rec. Trav. 22, 21; Ann. Serv. 18, 207 etc.
[3] Cf. Sethe, Bau- und Denkmalsteine p. 8.
[4] Urk. IV, 388; Beni Has. I, 26; Lansing 12, 2; Memphis V, 79–80, 15; Nauri 10; Dend. Zim. 13 [10].
[5] Eg. Urk. I, 272; IV, 43; LD. III, 71a, b; Rec. Trav. 22, 21 etc.
[6] Abyd. Mar. II, 3; Dend. Mar. I, 64a; Lansing 12, 2.
[7] Urk. IV, 294; 295. [8] Urk. IV, 388. [9] Beni Has. I, 26; Nauri 10.
[10] Urk. IV, 25; 214; 607; LD. Text I, 20; Ann. Serv. 18, 207; cf. Karn. Temp. A. II. [11].
[11] Not. Descr. II, 732; Karn. Temp. A. II. *ʿin ḥḏ nfr n rwḏt* (sic.).
[12] Harr. 25, 12; 45, 4; 8, 8; 57, 11; 57, 13; 58, 6. [13] Bau- und Edelsteine p. 85.
[14] Bau- und Denkmalsteine p. 12. [15] Fremdworte II p. 14 no. 248.
[16] Bau- und Denkmalsteine p. 13. [17] Urk. IV, 821.
[18] LD. III, 38d; LD. Text III, 15; Text IV, 98, 7; Rec. Trav. 16, 56; Mem. Miss. XV, 19, 1; Karn. ⟨251⟩. [19] Frequent, cf. Wb. II. 413.1.
[20] Daressy, Rec. Trav. 10 p. 143 is almost certainly wrong in distinguishing between *inr n rwḏt, inr nfr n rwḏt* and *inr ḥḏ nfr n rwḏt*.
[21] Urk. IV, 883; Dend. Mar. II, 10.
[22] *inr rwḏt*: Urk. IV, 843; Karn. Mar. 34; Borch. Baug. 40. *inr nfr rwḏt*: LD. Text III, 15; Mem. Miss. XV, 19, 1. Compare Sethe, 1, 73 [12] (*inr rwḏt*) with Luxor ⟨216⟩ (*inr n rwḏt*), and Sethe 3, 76 [13] (*inr rwḏt*) with Amada 14 (*inr n rwḏt*). [23] LD. Text IV, 9.
[24] LD. Text V, 3; Piehl Inscr. II, 75; Dend. Mar. I, 6 etc.
[25] Not. Descr. II, 732; LD. Text III, 100; Amada 14.

References to sandstone are rare before the New Kingdom[1], but from then onwards, *inr ḥḏ nfr n rwḏt* and its abbreviations occur with great frequency, chiefly as indication of the material of temples and other buildings[2] or parts thereof[3]. References to other uses of sandstone are few, which corresponds to the archaeological evidence. On at least one occasion, *inr n rwḏt* is used to indicate a stone other than sandstone, namely in the apparently unique expression *inr n rwḏt nt ḏw dšr*[4], which can only refer to the red crystalline quartzite sandstone of Gebel Aḥmar, later called *biꜣt*. That the stone was named in this way may perhaps suggest that the Egyptians recognised some connection between quartzite and sandstone, though it may well have been the notion of hardness that was uppermost. Another apparent example of the use of *inr n rwḏt* with reference to a stone other than sandstone is, however, susceptible of an alternative interpretation. This is the passage[5], already mentioned above (p. 23), where *inr n rwḏt* seems to mean alabaster, though it is perhaps more likely that the phrase simply means 'a block (*inr*) of hard stone (*rwḏt*)'. A third context in which the meaning of *inr n rwḏt* is doubtful may also be quoted, the *sbꜣw m inr n rwḏt* of the Amada stela being replaced by *sbꜣw m mꜣṯ rwḏt* in the Cairo duplicate[6]. Whether this must be taken to mean that the *inr n rwḏt* of the Amada copy refers to granite it is impossible to tell, since either may be mistaken, but the possibility cannot be ignored.

Apart from the expressions incorporating the element *rwḏt*, sandstone is also known by two other terms, which may, however, be confined to the Nubian variety. The first, noted by both Wendel[7] and Sethe[8] is *inr ḥḏ nfr n šꜣꜥt*[9], used to describe the sandstone temple of Thutmose III at Kummeh, the second *bnw ḥḏ*[10], referring to a stela of Nubian sandstone from Defenneh. The existence of the former might suggest that at an early date, sandstone, used only in small quantities, was loosely termed *inr ḥḏ*, more properly the name for limestone, and this might also account for the frequent mistaken references to sandstone as *inr ḥḏ nfr n ꜥin*[11] or *inr ḥḏ nfr n r-ꜣw*[12]. But definite evidence on this point is lacking, and it is at best a mere hypothesis.

Whether, with the few possible exceptions already noted, the various names for sandstone ever embraced other stones as well it is difficult to say. It must, however, be noted that similar terms such as *rwḏt*, *inr rwḏ* and *ꜥꜣt rwḏt*, discussed in a previous chapter, undoubtedly have a quite general meaning, and it therefore seems possible that in some instances *inr n rwḏt* at least may have been applied to other stones for which a more exact designation was lacking.

Granite

The original and most general term for all kinds of granite, and more specifically the common red variety, was *mꜣṯ* (Wb. II. 34.3f.), the name being perhaps derived as Sethe suggests[13], from that of a particular type of vessel of which granite was the characteristic material. Sethe[14] discusses *mꜣṯ* at some length, pointing out that although the word was in fact masculine, the change of *ṯ* to *t* led to its gender being mistaken, as is shown by the form

[1] Eg. LD. Text III, 129 (Senwosret I).
[2] Eg. Urk. IV, 818; 879; 821; Abu Simb. ⟨176⟩; Dend. Mar. I, 59c; Karn. ⟨251⟩ etc.
[3] Eg. Amada 14; LD. IV, 4b; Piehl Inscr. II, 38; Urk. IV, 856 etc.
[4] Urk. IV, 167. [5] Bersh. I, 14. [6] Urk. IV, 1295.
[7] Bau- und Edelsteine p. 80. [8] Bau- und Denkmalsteine p. 15.
[9] LD. III, 57a; 76b. [10] Tanis II, 42.
[11] Eg. Urk. IV, 607; 855; 858; 863; Dend. Mar. I, 64a. [12] Dend. Mar. I, 64b etc.
[13] Bau- und Denkmalsteine p. 18. [14] ib. p. 17.

of the adjective in the terms *mꜣṯ rwḏt*[1] and *mꜣṯ kmt*[2], and by the occasional loss of *ṯ*[3]. *mꜣṯ* was the earliest and most simple form of the name, *inr n mꜣṯ*[4] and *inr nfr n mꜣṯ* being later formations on the pattern of the names for limestone and sandstone. Not infrequently *mꜣṯ* has an indication of provenance added, the most common being *mꜣṯ ꜣbw*, granite from Elephantine, though *mꜣṯ rwḏt nt ꜥ-rsy* is known; moreover, in the 20th dynasty the expression *inr (n) ꜣbw* is also found. At first applied to granite of whatever colour, *mꜣṯ* later came to refer particularly to pink granite as opposed to black, which was represented by *inr km*[5], or *inr km n mꜣṯ*.

References to *mꜣṯ* are very frequent, and many provide direct evidence for the identification with granite[6]. Granite was the most common material for obelisks[7], and was often used for naoi and shrines[8], statues[9], vessels[10] and other objects, as well as in building[11], and in a few instances medicinally[12].

mꜣṯ is common in all periods, but the earliest examples of the term *inr n mꜣṯ* are of Middle Kingdom date[13], and it does not become at all frequent until the New Kingdom[14], when the form *inr nfr n mꜣṯ* also occurs[15]. That either of these variants has any distinct meaning is altogether unlikely, since they are merely formed on the model of the more common *inr n rwḏt* and *inr n ꜥin*. Whether the unique *ꜥꜣt n mꜣṯ* quoted by Brugsch[16] has any particular significance may also be doubted, though it is perhaps worthy of note that *mꜣṯ* and *biꜣt* alone of the building stones are referred to as *ꜥꜣt*. *mꜣṯ ꜣbw*[17], on the other hand, may be presumed to refer specifically to granite from Aswan, the most usual source of the material, unless *ꜣbw*, like *ḥwt-nbw*[18] was little more than an indication of quality. Moreover, *mꜣṯ rwḏt nt ꜥ-rsy*[19], describing an obelisk of Hatshepsut, in all probability indicates Aswan granite, and the expression *inr n ꜣbw*, which occurs from the 20th dynasty[20], must surely refer to granite from the same source, though it may be merely a synonym for granite in general[21].

mꜣṯ rwḏt may well be 'hard granite', with reference to some particular variety, though it must be admitted that it is used in a number of instances without any apparent distinction from simple *mꜣṯ*[22], and that the phrase *mꜣṯ nb mnḫt rwḏt*[23] does not seem to imply any specific kind. The apparent occurrence of *mꜣṯ rwḏt* and *inr n rwḏt* as variants in different versions of

---

1 Wb. II. 34.15; II. 412.17.
2 Wb. II. 34.16.
3 Eg. Harr. 5, 8; 26, 7; 45, 4; P. Tur. ⟨552⟩.
4 Wb. I. 34.13–14.
5 Wb. I. 97.15; V. 123.4.
6 Eg. Luxor ⟨710⟩; ⟨740⟩ (granite obelisk of Ramses II); Louvre D 29 (naos of red granite); Urk. IV, 362; 366 (granite obelisk of Hatshepsut); Leid. Granitnaos; Rom. Obelisk 19; Düm. T. I. I, III, 3, 6 (naos of black granite) etc.
7 Urk. IV, 362; 642; Luxor ⟨710⟩; ⟨740⟩; Münch. Bekh. Rück. 5; Rom. Ob. 19; Brugsch Wb. 159 etc.
8 Louvre D 29; Düm. Baug. 16; Edfou I, 10; Leid. Granitnaos; Düm. T. I. I, III, 3, 6.
9 Six Temp. 12, 4; Luxor ⟨669⟩; Abusir Mast. D. ⟨Var. 43⟩ etc.
10 Medum 13; Urk. IV, 23.
11 Urk. I, 107; IV, 295; Harr. 45, 4; LD. III, 210c; Lebens. 61; D. el B. 120; Abyd. Mar. II, 2c; Not. Descr. II, 150 etc.
12 Eb. 60, 11; 60, 13; 87, 20; Hearst 4, 11.
13 Ann. Serv. 24, 67–8; Lebens. 60–1.
14 Harr. 5, 8; 26, 7; Nauri 10; Abyd. Mar. II, 3, 4; II, 2c; Münch. Bekh. Rück. 5; Dend. Mar. IV, 35 etc.
15 Harr. 5, 10; LD. III, 210d.
16 Wb. p. 159.
17 Urk. IV, 843; Six Temp. 12, 4; Borch. Baug. 40; Rec. Trav. 20, 40, 4; P. Ch. B. IV vs. 7, 7 (*inr mꜣṯ n ꜣbw*).
18 Cf. Sethe, Bau- und Denkmalsteine p. 23.
19 Urk. IV, 362.
20 Harr. 45, 4; 47, 6; P. B. M. 10053 vs. 2, 15 (*inr ꜣbw*) etc.
21 NB. Sethe's interpretation of a passage in the Sehel, inscription (Sehel 14) as an identification of *mꜣṯ* and *inr ꜣbw* is invalid; cf. Barguet, Stèle de la Famine p. 22 n. 12.
22 Eg. Urk. IV, 366; 737; 849; 932; Cairo 34019, 16; cf. also Urk. IV, 362.
23 Urk. IV, 1046.

the same text[1] has already been mentioned in connection with sandstone, and although no very satisfactory explanation is forthcoming, it may be suggested that the adjective *rwḏt* was repeated automatically to indicate characteristic hardness.

mꜣt kmt, undoubtedly a specific term for black granite, seems to occur only once, as the material for a vessel[2], and cannot be regarded as a regular expression. The normal term for black granite in the New Kingdom is *ỉnr km*, which first appears in the reign of Hatshepsut[3], and is subsequently found in lists in contrast to *mꜣt*[4], red granite. References to *ỉnr km* are fairly numerous, and several provide direct evidence of the meaning[5], including the black granite naos from Saft el Henneh, which is described as *ỉnr km n mꜣt*[6], apparently the only example of this explicit term. *ỉnr km* is frequently cited as the material for statues[7], on one occasion[8] for composite statues of which the red crowns are *bỉꜣt*, and scarabs of *ỉnr km* are also mentioned[9]. In a few instances, the ▭ determinative is written after the adjective *km*[10], showing that the term had come to be regarded as a composite whole, and that the literal meaning 'black' or 'dark' stone had been forgotten. There can thus be little doubt that *ỉnr km* refers only to black granite, and not to other dark stones as well, and this is borne out by the direct evidence, and those instances where it is clearly opposed to *mꜣt*.

It may be noted in passing, that although *ỉnr km* existed as a specific term for certain shades of granite, from dark grey to black, the Egyptians did not at any time possess a precise name for the pink or red variety, and were merely content to adapt the common designation of all granite to this particular use. Moreover, since only this broad division between red and black granite was recognised, it is difficult to believe that fine grained grey granite (psammite gneiss) was ever intentionally referred to as *bḫn*, as Lucas seems to suggest[11], and it is more reasonable to suppose that in this single instance the stone was mistaken for greywacke.

The different writings of *mꜣt* as *mꜣt*, *mꜣ* and *mꜣiw* have already been mentioned, and it may also be observed that in some few cases the sign ⌂ or one of its later variants is apparently used as an ideogram[12]. What is probably a peculiar orthography of *mꜣt* occurs in the Sehel inscription[13], the reading being evidently *mtꜣy*[14], though it was mistaken by Brugsch and other earlier commentators. Granite is not otherwise mentioned in the text, which, since it was the most typical stone of this region, is strange, and it is therefore the more likely that this remarkable form conceals *mꜣt*.

That any other stones were ever confused with granite is altogether unlikely, since the stone was widely used from the earliest period, nor does it appear that it was any at time known by a name other than those here discussed.

[1] Urk. IV, 1295. [2] Urk. IV, 640.

[3] Urk. IV, 457.

[4] Normally *mꜣt, ỉnr km*: LD. III, 72a, 5; BIFAO. 13, 60; Luxor ⟨669⟩; ⟨594⟩; Sethe 2, 73 [14]; 15, 70 [15]. Once *ỉnr kṃ, mꜣt*: Not. Descr. II, 732.

[5] Eg. Rec. Trav. 19, 14 (a statue of black granite) = Cairo ⟨5⟩; Abyd. Mar. II, 11a, b (a door); Düm. T. I. I, 92, 12; 93, 20; Piehl Inscr. II, 92 (a naos).

[6] Goshen 6, 1.

[7] Rec. Trav. 19, 14 = Cairo ⟨5⟩; Luxor ⟨669⟩; ⟨594⟩; Sethe 2, 73 [14]; cf. Urk. IV, 457 (a sphinx).

[8] Ann. Serv. 38, 223, 12. [9] Harr. 4, 10.

[10] Harr. 4, 1; 7, 13 (); Sethe 15, 70 ().

[11] Materials p. 479.

[12] Eg. Düm. Baug. 16; Goshen 6, 1; ZÄS. 43, 8, III.

[13] Sehel 15.

[14] : Barguet, Stèle de la Famine p. 23.

Quartzite

The most common term for the characteristic red crystalline quartzite sandstone obtained principally from Gebel Aḥmar was undoubtedly *biꜣt* (Wb. I. 438.16f.), as is recognised by BRUGSCH[1] and WENDEL[2], both of whom note that the two 'Memnon' colossi are stated to be of this material, and suggest that it was probably so called by derivation from *biꜣ*, iron, on account of its extreme hardness. DARESSY[3], too, favours connection with *biꜣ*, but takes *biꜣt* as basalt, and *biꜣ km* (sic.) as black basalt. VARILLE[4], on the other hand, accepting the identification of *biꜣt* as red quartzite sandstone, thinks the name means literally 'une pierre merveilleuse', and this is also the view of SETHE[5], who tentatively suggests that the notion of 'wonder' may have originated with the 'Memnon' colossi themselves, since the word *biꜣt* is not previously known.

SETHE discusses quartzite at some length, noting that it was called *biꜣt*, *inr n biꜣt*, *inr nfr n biꜣt* and *ꜥꜣt biꜣt*, and that it was apparently valued by the Egyptians for its hardness and durability. That it is a species of sandstone seems to have been realised, since in one early instance an alternative term *inr n rwḏt nt ḏw dšr* is used, *ḏw dšr* being the common name of Gebel Aḥmar, the primary source of the stone, also called *ḏw n biꜣt*[6]. There is also evidence for another source of *biꜣt* near Elephantine, where SETHE thinks quartzite and black granite occurred side by side. In one or two instances, the stone is referred to as *bnwt*, evidently because it was the typical material for millstones, even as it is today.

The textual evidence for *biꜣt* wholly confirms the identification with quartzite sandstone, since in addition to the 'Memnon' colossi which are described as *biꜣt*[7], at least four other monuments provide a direct indication of the nature of the stone. These are a stela of 'grès silicieux'[8], a statue of 'grès ferrugineux'[9], a pyramidion of 'grès rouge'[10], and the base of a shrine of 'rotes Quartzkonglomerat'[11], three of which are said to be *inr n biꜣt*, and one *biꜣt ꜥꜣt*. Moreover, Gebel Aḥmar, usually called *ḏw dšr*[12], is also referred to as the *ḏw n biꜣt*[13].

The characteristic colour of *biꜣt* is specifically referred to in two passages in the Menshiyet es-Sadr stela of Ramses II, where the stone in compared to *mrw*-wood[14], a coniferous wood of red colour[15], and the red crowns[16] of statues of black granite are made of *biꜣt*[17]. This latter passage led SETHE to suppose that quartzite and black granite occurred side by side in a quarry near Elephantine, so that 'monolithic' statues could be made of two stones. But no actual instance of such a practice is known[18], whereas statues were not infrequently made with their crowns separate.

Quartzite was only rarely used in building, its chief employment being for statues and sarcophagi[19], and *biꜣt*, too, is mentioned mainly in connection with statues[20], stelae[21] and the

---

[1] Wb. Erg. p. 412.
[2] Bau- und Edelsteine p. 82.
[3] Rec. Trav. 10 p. 143; Ann. Serv. 13 p. 43 n. 3; ib. 22 p. 166.
[4] Ann. Serv. 33 p. 91.
[5] Bau- und Denkmalsteine p. 29f.
[6] *ḏw dšr* and Gebel Aḥmar have in fact the same meaning, lit. 'the red mountain'.
[7] LD. Text III, 144.
[8] Rec. Trav. 16, 56.
[9] Tell el Yah. p. 70.
[10] Ann. Serv. 23, 176f.
[11] Rec. Trav. 12, 4, III.
[12] Eg. Sin. R. 40; Urk. IV, 167, 3; Anast. I, 15, 6; Düm. G. I. IV, 127 etc.; cf. Onom. 2 p. 138.
[13] LD. III, 72, 4; Cairo 583, 16; cf. Sethe, Bau- und Denkmalsteine p. 30–1.
[14] Ann. Serv. 38, 220, 2.
[15] Ann. Serv. 16 p. 38; Lacau, Sarc. I p. 175; II p. 41.
[16]  not  as Sethe and Hamada.
[17] Ann. Serv. 38, 223, 12.
[18] But cf. B. M. 19, a monolithic statue of black and red granite.
[19] Lucas, Materials p. 79; p. 477.
[20] LD. Text III, 144; Harr. 4, 9; 26, 3; Six Temp. 12, 4; Luxor ⟨533⟩; Ann. Serv. 38, 219f. etc.
[21] Rec. Trav. 16, 56; Memphis I, 6.

like[1], though occasionally for buildings[2], and once for a model temple[3]. The word is usually written with the ideogram , but in the Harris papyrus it is written out fully, as also in the Chester Beatty mineral list[4]. As well as *bỉꜣt*, the stone is referred to more explicitly as *ỉnr n bỉꜣt*[5], *ỉnr nfr n bỉꜣt*[6], *ꜥꜣt bỉꜣt*[7], and even *bỉꜣt ꜥꜣt*[8], and in at least one context[9], *bỉꜣt* has the specific meaning of 'a block of *bỉꜣt*'. Whether there is any particular significance in the term *ꜥꜣt bỉꜣt* it is difficult to say, but it may be observed that of the building and monumental stones only *bỉꜣt*, and on one occasion *mꜣṯ*[10], are referred to in this way.

Before the first use of the term *bỉꜣt*, which appears to date from reign of Amenhotpe III, quartzite was known by two other names, neither, however, occurring before the beginning of the 18th dynasty. *ỉnr n rwḏt nt ḏw dšr* seems to be found only once[11], though *ỉnr n ḏw dšr* appears later[12], and can only refer to quartzite from Gebel Aḥmar, the form of the name strongly suggesting that the stone was recognised as in some way related to sandstone. The other word, *bnwt* (Wb. I. 458.12–14), though primarily the term for millstone[13], also indicates a specific kind of stone on at least two occasions, and perhaps more[14]. A doorway from Heliopolis, now in the Cairo Museum, is described as *ỉnr n bnwt*[15], and a stela from Defenneh is said to be *bnw ḥḏ*[16], while in a few Graeco-Roman texts, *bnwt* also seems to refer to a material[17]. The actual stone from which the Heliopolis doorway is made is not black granite as stated in Urkunden, but the red crystalline quartzite sandstone of Gebel Aḥmar[18], and there can be little doubt that the use of *bnwt* to describe it arose from its being the typical material of millstones[19]. The Defenneh stela, is, however, of Nubian sandstone[20], and the fact that this is described as *bnw ḥḏ*[21] may again indicate that some connection between quartzite and sandstone was recognised.

It is possible that *bnwt* also refers to quartzite in the phrase *ḥwyt nt bnwt* (Wb. III. 49.15), which occurs once in Ebers[22] for use in connection with the teeth, though it may equally well be used in the sense of millstone. The expression must then mean either 'crushed quartzite' or 'millstone-grindings', which in actual fact would amount to the same thing. But if this *bnwt* is to be taken in the sense of quartzite, it represents not only the earliest use of *bnwt* in this way, but also the earliest reference to quartzite under any name. What quartzite was called in the Middle Kingdom, when it was undoubtedly in use, cannot be determined, but either *ỉnr n bnwt* or *ỉnr n ḏw dšr* seems likely, though it is always possible that it was not strictly differentiated from sandstone at this time.

In short, it is clear that the earliest terms for quartzite were derived either from its use, as *ỉnr n bnwt*, or from its provenance, as *ỉnr n rwḏt nt ḏw dšr*, and that only towards the end of the 18th dynasty did a more specific word *bỉꜣt* come into use, perhaps as Sethe suggests, with reference to the wonder caused by the 'Memnon' colossi.

---

[1] Cf. Harr. 34a, 3; Rec. Trav. 12, 4, III.
[2] Harr. 4, 1; 7, 13; 25, 12.
[3] Rec. Trav. 8, 9.
[4] P. Ch. B. IV vs. 7, 6–7.
[5] Harr. 34a, 3; 26, 2; Rec. Trav. 12, 4, III, 16, 56; Memphis I, 6; Tell el Yah. 70; P. Ch. B. IV vs. 7, 6 (*ỉnr bỉꜣt*).
[6] Rec. Trav. 8, 9.
[7] LD. Text III, 144; Six Temp. 12, 4 (Wb. I. 165.18) both referring to statues.
[8] Ann. Serv. 23, 178; cf. ib. 179; 180 (the same).
[9] Ann. Serv. 38, 219, 3.
[10] Brugsch, Wb. p. 159.
[11] Urk. IV, 167.
[12] Hier. Ostr. 17, 1, 8.
[13] Eg. Prisse 5, 10; Harr. 65a, 9; 74, 7; Smith vs. 22, 2; P. Leid 343 vs. 2, 8 etc.
[14] Cf. Daressy, Ann. Serv. 13 p. 43; Rec. Trav. 10 p. 143.
[15] Urk. IV, 831.
[16] Tanis II, 42.
[17] Philae ⟨1998⟩ Phot. 254; Urk. VIII, 134 (twice).
[18] Sethe, Bau- und Denkmalsteine p. 32.
[19] Cf. Borchardt, Neferirkerēꜥ p. 11; Baedeker (8) p. 118.
[20] Sethe, op. cit. p. 33.
[21] Perhaps properly white quartzite.
[22] Eb. 89, 3.

Alabaster

The earliest reference to alabaster, at the beginning of the 4th dynasty[1] is by the common word *šs* (Wb. IV. 540.10f.)[2], but in the oldest graffiti in the alabaster quarries at Hatnub[3] there occurs a term *bit* (Wb. I. 433.1), which both Anthes[4] and Sethe[5] recognise as the name of the stone sought there. *bit* is, however, confined to these few instances, and the later graffiti[6] all refer to alabaster as *šs*, which, from the end of the Old Kingdom, is the common, and with one exception[7] the only name for the stone.

That *šs* refers to alabaster was first proved by Brugsch[8], who cites the dedicatory inscription of an alabaster shrine in the temple of Ramses II at Abydos, and this meaning has been universally accepted. The reading of the word has, however, been questioned, and various scholars including Wendel[9], Daressy[10], and latterly Jéquier[11], waver between *šs* and *ḳs*, the latter suggested by the Coptic ⲔⲰⲤ, ϬⲰⲤ, and by the compound *ḳs-ᶜnḫ*, which was thought also to refer to alabaster.

But the question of the reading is settled definitively by Gardiner[12], who points out that in the Old Kingdom ɤ was used only in bi-consonantal words, or tri-consonantals with a weak radical, and had the value *šs*. He then discusses in detail the translation of *šs* in the *ḥtp di nsw* formula, where it was first rendered 'cord', then 'linen bandages' (*šs mnḫt*) or 'clothing', and only recognised as alabaster by Sethe[13]. Gardiner shows that since 'clothing' was *šsr*, not *šs*, the latter can only be rope or alabaster, with a strong presumption in favour of alabaster, since vessels of this stone were frequently buried with the dead, whereas coils of rope were not. Brugsch's formal identification, and his comparison of *šs* with Hebrew *šēš* are then quoted, and references given to other monuments of alabaster directly described as *šs*[14]. Alabaster was the classic material for unguent vessels[15], and in some writings this is clearly shown, eg. [hieroglyph], where [hieroglyph] is an alabaster bowl, and cannot have the phonetic value *ḥb*. In the 18th dynasty, however, ɤ in [hieroglyph] was probably misinterpreted as the determinative for clothing.

Sethe[16] also discusses alabaster at length, pointing out that *šs* occurs as early as the beginning of the 4th dynasty[17]. It is often qualified as *wᶜb* or *bḳ*, just as limestone and sandstone are described as *nfr*, and the provenance *n ḥwt-nbw* is frequently added, *šs wᶜb n ḥwt-nbw* forming an exact parallel to *inr ḥḏ nfr n ᶜin*. *ḥwt-nbw* refers to the well-known alabaster quarries near Amarna, which do not, however, appear to have been worked in the New Kingdom, so that the phrase *n ḥwt-nbw* must at this period be a designation of quality rather than provenance[18]. The later source of alabaster was a quarry at Gauata near Siut[19], and this is probably referred to as *ḏw n šs wᶜb*[20]. Sethe deals with the occurrence of *šs* in the *ḥtp di nsw* formula, which springs from the very frequent use of the stone for unguent vessels.

---

[1] Medum 13.

[2] Cf. Hebrew *šayiš*, *šēš*, Syriac *šīšā*, Aramaic *šīš*; JAOS. 73 p. 155.

[3] Hatnub Gr. 3; 6; 7; 9 (reign of Pepi II). Both *inr*, blocks, and *ᶜȝt*, apparently irregular lumps, are mentioned.

[4] Hatnub p. 7.

[5] Bau- und Denkmalsteine p. 22.

[6] Hatnub Gr. 10; 25; 32; 43 etc.

[7] *inr n ḥwt-nbw*, Bersh. I, 14.

[8] Recueil I, 12, 3; cf. also ZÄS. 18 p. 6.

[9] Bau- und Edelsteine p. 86.

[10] Rec. Trav. 10 p. 143.

[11] BIFAO. 19 p. 93.

[12] BIFAO. 30 p. 161.

[13] Quoted by Gardiner, BIFAO. 30 p. 166.

[14] Ann. Serv. 4, 225; 23, 113; 24, 57.

[15] Cf. now Balcz, Mitt. Kairo 3 p. 65.

[16] Bau- und Denkmalsteine p. 22.

[17] Medum 13.

[18] Alternatively it may have been applied to other quarries in the region of Amarna which were worked during the New Kingdom; cf. Lucas, Materials p. 76.

[19] LD. Text II p. 157.

[20] Münch. Glypt. 32 (Dyroff-Pörtner II pl. 11) = Thes. 1451.

He compares the Greek ἀλάβαστρος, and tentatively derives it from *ʿ n Bȝstt* (* *ʿa-la-baste*), 'Gefäß der Göttin Ebáste (= Bubastis)'.

There is little to add to the discussions of Sethe and Gardiner, though the textual evidence for *šs* is considerable. It may, however, be noted that a writing *šst* (Wb. IV. 541. 6–10) is also known[1], and that an apparent reference to *rwḏt ḥwt-nbw*[2] is almost certainly *šst*. As pointed out by Sethe, *šs* is variously described as *šs n ḥwt-nbw*[3], *šs wʿb*[4], *šs wʿb n ḥwt-nbw*[5], and *šs bḳ n ḥwt-nbw*[6], and is mentioned in a number of the graffiti at Hatnub itself[7]. Moreover, in a single instance[8] what can only be alabaster is referred to simply as *inr n ḥwt-nbw*[9], parallel in form to *inr n r-ȝw*, *inr n ʿin*, *inr n ḏw dšr* etc. That Hatnub was the principal source of alabaster until the New Kingdom there can be no doubt[10], but there is ample evidence to show that some other quarries were worked both then and later.

Apart from its regular occurrence, in the *ḥtp di nsw* formula, *šs* is stated to be the material of buildings[11], stelae[12], statues[13], vessels[14], headrests[15] and other objects[16], and occurs in at least two contexts which seem to suggest that the Egyptians regarded it as one of the semi-precious stones[17]. *šs* also appears in the medical texts, either alone[18], or as *dḳw n šs*[19], or *imt n šs*[20], both of which must refer to alabaster in crushed or powdered form. *dḳw* is used of other mineral substances such as *didi* and *ḥsmn*, and also of vegetable products, but *imt* seems to be confined to this single context, where its meaning can only be guessed.

Except in the offering formula, where 'alabaster vases' is perhaps the more suitable rendering, *šs* is to be translated alabaster, though it must be borne in mind that geologically Egyptian alabaster is calcite, ie. calcium carbonate, not calcium sulphate, and that to this extent alabaster is an incorrect term[21]. *bḳs-ʿnḫ* has no connection with *šs*, and apart from the early term *bit*, the only other expression which refers to alabaster is *inr n ḥwt-nbw*. Whether *kfti* (Wb. I. 122.3), used once in the phrase *ḥs n kfti*, describing a vase of 'arragonite'[22], is another unique term for alabaster it is impossible to say, since it might equally well mean 'a vase of Keftiu workmanship'. That the Egyptians confused any other stones with alabaster seems somewhat unlikely.

bḫn

bḫn (Wb. I. 471.1–5) has been the subject of considerable controversy, and as yet no entirely satisfactory translation has been proposed. Brugsch[23] takes *bḫn* as porphyry, apparently from an incorrect identification of the material of the British Museum obelisks,

---

[1] Eg. Harr. 4, 8; 15b, 11; Cairo 20318; Louvre C 254; Flor. Sch. 1551; Urk. IV, 1044 etc.
[2] Urk. I, 107, ; cf. Montet, Kemi 3 p. 44.
[3] Urk. IV, 388.
[4] Recueil 12, 3; ZÄS. 18, 6; Abyd. Mar. II, 3; Cairo Wb. 219 ⟨438⟩; Cairo 42139 etc.
[5] Urk. IV, 424; Rec. Trav. 22, 118–9; Thes. 965 etc. [6] Urk. IV, 640.
[7] Hatnub Gr. 10; 25; 32; 43. [8] Bersh. I, 14.
[9] Cf. Sethe, Bau- und Denkmalsteine p. 23; Jéquier, BIFAO. 19 p. 93.
[10] Cf. Lucas, Materials p. 76 for later inscriptions.
[11] Recueil 12, 3. [12] Cairo Wb. 219 ⟨438⟩.
[13] Urk. IV, 457, 12; 457, 10; Harr. 41a, 4. [14] Medum 13; Urk. IV, 641; 639.
[15] Cairo 28083 (white); 28091 (red); 28092; 28123.
[16] Urk. IV, 458, 14; Medum 13; Urk. IV, 640 etc.
[17] Edfou VIII, 136–7; Philae Phot. 524–19; Thes. 41–2; Harr. 41a, 4. [18] Eb. 92, 15.
[19] Eb. 86, 3; 87,5; 87, 14; Smith vs. 21, 6; cf. Düm. G. I. II, 85b; *ḳȝw n šs* Hearst 10, 15; cf. Appendix I. [20] Eb. 69, 19; Wb. I. 76.2.
[21] Lucas, Materials p. 75. 'Arragonite' is also incorrect when used of Egyptian alabaster.
[22] Cairo 46082. [23] Wb. Erg. p. 444.

and Dümichen[1], noting that the stone came from Hammamat, regards it as 'ein dunkelgrüner, fast schwarzer, ungemein harter Diorit'. Wendel[2], too, interprets *bḫn* as diorite and diorite-breccia, perhaps also including basalt, and quotes Renouf's description of the British Museum obelisks as 'dark basalt'. He points out that Hammamat was the source of *bḫn*, and claims that only diorite, diorite-breccia, granite and porphyry are found there. Legrain[3], on the basis of the naos of Nectanebo said to be of *bḫn*, identifies the stone as the green breccia of Wadi Hammamat, while Daressy[4] calls *bḫn* schist and *biꜣt* basalt. Couyat and Montet[5] discuss in detail the rocks occurring in the region of Wadi Hammamat, which include breccia, schist, serpentine, granite and diorite, and point out that only one place in the Wadi bears evidence of extensive working, the stone there being black schist, which must be preferred as a translation of *bḫn*[6]. They then mention various monuments of *bḫn*, the actual material of which has been examined, and observe that although some cannot be checked, the naos described as 'green breccia' or 'black granite' is in fact schist from Wadi Hammamat. Sethe[7] notes *bḫn* (*bḫn*) and *inr* (*nfr*) *n bḫn* as the name of a stone of black or dark green colour, the *bḫn* of Wadi Hammamat being a kind of greywacke-schist of innumerable varieties[8]. The monuments said to be of *bḫn* are of a bewildering range of stones such as basalt, granite, breccia, schist etc. *bḫn* is the origin of Greek βασανίτης λίθος[9], which was clearly a black stone, and hence of Latin basanites, improperly basaltes[10]; Wadi Hammamat, the Egyptian *ḏw n bḫn*, is the βασανίτου λίθου ὄρος of Ptolemy. Sethe discusses the etymology and history of the word *bḫn*, and seeks to show that the stone was used as a touchstone in proving gold.[11]

More recently, the question of the exact identification of *bḫn* has been carefully re-examined on the evidence of the numerous monuments stated to be of the stone. Varille[12] points out that the Cairo naos described as *inr n bḫn ṯḫn* is schist[13], *ṯḫn* probably referring to the polish, or indicating a particular quality of the material; that the obelisks in the British Museum are not in fact basalt, but gneiss[14], as is proved by the Cairo fragment belonging to one[15]; that part of another obelisk in Marseilles is very hard serpentine rock[16]; and that the Cairo naos of Amasis from Benha is grey granite[17]. Thus scientific examination gives three distinct stones, gneiss, schist and serpentine-breccia, as equivalents of *bḫn*, and it becomes clear that colour was the ruling factor with the Egyptians.

Lucas and Rowe[18] also regard *bḫn* as the name of more than one stone, and give a chronological list of references to *bḫn*, with identification of the material wherever possible. From this it appears that objects of *bḫn* have been variously described as 'black basalt', 'black (basaltic) stone', 'diorite', 'basalt', 'green basalt', 'sandy schist', 'so-called basalt', 'dark hard stone', 'porphyry', 'diorite-breccia', 'green breccia', 'schist' etc. There are two ways of determining the real nature of the stone, – by examination of monuments stated to be *bḫn*, and by checking the statements as to provenance. Five monuments of *bḫn* in the Cairo Museum are then discussed, and Lucas and Rowe point out that re-examination revealed

---

1 Geschichte p. 124.
2 Bau- und Edelsteine p. 73; cf. p. 17.
3 Ann. Serv. 6 p. 122.
4 Ann. Serv. 13 p. 43 n. 3.
5 Hammamat p. 22f.
6 Goyon, Nouvelles Inscriptions p. 5f., mentions three principal areas in the Wadi where stone was quarried.
7 Bau- und Denkmalsteine p. 35.
8 Goyon, loc. cit., concludes that *bḫn* includes schist of three main colours, the classic *bḫn* being the black stone.
9 Cf. Wiedemann, Sammlung p. 17.
10 Cf. Pliny, HN. 36, 58; Isidorus 16, 5, 6.
11 For a probable reference to *bḫn* in the Bible, cf. Sethe, op. cit. p. 46; Lambdin, JAOS. 73 p. 148.
12 BIFAO. 34 p. 93.
13 Cairo 70019.
14 B. M. 523; 524.
15 Cairo 17030.
16 Marseille Ob. 1 [16] (Maspero, Cat. 1889 p. 1 no. 1).
17 Cairo 70011.
18 Ann. Serv. 38 p. 127; p. 677; 41 p. 347.

that 1) was not 'schist', but fine-grained grey granite, as ROEDER, better called 'psammite-gneiss'[1], 2) was 'schist', ie. greywacke,[2] 3) was also 'schist', ie. greywacke[3], 4) was probably 'schist', ie. greywacke[4], and 5) was greywacke[5]. Moreover the typical stone of the Wadi Hammamat is also 'schist' or greywacke, and the reference to *bḫn* in the Sehel inscription may indicate gneiss.

SHIAH[6], discussing the god *Bḫnw*, reviews the conclusions of LUCAS and ROWE regarding *bḫn*, pointing out that the reference in the Sehel text may be to mica-schist or hornblende-schist. He holds that *bḫn* originally referred only to greywacke and allied stones, and that the example which has been identified as psammite gneiss is a mistake, since it looks like schist, and was called *bḫn* simply because that was the name for the stone commonly used for naoi, whereas the term for psammite gneiss, if there was one, was unfamiliar. But SHIAH considers greywacke an unsatisfactory translation, because it may have either a very wide or a very restricted significance. Geologically it may be tuff, hornfels, greywacke 'proper', indurated mudstone, slate or schist, but for archaeological purposes, these distinctions, which cannot be perceived with the naked eye, may be ignored. Rather than adopt a term with precise geological connotations, SHIAH proposes that a new word should be created for the convenience of archaeologists, and suggests PETRIE's 'durite'[7], a sedimentary rock, varying in quality from mudstone to grit, derived from disintegration of other rocks or volcanic mud, indurated but unmetamorphosed. Different varieties of this stone could then be qualified as 'fine', 'coarse', 'green', 'black', 'grey' and so on.

There are thus two questions to be considered in relation to *bḫn*, first to what stone or stones the word refers, and second what is the most suitable term by which archaeologists can describe them.

Any detailed treatment of the textual evidence for *bḫn* would here be out of place, since all the important references have already been collected by LUCAS and ROWE, and although it may be possible to add one or two more from ostraca and other sources[8], none is of any great significance. It must, however, be observed that one of the examples quoted, text R, is not in fact relevant, since it contains no reference to *bḫn*, the word there being almost certainly *bḫnt*, 'pylon', as pointed out by SHIAH[9].

Of these texts, seven (B. C. D. F. I. J..K.) are either on the rock or on stelae in the Wadi Hammamat itself, while two others (A. G.) refer to it, and six (H. L. M. N. O. P.) are on objects either directly described as, or implied to be *bḫn*. Of the remainder, E. mentions two obelisks of *bḫn*, and Q. a shrine of *bḫn*[10], R. is irrelevant, S. refers to a plant receptacle of *bḫn*, and T. is the Sehel inscription which seems to indicate that *bḫn* was also found in the Elephantine region.

Now there can be no doubt that although more than one stone occurs in the Wadi Hammamat, the Egyptians were principally concerned with one in particular, namely the 'black schist' described by COUYAT and MONTET[11], which LUCAS[12] points out is in fact greywacke or tuff, and it seems likely therefore that this is the stone indicated by *bḫn*, and described by PLINY[13] as 'ferrei coloris atque duritiae'.

---

[1] Cairo 70011. [2] Cairo 70019. [3] B. M. 523; 524; Cairo 17030.
[4] Cairo 70031. [5] Cairo JdeE 40643; 49103.
[6] Ann. Serv. 41 p. 189. [7] Cf. Amulets p. 8; Scarabs p. 8.
[8] Eg. Ostr. Cairo 25262; cf. also graffiti published by Goyon, Nouvelles Inscriptions.
[9] Ann. Serv. 41 p. 197.
[10] Brunton, Ann. Serv. 40 p. 617 points out that this is not a piece of direct evidence, since the inscription in fact forms part of a statue, not a naos as described by Lucas and Rowe.
[11] Cf. Goyon, Nouvelles Inscriptions p. 5f. [12] Materials p. 478. [13] HN. 36, 58.

Moreover, the direct evidence would tend to support this conclusion, though not to the extent implied by LUCAS and ROWE, since two of their five examples from the Cairo collection (4 and 5, texts Q. and R.) are invalid, as has already been mentioned. Of the six objects said to be *bẖn*, the materials are identified as follows.

1) The Marseilles obelisk (text H.): 'black basalt' (MASPERO), 'serpentine rock, very hard' (REPELIN quoted by VARILLE) — no more recent opinion is available.

2) Statue of Horudja, at University College London (text L.): 'hardest and closest black basalt' (PETRIE) – no more recent opinion available.

3) Naos of Amasis, Cairo 70011 (text M.): 'grey speckled fine-grained granite' (ROEDER), 'psammite gneiss' ie. fine-grained grey granite (ANDREW).

4) Naos of Nectanebo I, Cairo 70019 (text N.): 'green breccia' (LEGRAIN), 'green basalt' (MASPERO), 'green schist (slate)' (ROEDER), 'sandy schist of Wadi Hammamat' (COUYAT-BARTHOUX quoted by VARILLE), 'greywacke' (ANDREW).

5) Part of a frieze, now in Bologna (text O.): 'basalt' (YOUNG), 'black basalt' (KMINEK-SZEDLO) – no more recent opinion available.

6) Obelisks of Nectanebo II in the British Museum, B. M. 523; 524 (text P.): 'black marble' (NIEBUHR), 'black basalt, of fine grain' (DESCRIPTION), 'basalt' (YORKE and LEAKE), 'dark-basalt' (RENOUF quoted by WENDEL), 'black basalt' (LONG, SHARPE, BUDGE, HALL), 'gneiss of micrograntic structure with amphibole' (KUENTZ), 'Hammamat greywacke' (ANDREW).

Thus, of the six objects, two are certainly greywacke, one is certainly psammite gneiss, two are described as basalt, and may therefore be greywacke, and one is described as a serpentine rock, which might possibly be greywacke. As a whole, therefore, the evidence seems to suggest that *bẖn* was the 'greywacke' of Wadi Hammamat, for which a suitable name must be found.

Basalt, although going back in origin to *bẖn*, is unsuitable, if only because it is more properly used to describe a fine-grained dolerite, quite distinct from the Hammamat greywacke[1]. 'Durite', invented by PETRIE, and favoured by SHIAH, though it might be a convenient way of avoiding the difficulty, seems altogether undesirable, since it may embrace certain stones, eg. slate, which were not included under *bẖn*. Schist, though frequently used to describe objects of greywacke, is an inaccurate term, as LUCAS[2] points out, and the only satisfactory solution seems to be to retain the term greywacke, while realising that in archaeological contexts it has a particular range of meaning.

That *bẖn* also embraced psammite gneiss, as LUCAS[3] seems to suggest, is unlikely, and the single example of this material may either be a genuine mistake, due to its superficially similar appearance, or a deliberate substitution. Since *bẖn* was the typical stone for naoi, any naos, whether of greywacke or not, could be described as such without undue difficulty. Similarly, the serpentine rock of the Marseilles obelisk, if such it is, may have been quite genuinely mistaken for *bẖn*, or simply stated to be such although its true nature was known. Lastly, the reference to *bẖn* in the Sehel inscription need not be interpreted as psammite gneiss, since both mica-schist and hornblende-schist occur near Elephantine, and it seems in any case extremely doubtful whether all the stones mentioned in the text, eg. *ḫsbḏ* and *mfkꜣt*, were in fact found in the area.

In conclusion, something must be said concerning a word  which occurs in some Ptolemaic texts[4], and is evidently a stone used for shrines. The obvious inference from the contexts would be that it is a writing of *bẖn*, but although  is frequently *ẖn*, 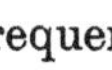 for *b* is

[1] Cf. Lucas, Materials p. 77. [2] Materials p. 477. [3] ib. p. 479.

[4] Dend. Chass. II, 24–5; II, 54–5; in all, 34 examples. I am grateful to Prof. Fairman for information on this subject.

altogether doubtful. 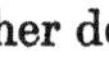-normally represents *ʿb*, and though *ʿ* can fall away in certain circumstances, it does not usually do so with *b*, the nearest parallel being 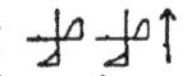, *ʿbb*, 'harpoon', which may, however, be due to false reduplication[1]. Neither Ptolemaic not Coptic gives grounds for believing that *ʿ* could be considered weak and disappear in these circumstances. On the other hand, if the word is not *bẖn*, it is difficult to see what it can be except *ʿb*, which occurs only three times elsewhere, and is unidentifiable. No very satisfactory conclusion is possible, but in spite of the philological difficulties, interpretation as *bẖn* cannot perhaps be ruled out entirely.

Concerning the lexicographical evidence for the six principal building stones as a whole, there is little to be said, though it may be observed that during the New Kingdom, expressions of the form *ỉnr n* . . . and *ỉnr nfr n* . . . were common, except in the case of alabaster. The term *ỉnr n ḥwt-nbw* is altogether unusual and apart from the few instances of *bỉt*, all the names for alabaster are variations of *šs*. What significance this may have it is not easy to see, though it may perhaps indicate that alabaster was not primarily regarded as belonging to the same class as the building stones, and this might be confirmed by its occurrence among precious stones in the texts of filling the sound eye. Again, only *bỉꜣt*, and on one occasion *mꜣṯ*, are compounded with *ʿꜣt*, a form of reference more usually associated with stones of some value, eg. *ʿꜣt bht*, *ʿꜣt ḥḏt*, the implications of which in the case of these two more common materials is obscure.

Only one stone used in building is apparently without a distinct name, though many materials used for statuary etc. are still unidentified, as will be shown in the following chapter. This is basalt (ie. fine grained dolerite, not the so-called basalt of Wadi Hammamat) which was employed in the Old Kingdom for pavements, and is widely distributed in Egypt[2]. That this stone was not identified by some specific term it is difficult to believe, and the probability is that there was in the Old Kingdom a word, not known to us through lack of documentary evidence, which fell into disuse in later times, when basalt was less generally employed.

[1] Cf. Ann. Serv. 43 p. 220 no. 135; p. 253 note XIX.

[2] Lucas, Materials p. 78.

CHAPTER 4

MISCELLANEOUS STONES

Beside the more important building stones treated above, and the semi-precious stones which form the subject of the next chapter, there are several, used by the Egyptians for statuary, vases etc., which cannot conveniently be classed with either, and are therefore grouped together here. With them may also be discussed a number of words which clearly refer to some species of stone, though exact identification is not possible, which precludes their being more satisfactorily classified. There are, of course, many other terms, occurring in lists of minerals, which may perhaps represent stones, but wherever doubt exists, such terms are better included among the miscellaneous minerals dealt with in chapter 7.

*ii*ꜥꜣꜥꜣ (?) (no Wb. ref.) appears to be known from only one example[1]. BALCZ[2] rightly refers to it as 'ein Materialname ... das ... unter anderen Gesteinsbezeichnungen vorkommt', though unable to suggest a satisfactory reading or meaning, while SELIM HASSAN[3] strangely regards it as a type of vessel, overlooking the parallelism with *tiiꜣꜣt* and *šs*, which shows that a material is indeed meant. There can be no doubt that the word refers to one of the many stones used for vessels in the early dynastic period, though the common form of the bowl shown does not assist in identification, and the precise meaning must remain problematic. The reading too is doubtful. BALCZ prints the two horizontal signs as maces (T. 1), which would require some such reading as *iimnwy*, but from the photograph the form seems more suggestive of ꜥꜣ (O.29), in which case the reading would be *ii*ꜥ*ꜣwy* or *ii*ꜥꜣꜥꜣ. In fact, without further examples no conclusion is possible, and both the reading and meaning of the word must remain uncertain.

inr ṯḥn (no Wb. ref.) is seemingly confined to a single unhelpful context[4], from which the meaning is not apparent. The term may possibly be an alternative for *ṯḥnt* or even *ds ṯḥn*, though it may equally well have a purely general sense such as 'glittering stone' or 'polished stone', without reference to any particular material. On the other hand, *inr ṯḥnt* also occurs as a place-name[5], probably corresponding to the 'Αλαβάστρων of PTOLEMY, which might suggest that *inr ṯḥn* is a word for alabaster[6], though this is perhaps unlikely.

inr n msdt (Wb. II. 152.15) is the material of which the two flagpoles of a model temple, or perhaps their bases, are made[7], the pylons of the same being of ꜥ*ꜣt ḥḏt*, so that no approximation to an actual building material need be supposed. Thus it is not impossible that

[1] Q. Saqq. VI p. 26.
[2] Mitt. Kairo 3 p. 75.
[3] Giza H. VI (2) p. 57.
[4] Mammisi 151.
[5] Kemi 6 p. 20; Onom. 1 p. 58; BIFAO. 12 p. 15; Gauthier, Dict. géog. I p. 85.
[6] Cf. Brugsch, Wb. Erg. p. 1374; Wendel, Bau- und Edelsteine p. 119, where *ds ṯḥn* is interpreted as alabaster.
[7] Rec. Trav. 8 p. 9.

msdt is a corruption of *msdmt*, though there do not appear to be any similar writings elsewhere, and no parallel example of *ỉnr n msdmt* can be quoted. On the whole, it seems more satisfactory to regard *ỉnr n msdt* as a ἅπαξ λεγόμενον, the meaning of which cannot be determined.

ỉnr n ḥꜣwt (no Wb. ref.) seems to occur only once, in a long list of minerals[1], where the context is of no assistance in determining the meaning. *ḥꜣwt* may be related to *ḥꜣw* 'neighbourhood' (Wb. II. 477.1), *ỉnr n ḥꜣwt* then meaning 'local stone' or the like, or to *ḥꜣyt* 'portico' (Wb. II. 476.4)[2] or 'ceiling' (Wb. II. 476.13)[3], when 'building-block' or 'roofing slab' would be more appropriate. On the other hand, there may be no connection with either, *ỉnr n ḥꜣwt* being a quite unknown species of stone.

ꜥꜣt ꜥnw (cf. Wb. I. 166.1; I. 187.12) has been the subject of some controversy. The phrase occurs only once[4], parallel to *ꜥꜣt ḫsbḏ*, as the description of a (red) vessel, and written 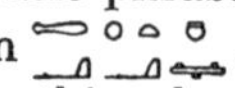. Balcz[5], regarding *ꜥꜣt* as 'wertvolle Steine', maintains that *ꜥꜣt ꜥnw* cannot be anything but a type of stone, the determinative of *ꜥnw* being therefore mistaken. He tentatively identifies *ꜥnw* with *ꜥỉn*, and takes the expression as equivalent to *ỉnr ḥḏ nfr n ꜥỉn*, with the slight difference that in this case it means 'hard limestone' and includes the coloured varieties. Varille[6], however, giving *ꜥꜣt* the wider sense of 'une matière dure', sees the phrase as *ꜥꜣt ꜥš*, quoting Loret[7] in support, and Montet[8] also thinks that *ꜥnw* is wood, not *ꜥš*, but an early writing of *wꜥn*. The Wörterbuch seems to point to a third interpretation, by which *ꜥꜣt* is a word for 'Steingefäß', and *ꜥnw* a type of vessel, so that *ꜥꜣt ꜥnw* would be a vessel of the type *ꜥnw*.

That *ꜥꜣt* in this instance is a word for (stone-)vessel is not altogether improbable, but to regard *ꜥnw* as the name of a particular class of vessel is to ignore the parallel *ꜥꜣt ḫsbḏ*, in which the latter element is clearly the name of a material. Now although the peculiar determinative of *ꜥnw* might suggest a wood, such an interpretation does seem somewhat unlikely in relation to what are fairly evidently stone vessels[9]. In any case, the pink or red colour would rule out *ꜥš*, which was a yellow wood, and possibly also the red-brown *wꜥn*, and no other wood suggests itself. An identification with *ꜥỉn*, limestone, is tempting, and a pinkish variety would then be referred to here, the determinative being erroneous.

In default of further evidence, no certain conclusion is possible, though it seems likely that *ꜥꜣt* is the term for stone-vessel rather than the commoner word for stone etc., and that *ꜥnw* is either (pink) limestone, or some totally unknown material.

ꜥb (no Wb. ref.) is known from two, possibly three, examples, from which, however, little can be deduced. It occurs among products of the Wadi Natrun[10], damaged but recognisable, and in the Chester Beatty mineral list[11], and a similar word ,*ꜥbw* or *ꜥb-ḫꜥw*, from the eastern desert, is found in the Edfu mineral list[12], the former reading being perhaps the more likely, and identification with *ꜥb* therefore possible. The nature of the mineral cannot be established,

---

[1] P. Ch. B. IV vs. 7, 12. [2] Cf. Chronique 28 p. 36.
[3] Not 'roof' as Wb.; cf. Rev. d'Ég. 8 p. 111 n. 3; Chronique 28 p. 35.
[4] Medum 13. [5] Mitt. Kairo 3 p. 70. [6] Ann. Serv. 33 p. 90.
[7] Ann. Serv. 16 p. 34 n. 4. [8] Scènes p. 308.
[9] It is perhaps possible that the expression refers to a dummy vessel of wood, which would explain the inconsistency. [10] Peas. R. 23. [11] P. Ch. B. IV vs. 8, 2. [12] Edfou VI, 202, 10.

though a stone seems likely from the determinative in the first two examples. The only other materials so determined in Peas. are *ꜥnw*, evidently pebbles of some kind, and *snt*, possibly a species of ochre, though there are many such in the Chester Beatty list.

The identification as a stone would be strikingly confirmed were the word, mentioned above (p. 81–2) in discussing *bḫn*, to be read *ꜥb*. The group occurs as the material for shrines[1], and is clearly the name of a monumental stone, which would seem to rule out *ḫsbḏ* as a possible interpretation. But the reading *ḫsbḏ* is in any case beset by almost insurmountable difficulties, and to interpret the word as *bḫn*, though tempting on archaeological grounds, is by no means easy from a philological standpoint, as has been shown above. Prof. Fairman thinks the reading may be *ꜥb*, and if this is accepted, identification with *ꜥb* under discussion would seem plausible.

But whether such an equation be accepted or not, the meaning of *ꜥb* remains obscure. It may be connected with *ꜥb* 'be pure' (Wb. I. 175.4), *ꜥb* (Wb. I. 176.8), *ꜥbꜣ* 'offering-table (Wb. I. 177.7), or *ꜥbꜣ* 'glitter' (Wb. I. 177.11), and some such translation as 'clear' or 'shining' stone may be suggested, though the stone to which this would refer cannot be established.

wtšy (Wb. I. 381.8) occurs once in the Sehel inscription[2]. Brugsch saw *wtšy ḥḏ*, and the Wörterbuch follows this, translating 'Art Stein von weißer Farbe, der bei Elephantine gefunden wird'. Barguet, however, sees rather than , and comments[3] 'le signe fait plutôt penser au peson *tḫ* avec lequel le mot *wtšy* (à lire *wtḫy*?) est peut-être apparenté'. From the photograph, the sign might be or , and the reading must remain in doubt. That the stone is to be identified with *tšy*, occurring in the same line, is not altogether impossible, but seems very unlikely, and there is no clue as to the meaning of either.

One further point must be mentioned. The '*wetesh*'-stone which appears in the index of Temple of Hibis III, and is described in the text[4] as 'beautiful white blocks of '*wetesh*'-stone', is a gross error. The writers have unaccountably misconstrued the common word *wmt*, 'pylon', as an unknown stone '*wetesh*', ignoring the fact that *inr ḥḏ nfr* immediately precedes a lacuna in which *n rwḏt* almost certainly stood. In fact, the passage refers to 'a pylon of beautiful white stone (of sandstone)', and no connection with the *wtšy* of the Sehel text can be maintained.

bḫs (Wb. I. 469.12) is apparently ἅπαξ in a context in Harris[5], referring to 'large statues like mountains of alabaster and *bḫs*', from which it is clear that *bḫs* is a monumental stone. The Wörterbuch reading *bḫs* is not the only possible interpretation; Piehl prefers *ḥws*, and *ḥts* must also be considered. The latter reading might suggest an identification with *ḥts* (Wb. III. 203.11), though there are objections, particularly if *ḥts* is to be regarded as an aberrant writing of *ḥrst*[6]. Without further evidence no solution seems possible.

pꜣḳt (no Wb. ref.), written, is known only from a damaged passage in the Menshiyet es-Sadr stela[7] running as follows: *kt ẖꜣt m pꜣḳt inm.s mi ḥḏ iꜥ*. The reading is problematic, but *pꜣḳt* seems likely. The second sign is in lacuna, though traces remain which suggest that (H. 3) stood there, perhaps reversed, and this is confirmed by philological considerations. That *pꜣ* is not the article is shown by the feminine *inm.s*, so

---

[1] Dend. Chass. II, 24–5; II, 54–5; 34 examples in all. [2] Sehel 15.
[3] Stèle de la Famine p. 23. [4] Temple of Hibis III p. 31, re pl. 58.
[5] Harr. 4, 8. [6] Cf. below p. 123. [7] Ann. Serv. 38, 223, 13.

that the word must have *pꜣ* as its first syllable, and end in *t*, which again points to *pꜣḳt* as almost the only possibility.

The context is suggestive. 'Another quarry of *pꜣḳt*, its colour like washed silver', points to a stone of a whitish or silvery colour, and since the quarry is at Elephantine, the source is also known. Comparison with other words of similar form yields several possible interpretations. *pꜣḳt* (Wb. I. 500.1–3) bears the meaning 'potsherd' in at least one instance[1], and Pliny[2] mentions a stone ostracitis of which samples are like potsherds, and are used like pumice for rendering the skin smooth; Dioscorides[3] gives much the same information as Pliny. The identity of Pliny's ostracitis is not certain, but Bailey[4] points out that 'some sandstones would answer the description fairly well', and sandstone does in fact occur at Aswan.

Alternatively, if *pꜣḳt* is connected with *pꜣḳt* in its meaning of shell[5], it may perhaps represent a limestone containing ammonites and other shells, since limestone also occurs near Aswan. A nummulitic limestone would also be suitable if *pꜣḳt* were related to *pꜣḳ*, 'fine, thin' (Wb. I. 499.6–7), and *pꜣḳt*, 'plate' (Wb. I. 499.8), the latter being a possible description of the flat nummulites. On the other hand, derivation from *pꜣḳt*, 'thin plate', and comparison of the colour to that of washed silver, might suggest a mica-schist such as in fact occur in the Aswan region[6].

The problem is a difficult one, and no satisfactory translation of *pꜣḳt* is possible. The most that can be said is that the word represents a stone quarried at Elephantine and used for statuary, the identification with mica-schist being perhaps the most satisfactory.

pꜣgꜣw (no Wb. ref.) is known from two examples, occurring once in a list of materials, mainly precious stones, over which a spell is to be recited[7], and once in the Chester Beatty mineral list[8]. In the first example, *pꜣgꜣw* is quite clearly a material, since it has the 𓈒𓏥 determinative, and comes between *mꜣṯ* and *biꜣ n pt*. In the second, however, *pꜣgꜣw* occurs in the phrase *ḫsbḏ pꜣgꜣw šbnw ḥr wꜣḏ šmꜥ*, and has 𓈅 as determinative, giving the appearance of being a place name, the provenance of *ḫsbḏ*. This is quite possible, since a place *pꜣgꜣw* is otherwise known[9], evidently a region in the delta, but such a location would hardly be suitable as a source of lapis, unless it were a trading station. But even if *pꜣgꜣw* is to be interpreted as the name of a locality in the latter context, it can hardly be such in the former, and the only satisfactory compromise would be to regard *pꜣgꜣw* as the name of a place, and of a mineral, probably a precious stone, coming from it (cf. *tfrr*). In effect, no conclusion is possible on the present evidence.

mnhyw (no Wb. ref.) is apparently ἅπαξ in a list of statues and their materials[10]. The substances mentioned include *šndt*-wood, ebony, ivory, *bḫn*, *ss(n)ḏ(m)*-wood, *mnhyw* and granite, and the determinative of *mnhyw* shows it to be a monumental stone. No identification suggests itself, and confusion with any other more common name seems unlikely.

---

[1] Eb. 78, 17. P. Kah. Vet. 55 is probably linen; Westc. 6, 10 would be better translated 'shell' – cf. 'tortoiseshell', Eb. 67, 5–6; 71, 15; 86, 11; 87, 18.

[2] HN. 36, 31, 139; cf. ib. 37, 177; 37, 151; Dioscorides 5, 146 (W); Isidorus 16, 4, 20, 25.

[3] 5, 146 (W); cf. also 5, 74 for an ostracitis which is a form of cadmia.

[4] Elder Pliny 2 p. 256.

[5] Cf. above note. 1.

[6] Cf. Hume, Geology 2 pt. 1 pp. 47; 49; 51; 104–5.

[7] P. Ch. B. IX B, 18, 8.

[8] P. Ch. B. IV vs. 7, 5.

[9] Gauthier, Dict. géog. II p. 41; cf. also Onom. 2 p. 77f. re *pr-šs*.

[10] ZÄS. 37, 96.

mntt (Wb. II. 91.10–12) is represented by only a very few examples, from which, however, the nature of the stone can be deduced with some certainty. The word was originally translated as black granite[1], on the basis of LEGRAIN's description of the material of the Cairo statue 42004 as 'granit noir avec taches rouges'[2], and the same view is also taken by BALCZ[3], who regards *mntt* as a form of *mnw*, and renders 'schwarzer und schwarzweiß gefleckter Granit (?)'. The correct interpretation is outlined by SETHE[4], who notes that an inscription in the diorite quarries in the western desert mentions *mntt* as the stone brought thence, and that the word is also known from two statues, B.M. 870 and Cairo 42004, both erroneously described as black granite. SETHE questions this description, pointing out that the London statue has yellowish-green blotches 'rather like a breccia', and that the Cairo one is diorite-breccia according to BORCHARDT. ENGELBACH[5] also refers to *mntt* on the diorite stela of Amenemhet II from the Tushkah quarries, mentioning that both 'Chephren-diorite' and the white speckled variety used for bowls in the predynastic period were found there. The Cairo statue, however, he wrongly describes as 'speckled dark granite'. ROWE[6], too, takes *mntt* as the word for diorite, but also for a type of dark red granite with prominent spots of red felspar, presumably from a mistaken analysis of the Cairo statue.

The most significant reference is unquestionably that on the stela from the Tushkah quarries[7], where *mntt* and *mḥn(m)t* are mentioned as being fetched thence. The quarries were undoubtedly the source of the peculiar variety of diorite from which the well-known statues of Khafrēᶜ are made, and also, according to ENGELBACH, of the commoner white speckled kind[8].

The two statues[9] described a *mntt* are of considerable interest. That in Cairo has been variously described as 'black granite', 'speckled dark granite', 'dark red granite', 'dark grey granite' and 'diorite-breccia', and reference has been made to the spots or crystals of red felspar which are present. No exact petrographic analysis seems to have been made, but from the general description it is apparent that the stone is in fact the distinctive diorite from Tushkah. The London statue, also described as 'black granite'[10] is said to be spotted with 'irregular yellowish green blotches; rather like a breccia'[11], and there seems good reason to suppose that it too is of Tushkah diorite, or something so similar that it was thought to be such by the Egyptians.

mntt is also the material of certain stone vessels[12], for which both the Tushkah diorite and the coarse-grained black and white speckled variety were used in early dynastic times[13]. From this it might appear that *mntt* can refer to both varieties, and ENGELBACH's statement that the coarser species is also found at Tushkah would tend to confirm this. In favour of such an extended meaning is also the fact that there does not otherwise seem to be a term for the commoner diorite.

If *mntt* is then translated diorite, it must be borne in mind that this is not the most satisfactory name for the characteristic Tushkah variety. LUCAS points out that the rock has a

---

[1] Eg. Sethe, Sahurēᶜ II p. 125.
[2] Statues I p. 3.
[3] Mitt. Kairo 3 p. 68.
[4] Bau- und Denkmalsteine p. 49.
[5] Ann. Serv. 33 p. 65.
[6] Ann. Serv. 38 p. 395; p. 682.
[7] Ann. Serv. 33, 70 fig. 2.
[8] Cf. Lucas, Materials p. 465–67.
[9] Cairo 42004, Rec. Trav. 26, 222; B. M. 870, Auswahl IX b, c.
[10] Budge, B. M. Guide, Sculpture p. 17.
[11] Glanville quoted by Sethe, loc. cit.
[12] LD. II, 28; Sahurēᶜ pl. 61 (a fluted (?) bowl, but cf. also Ann. Serv. 34 p. 110 fig. 18).
[13] Cf. Orientalia Suecana 5 p. 7f. for a bowl of Tushkah anorthosite quite unlike the characteristic variety of the Khafrēᶜ statues.

gneissic structure, and that diorite-gneiss or even anorthosite-gneiss would be more suitable descriptions. While, however, the latter may be more scientifically correct, it seems necessary to retain the diorite connection for archaeological purposes, and diorite-gneiss is therefore to be preferred as a name for the so-called 'Chephren diorite'. *mntt* therefore is to be translated 'diorite-gneiss', and perhaps also simply 'diorite'.

mḫtbtb (no Wb. ref.) appears to be ἅπαξ in the Sehel inscription[1], where BRUGSCH was unable to suggest a satisfactory reading. BARGUET's *mḫtbtb* seems from the photograph to be accurate, and his suggested connection of the word with *mḫtbt* (Wb. II. 133.1) is plausible[2]. Following GRAPOW[3] and SETHE[4], he regards *mḫtbtb* as 'un substantif à préformante *m*, formé sur le verbe *ḫtb* (*ḫꜣtb*) dont les deux dernières radicales ont été redoublées', which is no doubt correct. His contention, however, that the related *mḫtbt* is '(le) nom d'une amulette de couleur verte', is puzzling, and cannot be supported. GRAPOW[5] regards it as 'Art goldenes Schmuckstück', and SETHE translates 'Platten'[6], taking it as an amulet of gold[7]; RANSOM-WILLIAMS[8] suggests 'ingot of gold'. Whatever the nature of the object, the WÖRTERBUCH references make it quite clear that it was of gold, and not, therefore, 'de couleur verte'. From connection with this amulet, *mḫtbtb* might be translated 'the golden stone', and compared with Greek χρυσόλιθος, though the related verb *ḫꜣtb*, 'schonen', might suggest a more general meaning such as 'protective' or 'amuletic' stone, which cannot be related to any particular material.

nmtt (Wb. II. 272.2) and *nmti* (Wb. II. 272.1), separated by the WÖRTERBUCH, are identified by BALCZ[9], and this seems justified by comparison of the single example of *nmti*[10] with the two early writings of *nmtt*[11]. *nmtt* is translated as red granite by the WÖRTERBUCH, no doubt from the occurrence of the word on a naos described by GAUTHIER as being of that material[12]. BALCZ, however, takes *nmtt* to be white quartzite, and possibly also the red variety, quoting verbal information from LACAU concerning a Cairo naos, stated to be of *nmtt*, which is in fact white quartzite from Gebel Aḥmar. SELIM HASSAN[13], who may have been in a position to check this information, seems also to regard *nmti* as quartzite.

The most important clue to the meaning of *nmtt* is its occurrence on the naos or naoi in Cairo. It is surely not without significance that both these naoi are said to be in Cairo, and to belong to the Saite period, and it may be suspected that they are in fact one and the same. In this case, LACAU's 'white quartzite' is more likely to be correct, both because it is a later identification, and because several superficially similar stones are frequently mistaken for granite. Furthermore, *nmti* occurs together with *mꜣt* in the Medum inscription, and to have two different words for red granite in the same context, and at such an early date, is unreasonable. The use of *nmtt* as a material for stone vessels is of little assistance in determining the meaning since most varieties of stone were so employed during the early dynasties.

---

1 Sehel 15.
2 Stèle de la Famine p. 23.
3 Wortbildung p. 27.
4 ZÄS. 48 p. 143; Dram. Texte p. 187.
5 loc. cit.
6 ZÄS. 48 p. 143.
7 Dram. Texte. p. 187.
8 Jewelry p. 2.
9 Mitt. Kairo 3 p. 66.
10 Medum 13.
11 Medum 20; Giza J. I, Ab. 53.
12 Ann. Serv. 23 p. 171.
13 Giza H. VI (2) p. 57.

nmtt, therefore, is probably to be taken as white quartzite, for which no other specific term exists[1]. That it also refers to the red variety is not impossible, since the regular *biꜣt* is not known from the Old Kingdom, and nor are the other terms *bnwt* and *inr n ḏw dšr*. A precise petrographic analysis of the Cairo naos or naoi would, however, be desirable.

nn (no Wb. ref.) seems to be confined to a single inscription[2], in which it occurs four times, in each case as the material of a divine statue – Isis, Nephthys, Isis and Selkis –, twice gilded. DARESSY[3] comments, 'Je ne connais pas d'autres exemples de ce mot . . . Il est probable qu'il désigne le schiste ou la pierre saponaire, si usités pour les statuettes de divinités'. This may well be correct, since new names for materials are not infrequent in late texts[4], and the statues of goddesses call to mind the green 'schist' figures of Osiris, Isis and Tewosret in the Cairo Museum, belonging to the 26th and 30th dynasties. A somewhat similar 'green stone' figure in the British Museum has the remains of a gold mask[5]. It may perhaps be objected that 'schist' is properly *bḫn*, but although this may be true, it is not impossible that *nn* should refer to a specific variety of a characteristic colour, and used for a particular purpose. However, the evidence is by no means conclusive, and no valid translation is possible.

rꜥgs (Wb. II. 403.6) appears to be ἅπαξ in the Sehel inscription[6], where neither BRUGSCH nor BARGUET comments upon it. The reading is certain, though the third sign might conceivably be *nst* rather than *gs*. Philologically, the name suggests nothing, unless it is in some way connected with the preposition *r-gs*, and no possible interpretation can be offered[7]. The meaning must remain obscure, though it is probable that a semi-precious or monumental stone is meant.

ḥsg (Wb. III. 169.8) occurs only once, as a 'label' beside two tall jars given by Thutmose III to Amūn[8]. That *ḥsg* is the name of a stone is evident from the determinative, and from the fact that the other parallel 'labels' are the names of materials and not of different types of vessels. The word may represent one of the rarer Egyptian stones, in which case no identification is possible. On the other hand, a connection with Akkadian *ᵃḫusigu*, tentatively translated 'amazon stone' by CAMPBELL THOMPSON[9] is just feasible.

ẖt (Wb. III. 359.5) is found in three versions of the same passage in the 'Book of the Dead'[10], as material of a temple, in place of the older *ḥrst*. Ten texts are collected by SETHE[11], four having the phrase *ḥwt-nṯr.f m ḥrst*, and two corruptions of *ḥrst*, three replacing *m ḥrst* by *m ẖt*, and one by *m šꜣ* . . . The significance of *ẖt* in these instances is difficult to determine. It can scarcely, as SETHE points out[12], have anything to do with *ẖ(ꜣ)t*, used like *ꜥꜣt* in late

---

[1] *bnw ḥḏ*, Tanis II, 42, refers not to white quartzite, but to Nubian sandstone; cf. above p. 76.
[2] Rec. Trav. 30, 2. [3] loc. cit.
[4] The text is of Saite date, but purports to be a copy of a 4th dynasty original. Whether such an original ever existed is doubtful.
[5] B. M. 1482, said to be of 20th dynasty date. [6] Sehel 15.
[7] Cf. perhaps Demotic *gs*, P. Ryl. IX, 24, 3, which, however, seems to be a generic term for stone.
[8] Urk. IV, 641. [9] Dictionary p. 178. [10] Totb. Nav. 108, 3.
[11] Totb. Stud., ZÄS. 59, 45*. [12] ZÄS. 59 p. 81.

texts[1], nor does it seem likely that it is merely a new word for carnelian, used instead of *ḥrst* on these occasions only. The double determinative 𓈇 and ∘∘∘ might perhaps point to an earthy substance, but no suitable rendering suggests itself. In default of further evidence, no conclusion can be reached, and *ẖt* must be regarded as a quasi-artificial term of doubtful meaning, possibly some kind of stone.

ḳḥ (Wb. V. 67.1) seems to be confined to a single text[2], in which it occurs a number of times in the phrase *r šꜥd ḳḥ*. It is evident from the context that *ḳḥ* is a building material of some kind, and Gardiner[3] is inclined to emend to *ḳḏ*, which would not be inappropriate. The text is apparently concerned with only one material, and since gypsum-workers, (*ḳḏy*), are mentioned[4], it seems logical to suppose that the material was gypsum, and that *ḳḥ* is for *ḳḏ*. On the other hand, it is difficult to see why corruption has occurred only in the writing of *ḳḏ*, and not of *ḳḏy*. That gypsum should be used as a building material is not impossible, since it does sometimes occur in a rock-like formation[5].

Two possible examples of *ḳḥ* outside the Turin text must here be noted, both being more fully discussed below sv. *ḳꜣḥ*. An ostracon from Deir el Bahari[6] mentions *nḏw ḳḥ* in a list of workers, and although this may mean 'kneader of clay' or the like, it could equally well refer to the crushing of *ḳḥ*, the material under discussion. More significant perhaps is the *ḳḥ* found in an erotic poem[7], to which a part of the girl's body is compared. If the comparison is with teeth, then *ḳḥ* must be a white stone, and might be the *ḳḥ* of the Turin text. But if such an identification were accepted for either of these two references, *ḳḥ* could hardly be regarded as a corruption of *ḳḏ*, and would have to be considered as a separate word.

A possible explanation is that *ḳḥ* refers to (gypsum) plaster, which would suit the white colour, and account for the presence of gypsum-workers in the Turin text[8]. On the other hand, the determinative and the use of the verb *šꜥd* suggest something more solid, and in default of further evidence such an interpretation can be no more than tentative[9].

ḳḏ (Wb. V. 82.7), originally regarded as a word for rubble or the like[10], is clearly the regular name for gypsum in the New Kingdom. This is proved by the two lumps of gypsum, found by Wiedemann and published by Spiegelberg[11], which bear hieratic inscriptions identifying them as *ḳḏ*. To these original examples may now be added a number of others, gypsum cones from Amarna bearing the date and the description *ḳḏ*[12]. No further proof is required, though it is interesting to note, as Gardiner points out[13], that the use of *ḳḏ* in the Leopold-Amherst papyrus[14] can be parallelled by the actual employment of gypsum to protect the entrance to the burial-chamber of Tutankhamūn. Worthy of note, too, is the use of gypsum plaster to cover wood[15].

---

[1] Wb. III. 360.15; cf. above p. 23.

[2] P. Tur. PR. 61 = RAD. 22, 1, 3; 1, 15; 1, 18; 2, 3; 2, 6; cf. however, the *inr ḳꜣḥyt* of Coff. Texts IV, 100e, which may be this word.

[3] RAD. 22 note on 1, 15.

[4] RAD. 22, 1, 10; 1, 11; 2, 3.

[5] Lucas, Materials p. 97; p. 470; Theophrastus § 65 – but no use in building is quoted.

[6] Ostr. D. el B. ⟨var. 198⟩.

[7] Louvre C 100.

[8] The *nḏw ḳḥ* would then be a workman engaged in preparing plaster; cf. below p. 205f.

[9] The verb *sḳ(ꜣ)ḥ* might be related to *ḳḥ* rather than *ḳꜣḥ*, which would be more consistent with its occurrence in connection with wood; cf. below p. 205.

[10] Goodwin, ZÄS. 14 p. 106; Brugsch, Wb. Erg. p. 1270.

[11] ZÄS. 58, 51.

[12] Cit. of Akh. III nos. 331–5.

[13] JEA. 22 p. 178.

[14] P. Leo. Am. 2, 11.

[15] Hor. & Seth 13, 5–6.

The orthography of *ḳḏ* shows quite clearly that it is a loan-word, such aberrant writings as the *ḳꜣḏꜣḏꜣ* of a Cairo ostracon[1] being particularly significant. The word is doubtless to be derived from Akkadian *gaṣṣu*, from which Greek γύψος, Latin gypsum, and hence English gypsum, German Gips etc. all stem. But if *ḳḏ* is an adopted name for gypsum in the New Kingdom, the problem arises of what was the older word for gypsum which the new term *ḳḏ* replaced. Gypsum was in use in predynastic times, and was widely employed as mortar and plaster, and for vases, in the Old Kingdom[2], so that there must have been a name for it before the introduction of *ḳḏ*. It is suggested below (p. 190) that this may perhaps have been *bsn*, but there is room for doubt. As material for vessels, it could possibly have been confused with limestone or even alabaster, and may have been referred to accordingly.

kꜣ (Wb. V. 93.10) is known from only a few examples, all from mastabas of the Old Kingdom[3], where it occurs with e.g. *tiiꜣꜣt* and *nmtt* as the material of certain stone vessels. Balcz[4] notes the various examples, pointing out that nothing is to be gained by comparison of the types of vessels represented, and is unable to offer any suggestion as to the meaning. Selim Hassan[5] strangely refers to *kꜣ* as 'heading for a class of vessels', though it is quite clearly a material. Like *iiꜥꜣꜥꜣ* and *tiiꜣꜣt*, *kꜣ* presents an insuperable problem. There can be no doubt that it refers to one of the stones used for making vessels in the early dynastic period, but there is insufficient evidence to support an identification. That *kꜣ* is connected with the semi-precious stone *kꜣ* discussed in the next chapter (p. 133) is possible, but unlikely, since there does not seem to be any translation equally suitable to both.

tiiꜣꜣt (no Wb. ref.) is known only as the material of stone vessels, and is apparently confined to inscriptions of Old Kingdom date. The meaning is problematic, and the reading by no means certain. Balcz[6] rightly points out that 'es muß ein äußerst geschätztes Material für Vasen gewesen sein', and tentatively reads *tiꜣt* or *tiiꜣꜣt*. Murray and Griffith[7] read *ti-šps* or *ti-iꜣm*, the former of which Balcz rejects out of hand, while conceding that the latter is possible in so far as [8] clearly represents a tree. But *ti-iꜣm*, though possible, is unlikely, owing to the existence of the variant [9]. The nature of the bird is not certain, but Balcz inclines to regard it as *ꜣ*, reading *tiꜣt* or *tiiꜣꜣt*, and pointing out that there is a word *iꜣꜣt* (Wb. I. 27.10) meaning 'Art Keule oder Szepter', and another (Wb. I. 27.11) meaning 'Rute, Zweig', so that may be a compromise between and , with the value *iꜣꜣt*. The suggestion seems reasonable, and no more satisfactory reading can be proposed.

The vessels represented as being made of *tiiꜣꜣt* are of various forms, and are of little assistance in determining the meaning, though Balcz suggests reasons for thinking that the stone was red. Selim Hassan[10], who accepts the reading *tiiꜣꜣt*, refers to it as 'a general heading for a class of vessel', and, more reasonably, as 'vessels of an unknown stone'. The nature of the stone cannot be determined, though if it was in fact red, it may have been a breccia[11], or, less probably, imperial porphyry[12]. However, not even the colour is certain, and the meaning of *tiiꜣꜣt* must remain obscure.

---

1 Ostr. Cairo 25521 rt.
2 Lucas, Materials p. 96; p. 470 etc.
3 Cairo 1385 = Mar. Mast. A 2; Giza J. I, Ab. 53; Medum 19.
4 Mitt. Kairo 3 p. 73.
5 Giza H. VI (2) p. 57.
6 Mitt. Kairo 3 p. 71.
7 Saqq. Mast. 1 p. 34.
8 Medum 13; 19; 20; cf. also the det. of Giza J. I, Ab. 53.
9 Q. Saqq. VI pl. 26; ? also ib. pl. 19.
10 Giza H. VI (2) p. 57.
11 Lucas, Materials p. 464.
12 ib. p. 474.

tỉ(ḥ) (Wb. V. 242.10) is virtually ἅπαξ, occurring in some dozen versions of the same passage. The Wörterbuch reading *tỉḥ* is derived from only one of these[1], and none of the other examples[2] has an *ḥ* written. The writings in the Coffin Texts include , , , , and , the last perhaps indicating that the change in interpretation to *ỉnb* had already begun. This change is exemplified in the parallel texts of the 'Book of the Dead'[3], where beside and are three examples with , two with *ỉnb* written fully, and three with *m ỉ3btt pt*. The context runs: 'I know this mountain of *bḫ*, upon which the heaven rests; it is of *tỉ(ḥ)*, 300 ells long, 150 broad'. Thus *tỉ(ḥ)* is the substance of which the mountain is made, possibly an archaic term, later misunderstood and corrupted to *ỉnb*, probably because of the sign. That some kind of stone is meant seems certain, since in every case but one the determinative is . The writing which might be *ḥsmn*, is probably to be regarded as an attempt at rationalisation by a scribe faced with a word he did not understand. The exact significance of *tỉ(ḥ)* cannot be determined but it seems likely that it represents a building material rather than a specific stone or mineral. That the true reading is *tỉ(ḥ)* may be doubted, and *tỉ* or even *tỉt* would appear preferable from the Coffin Text writings. On the other hand, the possibility cannot be ignored that is a corruption of , and that the reading is therefore *tỉỉ33t*, the word being identical with that discussed above.

A similarly written word is found in a fragment of the unpublished Abusir papyrus[4], where it is parallel to *šndt*-wood and other substances as the material for various objects. The writing alone would seem to point to an identification with the word under discussion, but the context appears rather to suggest interpretation as *tỉỉ33t*, so that this one example may possibly provide a connection between the two. However, the point cannot be pressed, and in view of the uncertainty of the evidence, the terms are best considered apart.

tm-ỉḳr (Wb. V. 306.11) occurs only in the Sehel inscription[5] ,where Brugsch saw only *tmy*, the reading which appears in the Wörterbuch. Barguet[6], however, reads , which is confirmed by the photograph, the *r* alone being damaged. He transliterates *tm-ỉḳr*, and comments, 'il faut peut-être comprendre, lit. "celle qui est absolument irréprochable", et l'on pourrait alors penser à un cristal de roche'. The translation and literal interpretation are plausible, but it is difficult to believe that *tm-ỉḳr* is yet another name for rock crystal, for which several terms are known. It is more likely to represent some rare or newly discovered stone, and may perhaps be a loan-word, or an attempt to render a foreign, perhaps Greek, word in Egyptian.

tšy (Wb. V. 330.1) occurs once in the Sehel inscription[7], where the reading is certain. Barguet[8] regards it as perhaps identical with *wtšy*, which seems most unlikely, since the two words are found in the same line. *tšy* is said to be in the middle of the river, *wtšy* to the west and in the river, so that some distinction must be implied. Identification of *tšy* with Akkadian *ªdušû*[9] is tempting, but unlikely, in that another word for crystal or the like would be superfluous, and no translation is possible.

[1] Totb. Stud. ZÄS. 59, 44*.
[2] Coff. Texts II, 376b; Totb. Stud. ib.
[3] Totb. Stud. ib.
[4] P. Abusir frag., unpublished [17].
[5] Sehel 16.
[6] Stèle de la Famine p. 24.
[7] Sehel 15.
[8] Stèle de la Famine p. 24.
[9] Campbell Thompson, Dictionary p. 175.

In addition to the terms discussed above, there are a few, known from Demotic texts, for which no hieroglyphic original can be traced. Of these, one in particular deserves some mention here.

tmgy occurs three times in P. Rylands IX, where it is evidently the name of a stone employed for statuary and the like. The first passage[1] runs as follows: 'He reached Elephantine, he caused a tablet of stone of Elephantine to be quarried, and the blocks for two statues of *tmgy*', in the second[2], details of the statues are given, and the third[3] refers to the statues (*ꜣny tmgy*) being cast into the river. GRIFFITH[4] tentatively connects *tmgy* with  and 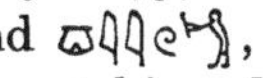, hardstone mason, and points to *mꜣky* (sic.) of the Sehel inscription[5] as the nearest hieroglyphic equivalent.

Neither of these suggestions seems at all plausible, and identification with *mꜣky* (sic.) is ruled out altogether, since *mꜣky* (sic.) is certainly *ḥmꜣky*, for *ḥmꜣgt*. But no more satisfactory interpretation can be offered. Corruption of *ṯmḥy*[6] to *tmgy* is practically impossible, although the Nubian provenance of the former might favour a connection, and derivation of *tmgy* from the *tm-iḳr* of the Sehel inscription[7] seems equally unlikely, though again the provenance would be appropriate. In fact, no likely hieroglyphic form of *tmgy* can be traced, and since the Demotic contexts are unhelpful, the exact meaning of the word must remain in doubt, though it quite clearly represents some kind of monumental stone.

Reference may also here be made to a word *waṯha*, occurring several times in the Tell el Amarna correspondence, and evidently regarded by CAMPBELL THOMPSON[8] as a white stone of which the Egyptian name was something like *waṯha*. But the word does not refer to a stone, as is shown by letter 14, line 53, the only significant reference, from which it is clear that *waṯha* is the name of an oil vessel[9], and this is apposite in every other instance.

Of the various terms discussed in the present chapter, relatively few have been identified with any degree of certainty. The majority remain in doubt, and no systematic treatment is therefore possible. It is, however, probable that among these words are hidden the names of a number of the secondary building and monumental stones used by the Egyptians, and to which reference cannot otherwise be traced. These include anhydrite, basalt, breccia (both green and red), diorite (other than the Tushkah variety), dolerite, dolomite, gypsum (before the New Kingdom), marble, obsidian, imperial porphyry, porphyritic rock, and slate. That so many materials were not represented in the language is unlikely, and although some may perhaps have been confused with other, more common, stones, the majority must surely have had distinctive names.

Some tentative identifications have been suggested in this and other chapters. Green breccia and slate were perhaps confused with greywacke, and called *bḫn*[10], the commoner varieties of diorite may have been embraced by *mntt*[11], *ʿꜣt ḥḏt* may refer to dolomite or marble, or both[12], gypsum was probably confounded with limestone or alabaster, or included

---

[1] P. Ryl. IX, 7, 14f.
[2] ib. 7, 17.
[3] ib. 18, 20.
[4] Cat. Ryl. III p. 229.
[5] Sehel 16.
[6] Cf. below p. 154.
[7] Sehel 16.
[8] Dictionary p. 142.
[9] Cf. Lambdin, Orientalia 22 p. 368, though the explanation is incorrect. The word is probably Egyptian *wḏḥ*, cf. Ostr. DM 318,5; 434, II, 8.
[10] Cf. above p. 78f.
[11] Cf. above p. 87f.
[12] Cf. below p. 100f.

under *bsn*[1], and *tiiꜣꜣt* may be either red breccia or imperial porphyry[2]. But none of these suggestions can be accepted without the utmost reserve, and all may be wide of the mark. Basalt and dolerite, which are fundamentally the same, may well have been regarded as such by the Egyptians[3], but even so no name is known, and neither porphyritic rock nor the very characteristic anhydrite have been identified from textual references. The obvious conclusion is that these, and perhaps other stones, are among those occurring in Old Kingdom inscriptions and elsewhere, and discussed in this chapter, though no identifications are possible.

---

[1] Cf. above p. 90f. [2] Cf. above p. 91.

[3] NB. The Assyrians apparently confused dolerite with diorite. Campbell Thompson, Dictionary p. 163, gives [a]*ušû*, [a]*ESI*, both meanings.

CHAPTER 5

SEMI-PRECIOUS STONES

The various substances grouped together here are those used by the Egyptians in making jewellery and for other similar purposes, and generally regarded by them as next in value to the noble metals. The majority are what may loosely be termed semi-precious stones[1], though some are hardly such by present-day standards; the remainder are faience and glass imitations, equally esteemed by the Egyptians, and never clearly distinguished from the true minerals. Not only could the name of any stone be applied to a similarly coloured glass, but the more general terms for faience and the like, eg. *ṯḥnt*, are listed together with semi-precious stones on an equal footing.

In distinguishing between the lesser gem stones, the Egyptians must have been guided to a large extent by colour and provenance, and evidence of the former is perhaps the greatest help towards lexicographical identification. In several instances where only the colour of a stone is known, it is possible by a simple process of elimination to suggest a probable meaning, and indeed it was by colour alone that the nature of *ḫsbḏ* was originally established.

ꜣst (Wb. I. 20.16) occurs only once[2], in the following context: 'Cause to be brought the beads of faience (*ṯḥnt*), concerning which I wrote to you, as well as the requisites of *ꜣst*'. The parallelism with *ṯḥnt* would suggest a semi-precious stone or kindred mineral, and *dbḥt*, 'requisites' may then be 'Art Schmuckstücke', as the Wörterbuch translates (V. 442.4), though Caminos[3] prefers to render etymologically 'the (things) which are required'. In editing the text, Gardiner[4] noted *ꜣst* as an unknown word, probably a corruption of *ḥrst*, and Caminos accepts this view. The identification is tempting, since *ṯḥnt* and *ḥrst* occur together elsewhere[5], but the possible existence of a distinct word *ꜣst* cannot entirely be discounted.

iirrt (no Wb. ref.) is apparently ἅπαξ in Harris[6], where it stands between analogous items of *ḫnmt*, *wꜣḏ*, *ṯḥnt* and *ḥrst* as the material for *smdt*. The context thus suggests a gem stone or some species of faience, and the word is not to be confused with similar writings of *iꜣrrt*, as Piehl appears to do. On the other hand, *iirrt* may be derived from *iꜣrrt*, with some such meaning as 'the grape-coloured stone', which might suggest amethyst. This is assumed by Iversen[7], but the existence of the word *ḥsmn* makes such an interpretation extremely

---

[1] Ball, Roman Book p. 13, objects that 'semi-precious' is a meaningless term, since a stone is either precious or it is not. He therefore prefers to describe such stones as 'lesser gems', following Pliny, who divides gems into the more noble, and those of inferior quality. None of the true precious stones was known to the Egyptians.

[2] P. Bol. 1094, 2, 3.

[3] LEM. p. 8.

[4] LEM. p. 2a.

[5] Harr. 64b, 15.

[6] Harr. 64b, 14; cf. however, Harr. 64c, 13, which may be the same word.

[7] Paints and Pigments p. 10.

doubtful. Alternatively, *iirrt* may be connected with the verb *iri*, which would point to an artificial material, perhaps a specific variety of faience. Without, however, further examples of the word, no positive conclusion is possible.

ibhty (Wb. I. 64.1) and *bht* (no Wb. ref.) are undoubtedly the same word, as was realised by Brugsch[1] and Sethe[2], though the identification has been implicitly denied by Wendel[3], and questioned by Jéquier[4] and the Wörterbuch, whose comparison with *bht*, for which no entry in fact appears, is only tentative. That the two words are in fact one and the same is proved conclusively by the texts of filling the sound eye, in two versions of which *ibht(y)* appears[5], while a third has *bht*[6].

From the contexts in which it occurs parallel to other semi-precious stones[7], it is clear that *ibhty* was regarded as such by the Egyptians, and this is confirmed by the use of *ꜥꜣt* in reference to it[8]. The precise meaning is, however, a matter of some doubt, and various views have been advanced. Brugsch[9] favours derivation from *ibḥꜣt*, and hence renders as red granite, though elsewhere[10] he quotes Hebrew and Arabic parallels and translates 'unechter Marmor', 'marmorähnlicher Stein'. Wendel[11], on the other hand, follows Dümichen[12] and Mariette[13] in regarding *bht* as porphyry, and seeing *ḏw nw bht* as the Egyptian name for the mons porphyrites, Gebel Dokhân, which was exploited at a late period. Jéquier[14] offers no definite translation, but seems to favour a type of marble or false emerald (σμαραγδίτης), and the latter meaning is accepted by Barguet[15] for the Sehel passage[16]. Sethe[17], however, supports derivation from *ibḥꜣt*, takes the *ḏw nw bht* to have been in the southern, Nubian, mountains on the west of the Nile, and regards *ibhty* as amethyst from Tushkah. None of these translations is altogether satisfactory, and the evidence requires careful reassessment.

The most significant point is undoubtedly the location of the *ḏw nw bht*, situated in the region *št*, and mentioned together with the granite mountain[18], which must be the mons Claudianus, Gebel Fatireh. From this connection it seems likely that *št* is the Antaeopolite nome, and that the *ḏw nw bht* is the other great mountain on the Red Sea coast, the classical mons porphyrites[19], which is the view of Wendel. But Sethe notes that the region *št* is included among those of the western hill-country, *ḏw mꜣnw*[20], which would preclude identification of *ḏw nw bht* with Gebel Dokhân on the east coast. The problem is indeed complicated, and it may well be doubted whether the Egyptians were altogether certain of the meaning of *mꜣnw*. The Wörterbuch (V. 542.10–11) describes *ḏw mꜣnw* as 1. Berg des Westens[21], and 2. östlicher usw. Berg[22], with evidence for both interpretations, and it seems at least possible, therefore, that there is some confusion in the case of the *ḏw nw bht*, particularly since the *ḏw nw mꜣṯ* is similarly classified.

[1] Wb. Erg. p. 438.
[2] Bau- und Denkmalsteine p. 50.
[3] Bau- und Edelsteine p. 77.
[4] BIFAO. 19 p. 26.
[5] Edfou VIII, 136–7; Thes. 41–2.
[6] Philae Phot. 522.
[7] loc. cit. and Dend. Mar. IV, 87; Düm. G. I. II, 67; IV, 163; IV, 176; Admon. 3, 2.
[8] Düm. G. I. IV, 176.
[9] Hungersnoth p. 130.
[10] Wb. Erg. p. 438.
[11] Bau- und Edelsteine p. 77.
[12] Geschichte p. 167.
[13] Dend. Supp. pl. I.
[14] BIFAO. 19. p. 26.
[15] Stèle de la Famine p. 25.
[16] Sehel 17.
[17] Bau- und Denkmalsteine p. 50.
[18] Düm. G. I. II, 67, 9.
[19] For the two mountains cf. Fitzler, Steinbrüche p. 94f.
[20] Düm. G. I. II, 67; cf. Gauthier, Dict. géog. VI p. 118.
[21] Hamm. 19, 4; Mem. Miss. 15, 12, 3; Med. Hab. ⟨437⟩; Theb. Grab. 65 ⟨682⟩.
[22] Totb. Budge 189 (Nu 19, 27); Amarna V, 28; Goshen 1 right.

The translation porphyry has, however, serious objections, and can hardly be accepted. Imperial porphyry is confined to three localities in the eastern desert, and was used but rarely before the Roman period, when it was largely exported[1]. *ỉbhty* on the other hand, is said to come not only from the *ḏw nw bht*, but also from the neighbourhood of Elephantine[2], and is, further, brought as tribute from an Asiatic region[3]. Nor is it confined to texts of Graeco-Roman date, but occurs in the 18th dynasty and even earlier[4]. Thus, whatever view may be taken of the geographical question, porphyry seems a somewhat unlikely interpretation.

That *ỉbhty* is amethyst is even less likely, since *ḥsmn* is undoubtedly the regular word for the stone, and the two are juxtaposed in one passage[5].

The third translation, based on the Hebrew parallel *bahaṭ* rendered σμαραγδίτης in the Septuagint, is perhaps more satisfactory, though marble is to be preferred to false emerald, since the latter was in all probability *wꜣḏ*, and *ḏw nw bht* can scarcely be the mons smaragdites[6]. Various kinds of marble are found in Egypt, chiefly in the eastern desert, and it is found in use as early as predynastic times, and is employed extensively for statuary in the New Kingdom and in Roman times[7]. The chief objection to translating *ỉbhty* as marble seems to be that it can hardly be regarded as a semi-precious stone, and is not suitable for jewellery etc., though Pliny[8] does mention a species called memphites which was akin to the precious stones.

The references to *ỉbhty* and *bht* are not in themselves of any great significance, and provide no clue as to the nature or colour of the stone. *ỉbhty* occurs together with other semi-precious stones[9], and also with more miscellaneous mineral substances[10], being determined both with □[11] and ∘∘∘[12]. It is the material of amulets[13], and jewellery[14], but the two objects in the Dendera amulet list are made in a variety of materials, including 'basalt', lapis lazuli, serpentine, haematite, carnelian, steatite and porphyry, so that nothing can be deduced. The possible etymological connection with *ỉbhꜣt* is of little assistance in determining the meaning, since the precise location of that region is uncertain. Sethe[15], however, seems to see *ỉbhꜣt* as the source of diorite-gneiss, and there may perhaps be some significance in this.

In fact, the identification of *ỉbhty* presents a practically insoluble problem, since none of the possible suggestions appears to be consistent with all the evidence. Amethyst may be rejected, porphyry and 'false emerald' seem unlikely, and marble is not altogether satisfactory. It may be that *ỉbhty* is none of these, but some unusual species of diorite or the like. The most that can be said is that it represents a more or less rare stone, found in the southern hills and elsewhere, and used for amulets, beads and other similar objects.

ỉmyt (Wb. I. 78.4) is known only from a Louvre stela[16], where the reading is somewhat uncertain. The second sign, taken as ⊂ by the Wörterbuch, appears as 〜, and is in all probability a misunderstanding of a hieratic sign. Maspero, Madsen, and Sottas all

---

[1] Lucas, Materials p. 475.
[2] Sehel 17.
[3] Urk. IV, 715.
[4] Urk. IV, 715; M. u. K. vs. 4, 7; 6, 6; Admon. 3, 2.
[5] Admon. 3, 2.
[6] Cf. below p. 103f.
[7] Lucas, Materials p. 472.
[8] HN. 36, 11, 57.
[9] Edfou VIII, 136; Thes. 41; Philae Phot. 522; Admon. 3, 2. Dend. Mar. IV, 87 etc.
[10] Urk. IV, 715; Sehel. 17.
[11] Urk. IV, 715; Admon. 3, 2; Sehel 17.
[12] Dend. Mar. IV, 87; Düm. G. I. II, 176; M. u. K. vs. 4, 7 etc.
[13] Dend. Mar. IV, 87; M. u. K. vs. 4, 7.
[14] Admon. 3, 2; M. u. K. vs. 6, 6 (beads).
[15] Bau- und Denkmalsteine p. 49.
[16] C 14, 11.

read it as , the last specifically rejecting . 'Le monument', he maintains, 'ne porte certainement pas ; si le  est légèrement déformé, cela tient en grande partie à l'exiguité de l'espace à lui réservé au-dessus de '[1]. It must be agreed that the dubious sign is not , but to describe it as a slight deformation of  requires considerable imagination. But the value of 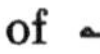 is not an insurmountable difficulty, since ?  is probably to be read *im* in any case; the real problem lies in the hopelessly obscure context. A brief summary of previous translations will serve to illustrate the point.

Maspero[2]: 'I know the making of amulets, that we may go without any fire giving its flame, or without our being washed away by water'.

Madsen[3]: 'Je savais fabriquer des faiences qui . . . sans les faire brûler et dont pourtant on ne pouvait pas délaver (les couleurs) avec l'eau'.

Sottas[4]: 'Je savais faire des enjolivures de matières incrustables, sans les brûler au feu, et, en même temps, non délavables à l'eau'.

Spiegelberg[5]: 'Ich kenne das Machen von *imyt*, einer eingelegten Masse, die nicht vom Feuer verbrannt wird, die auch nicht vom Wasser abgewaschen wird'.

Baud[6]: 'C'est ainsi que je savais exécuter des décorations qui s'incrustent (*ḥꜣꜣ* exactement est "qu'on fait entrer dans une alvéole") sans qu'elles soient fondues au feu et qui ne sont pas délavables à l'eau non plus'.

Wilson[7]: 'I know how to make [things of paste and inlaid things] without letting the fire melt them, nor do they wash off in water either'.

The tendency of all these translations, save that of Maspero, is to regard *imyt* as 'paste' or faience, and this is also the view of Erman[8] who renders 'Tonwaren', and Brugsch[9] who favours 'Farbe' or 'Farbenschmelz'. Spiegelberg and Baud, however, go even further. The former, commenting on *imyt*, translates 'Farbenpaste', and refers to the famous boast 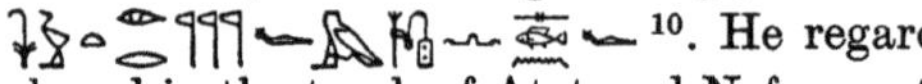[10]. He regards *imyt* as a word for the kind of decoration employed in the tomb of Atet and Nefermaat, and derived from *im. iḫt ḥꜣꜣt.n* would then mean 'eine Masse die hineingeht', or 'die hineingedrückt wird'. The chief objection to this view seems to be that the type of decoration Spiegelberg has in mind is unique, and this reference to *imyt* is a millenium later than the Medum tomb. More convincing is the theory of Baud, who sees the passage as a reference to 'ces pâtes ciment qu'on incrustait dans les alvéoles creusées dans le calcaire', the technique used for the hieroglyphs on the base of the statue of Hemiun at Hildesheim, and the sarcophagus of Seti I. At all events, it seems that *imyt* was a form of decoration the virtue of which lay in its resistance to heat and damp, in all probability an inlay, either of solid pigment or of specially prepared frit[11]. If this is indeed the meaning of *imyt*, a connection with the somewhat obscure *im* mentioned in relation to tomb decoration[12] is tempting.

In conclusion, however, it must be pointed out that *imyt* may not represent a material at all. It is just possible that the passage bears some such meaning as 'I knew how to make the outlines of things descend deep etc.', and is a simple reference to the technique of relief en creux, the durability of which was so superior to that of ordinary relief.

---

[1] Rec. Trav. 36 p. 162.
[2] TSBA. 5 p. 562.
[3] Sphinx 12 p. 248.
[4] Rec. Trav. 36 p. 162.
[5] ZÄS. 64 p. 94.
[6] Chronique 13 p. 21.
[7] JNES. 6 p. 245.
[8] ZÄS. 46 p. 104.
[9] Aegyptologie p. 415.
[10] Medum 24 = Urk. I, 7.
[11] The fact that such an inlay would eventually contract and fall out does not seem to have been realised by the Egyptians.
[12] Louvre C 12.

inf (no Wb. ref.) is apparently confined to two certain examples[1], which the WÖRTERBUCH seems to confuse with *inf* (Wb. I. 96.10), 'Ausfluß aus den Augen'[2]. In both cases the word occurs in lists of tribute, as a variant for *w3d*, malachite, and seems to be an artificial name of the same kind as *hb* and *hkn* discussed below. There does not, however, appear to be any pun or the like from which the word could be derived, and its precise standing is somewhat obscure. But it is unlikely that *inf* has any exact meaning, and it is best regarded as a simple variant of *w3d*, perhaps related to *inf*, 'Ausfluß', either through connection with eye-paint and the eyes, or for some more obscure religious reason.

inr n wdh (Wb. I. 98.5) is evidently a term for artificial stones of some kind, and the WÖRTERBUCH translation 'Glasflüsse' is in all probability correct. The verb *wdh* is applied to melting copper[3], and PIEHL[4] describes it as 'verbe qui désigne un certain traitement appliqué à des pierres et des métaux'. In the case of stones this can only be the production of faience or glass imitations. The passage in which *inr n wdh* occurs[5] lists *ʿ3t nbt nt h3st tn hnʿ inrw ʿš3w n wdh* among Asiatic tribute, and it seems possible that the reference is to some type of artificial gem produced in Syria which the Egyptians did not recognise as faience. The use of *inr* in this phrase is perhaps a little unexpected, but it is merely another case where *inr* has usurped the function of *ʿ3t*[6].

inr n db3 (no Wb. ref.) occurs only in the Dendera amulet list[7], where it is the prescribed material for two scarabs and a *wd3t*-eye. Both these amulets are found in such a variety of materials that no suggestion as to the meaning of the term is possible. It is, however, clear that *inr n db3* is a semi-precious stone, and it is possible that *db3* is derived from the verb *db3* (Wb. V. 556.11) meaning 'to adorn' or 'decorate'. *inr n db3* would then be 'decorative stone', though to what this would refer is not clear. On the other hand, *db3* may be the name of Edfu, and *inr n db3* a name of the same type as *inr n 3bw*, *inr n hwt-nbw* etc., but the identity of 'Edfu-stone' cannot be determined either. In fact, no conclusion seems justified, and *db3* might equally be connected with *db3* (Wb. V. 561.2–7), 'palace', or *db3* (Wb. V. 561. 8–12), 'shrine' or 'sarcophagus', or even *db3* (Wb. V. 558.9f.), 'to block'.

irh (no Wb. ref.) was first recognised by DÉVAUD[8], who saw that an obscure group[9], regarded by BURCHARDT[10] as 'Verderbter Name eines Edelsteines', in fact concealed *ismr* and another word which he read *ilh*. The writing suggests a Semitic loan-word, and this is supported by the orthography of what is clearly the same word in the Toronto duplicate of the Chester Beatty mineral list[11] where, however, the reading *irh* is to be preferred. KUENTZ[12] compares Akkadian *arhu*, 'aeris squama'[13], *iarahu, uriahu*, 'colcothar, red vitriol'[14], and *airahi*, a name for gold[15], none of which is altogether convincing, though the second might

1 Dend. Mar. I, 71; Düm. G. I. II, 67, 7 = Edfou II, 290, 15.
2 Edfou IV, 333, 12; III, 154, 4–5; cf. also V, 191, 16; Dend. Chass. III, 183, 7–8.
3 Wb. I. 393.11. 4 Dict. p. 21. 5 Urk. IV, 695. 6 Cf. above p. 20.
7 Dend. Mar. IV, 87. The ▭ of Mariette's copy is evidently ▭.
8 Kemi 1 p. 137. 9 P. Tur. PR. 67, 11. 10 Fremdworte II p. 5 no. 81.
11 Ostr. Tor. A. 4 = P. Ch. B. IV vs. 7, 11 (fragmentary). 12 BIFAO. 36 p. 185.
13 Campbell Thompson, Dictionary p. 67–9. 14 ib. p. 97–8.
15 ib. p. 61; p. 97.

be suitable in the Chester Beatty context. But a word *ylḫ* also occurs in the Demotic magical papyrus[1], in a passage where *sꜣ-wr* stone and stone of *ylḫ* (*ꜣny n ylḫ*) are to be put on a brazier. The use of the phrase *ꜣny n ylḫ* would seem to establish *ilḫ* as a stone, and it may perhaps be compared with the Demotic *yl*, 'glass', which occurs in the same papyrus[2], *ilḫ* then being a form of natural glass or crystal. There are, however, objections to such a view, and the nature of *irḫ* cannot satisfactorily be determined.

irḳbs (Wb. I. 116.4) is apparently ἅπαξ in a list of Nubian tribute[3], the orthography suggesting a loan-word. BURCHARDT[4] compares Akkadian *ᵃalgamišu*, and translates 'Bergkristall', which is accepted by GARDINER[5] and CAMINOS[6]. But the meaning crystal depends on identification of *ᵃalgamišu* with Hebrew *algabiš*, Arabic *algibšu*, 'hail', which CAMPBELL THOMPSON[7] rejects in favour of a connection with ᵃAŠ.MUR, discussed below sv. *ismr* (p. 163f.), and Arabic *ǰmst*, 'amethyst'. He gives the word a fairly wide range of meaning, including corundum, emery and oriental amethyst, and suggests that it may possibly be the origin of Greek ἀδάμας[8]. Since *ᵃalgamišu* was evidently used as a drill, it is likely to have been a hard stone, and some species of corundum seems probable. On the other hand, quartz is also possible, and the Nubian provenance of *irḳbs* would rather suggest this. Without more precise evidence no certain conclusion can be reached, and a definite translation such as amethyst cannot be justified.

ꜥꜣt wḏḥ (Wb. I. 165.21), like *inr n wḏḥ* discussed above, is undoubtedly a term for artificial gem stones, and again the WÖRTERBUCH rendering 'Glasfluss' is perhaps adequate. The expression occurs only once[9], and is there contrasted with *ꜥꜣt nbt*, the genuine stones, which are given precedence over the base metals, while the *ꜥꜣt wḏḥ* are not. The artificial, 'molten' stones were probably globules, blocks, or thin plates of faience or coloured glass prepared ready for use in jewellery, particularly for inlay work, and this is one of the few instances where any distinction is drawn between the genuine stones and their cheaper imitations.

ꜥꜣt ḥḏt (Wb. I. 165.17), obviously a stone white or clear in colour, and regarded as precious by the Egyptians, is not easily identifiable. BRUGSCH suggests onyx[10], and white jasper[11], and DE ROUGÉ[12] proposes 'cristal de roche', though both are tentative in their identification, and neither WENDEL[13] nor BISSING[14] considers that the evidence warrants a definite translation.

The references to *ꜥꜣt ḥḏt* are not, indeed, very explicit. The word occurs among the 24 precious stones enumerated at Dendera[15], and with other valuable minerals brought as tribute[16], as well as in the Chester Beatty list[17], from which its general nature is sufficiently evident.

---

[1] P. Dem. Mag. 3, 25. [2] ib. vs. 3, 18 etc.
[3] Koller 4,2; cf. however, the broken *irg* . . . of P. Ch. B. IV vs. 8, 1, which may be the same word.
[4] Fremdworte II p. 7 no. 113. [5] Hier. Texts I p. 41. [6] LEM. p. 442.
[7] Campbell Thompson, Dictionary p. 167; p. 134; cf. also Quiring, Sudhoff's Archiv. 38 p. 205.
[8] ib. p. 167; cf., however, p. xlvi. [9] Harr. 73, 13.
[10] Wb. p. 159. [11] Wb. Erg. p. 178. [12] Rev. Arch. N. S. I (2) p. 303.
[13] Bau- und Edelsteine p. 111. [14] Stat. Taf. p. 41.
[15] Dend. Mar. IV, 36, 50; 39, 141. [16] Dend. Mar. I, 70; I, 71; II, 12.
[17] P. Ch. B. IV vs. 7, 7.

Two blocks of *ꜥꜣt ḥḏt* are mentioned as tribute from Ḫatti[1], and it is found, with other semi-precious stones, among the tribute of Retenu[2]. Native sources of the stone include *ḫt(t)* or *ḫtšd*[3], apparently a region to the south of Egypt[4], though thought by WENDEL to be in Asia, *ḏw n tp-iḥt*, the northern range of the eastern hills between Turah and Atfiḥ[5], and the *ḏw n ꜥꜣt ḥḏt* cited by BRUGSCH[6]. Concerning the use of *ꜥꜣt ḥḏt* we know only that it was the material from which the pylon of a model temple was made[7], while a figure of Rēꜥ-Ḥarakhti is also said to be of this stone[8].

Not one of the white or colourless semi-precious stones known to the Egyptians seems suitable as a translation of *ꜥꜣt ḥḏt*. Chalcedony, which when pure is white or greyish, would seem a possible rendering, but it will be shown below that it was known as *ḥrst ḥḏ*, and so cannot be considered here. For the same reason, milky quartz, *mnw ḥḏ*, may be disregarded as a possible translation. Iceland spar, a very fine alabaster, pure and transparent, was probably regarded as a natural glass, and the same may be true of rock crystal, which, however, may have been recognised as a form of quartz. White coral is out of the question, because it was not mined as *ḏw* implies, and white jasper is unlikely, since it was only very rarely, if at all used in Egypt. Lastly onyx, favoured by BRUGSCH, seems to be ruled out, as it not white, but banded white and black, and is, in any case, of some rarity. Concerning all these stones, it must also be pointed out that none of them, except perhaps quartz, occurs in large enough aggregates for sizeable blocks (*inr ꜥꜣ*)[9] to be extracted.

It seems therefore probable that *ꜥꜣt ḥḏt* is not what would now be regarded as a semi-precious stone, but may be a species of marble or fine limestone, or even the compact rock-like form of gypsum which resembles alabaster[10]. Now PLINY[11], speaking of types of marble, mentions one, called memphites from its place of origin, which is of the nature of a precious stone, and concerning which LUCAS[12] remarks that it was probably some sort of limestone[13]. This seems likely in that Memphis was the source, and it is interesting that *ḏw n tp-iḥt*, from which *ꜥꜣt ḥḏt* was obtained, is in that area. It is thus possible that the Egyptians confused fine limestone and marble, a crystalline form of limestone, and perhaps also dolomite, though the latter has a somewhat characteristic appearance[14], and it may be to these that *ꜥꜣt ḥḏt* refers. Marble was employed for statuary in the New Kingdom, and again in Roman times[15], and was also imported[16], which conforms to what is known of *ꜥꜣt ḥḏt*, and both marbles and hard crystalline forms of limestone occur in Egypt. Concerning marble EPIPHANIUS says that 'they also call the marble stone falsely onyx, because of the pure whiteness of the stone'[17], which would account for its inclusion among semi-precious stones.

In default therefore of further evidence, it may perhaps be assumed that *ꜥꜣt ḥḏt* represents white and pale grey marble, and analogous forms of hard crystalline limestone, though perhaps including other similar stones of rarer occurrence.

---

1 Urk. IV, 701. 2 Urk., IV, 722. 3 Dend. Mar. I, 70; I, 71; II, 12.

4 Gauthier, Dict. géog. IV p. 188; p. 194.

5 Rec. Trav. 16, 52; Gauthier, Dict. géog. VI p. 123.

6 Wb. Erg. p. 178. 7 Rec. Trav. 8, 9. 8 Goshen 5, 3.

9 Urk. IV, 701. 10 Lucas, Materials p. 470.

11 HN. 36, 11, 57; cf. Isidorus 16, 4, 14, and contrast Dioscorides 5, 140 (W).

12 Materials p. 473.

13 Cf. the 'very hard marbly white limestone' used for sculptures in the Labyrinth, Petrie, Labyrinth, Gerzeh and Mazguneh p. 31.

14 Lucas, Materials p. 467. It was, however, chiefly used in dynastic times.

15 ib. p. 472.

16 Fitzler, Steinbrüche p. 6.

17 Blake & De Vis, De Gemmis p. 211.

w3d (Wb. I. 267.3–8 & 9–15) has been discussed in some detail by a number of scholars, though with little positive result, as is shown by the uncertainty which remains in the WÖRTERBUCH. DÜMICHEN[1] recognises three distinct species, 1. *w3d n b3h*, apparently a Ptolemaic variant for *mfk3t*, 2. a green pigment and eye-paint, and 3. a green stone, occurring as *w3d šmᶜ* and *w3d mh*, and observes that in the lists of the 24 *ᶜ3t* at Dendera[2], *w3d*, *w3d mh*, and *w3d šmᶜ* are clearly differentiated. BRUGSCH[3] similarly regards *w3d* as both a precious stone and a pigment or eye-paint, and notes the two varieties *w3d mh* and *w3d šmᶜ*, 'der asiatische und der afrikanische Grünstein', together with the third, *w3d n b3h*, which is quite separate as 'eine besondere Bezeichnung des sonst *mfk3t* genannten Minerales'.

A more definite attempt at identification is made by WENDEL[4], who takes *w3d* to be the name of the native Egyptian green stones as opposed to the imported *mfk3t*. Of these he claims there were only two, green feslpar and green jasper, from both of which small objects were made. He further compares *w3d* with classical smaragdus, and discusses the evidence for the latter.

The translation green felspar, proposed by LEPSIUS[5], and quoted by DÜMICHEN, has been accepted by several authorities, including the WÖRTERBUCH[6], but others, among them JÉQUIER[7] and NEWBERRY[8], have inclined towards malachite, the latter regarding *w3d* as the commoner name for *šsmt*. A similar view is also taken by IVERSEN[9], who sees *w3d* as a term which could theoretically represent any green stone or green mineral, though it did in fact refer to one in particular, namely malachite.

That this is essentially correct can scarcely be doubted, and it will be shown below[10] that the frequent references to *w3d* as an eye-paint, pigment, or medicament, can only indicate malachite, the analogous verdigris and chrysocolla, or the green frit made in imitation of them. But the Egyptians do not seem to have drawn a distinction between these instances, and those where *w3d* is clearly a semi-precious stone, and it is therefore reasonable to suppose that the same meaning malachite generally applies to the latter also, though with certain reservations.

Now it is a well-known fact that apart from its regular use as an eye-paint, malachite is comparatively rare in connection with ancient Egypt, and few instances of its occurrence in jewellery or as the material for small objects such as beads and scarabs can be cited[11]. It is not therefore surprising that *w3d*, although frequently mentioned in lists of minerals[12], was not regarded as one of the more important stones[13], or that specific references to objects of *w3d* should be few, and confined to such things as collars and necklaces[14], a statuette of the Nile god[15], and a sistrum[16]. Indeed, it seems likely that the majority of examples of *w3d* brought as tribute or offered with other valuable materials[17] are to be interpreted as the raw material for green eye-paint rather than precious stones destined for use in jewellery.

Malachite was obtained largely from two localities, Sinai and the eastern desert[18], the former being perhaps the only source in the Old Kingdom, and these may well correspond

[1] ZÄS. 10 p. 100. [2] Dend. Mar. IV, 36, 49–50; 39, 140–2.
[3] Wb. p. 361; Erg. p. 404. [4] Bau- und Edelsteine p. 98.
[5] Métaux p. 44–5. [6] Cf. Paton, Records III, 2 p. 54 n. 78.
[7] Frises p. 154. [8] Griff. Stud. p. 320. [9] Paints and Pigments p. 9.
[10] p. 143f.. [11] Lucas, Materials p. 456.
[12] Eg. P. Ch. B. IV vs. 7, 5; 7, 13; Edfou VI, 201f.; Dend. Chass. IV, 159; Dend. Mar. IV, 36, 49–50; 39, 140–2.
[13] Düm. G. I. II, 97, 13–14; Dend. Mar. I, 22. [14] Cairo 28027; 28083; Dend. Mar. I, 31.
[15] Harr. 40b, 15. [16] Dend. Mar. II, 53b.
[17] Eg. Dend. Mar. I, 70; I, 71; II, 12; Düm. G.I. II, 63–4; 66–7; 69–70; IV, 102; 170; 180; 186; Harr. 64b, 13 etc. [18] Lucas, Materials p. 457.

to the *wꜣḏ mḥ* and *wꜣḏ šmꜥ* of the texts[1]. Moreover, *ršt* (*ršꜣwt*), which is probably Sinai or a part of it[2], is actually mentioned as a source of *wꜣḏ*[3], and *wꜣḏ* also occurs in a list of stones from the eastern desert[4]. Other localities from which *wꜣḏ* was obtained are less readily identified, and include several which are probably regions in Nubia[5].

Of particular importance is the expression *wꜣḏ n bꜣḫ*, which has been interpreted as an alternative name of *mfkꜣt*, since it occurs as a variant of the latter in the Dendera tribute tables[6]. That *bꜣḫ* did in fact produce *mfkꜣt* is apparent from the mention of *mfkꜣt mꜣꜥ n bꜣḫ*[7], and Prof. Černý suggests that it may perhaps refer to Sinai in some instances. This would find confirmation in the citation of *ršt*, probably Sinai, as a source of *wꜣḏ n bꜣḫ*[8], and an identification of *wꜣḏ n bꜣḫ* and *mfkꜣt* would then seem likely.

On the other hand, *bꜣḫ*, originally the name of a region to the west of Egypt, later normally refers to the eastern hill-country out of which the sun rises[9], and may in relation to *wꜣḏ* be a term for the classical mons smaragdites, Gebel Zebara, on the Red Sea coast. In this case, *wꜣḏ n bꜣḫ* refers not to *mfkꜣt* itself, but to some stone which had a somewhat similar appearance and was perhaps confused with it. Now the chief green stone that occurs in the eastern hills, and specifically at Gebel Zebara, is a green beryl or 'emerald', which is otherwise unidentified lexicographically. It was worked extensively in Ptolemaic times, though almost certainly not before[10], which would coincide with references to *wꜣḏ n bꜣḫ*, and in general appearance and value would be sufficiently akin to turquoise to be identified with it.

But if *wꜣḏ n bꜣḫ* is to be regarded either as beryl or as turquoise, it becomes apparent that *wꜣḏ*, although primarily malachite, also has a more general meaning, embracing other green stones, and this is also suggested by certain other examples. Thus, in the Pyramid texts[11], *wꜣḏ*, *šsmt* and *mfkꜣt* occur together, and if *šsmt* is malachite, *wꜣḏ* must surely have some other meaning. Similarly, in another text[12] *wꜣḏ šmꜥ n mꜣꜥ* and *šsmt n mꜣꜥ* are juxtaposed, and although they may refer to malachite from the south and from Sinai, some distinction seems implicit, and *wꜣḏ šmꜥ* may well indicate another green stone such as jasper. The same is perhaps true also of *wꜣḏ mꜣnw*, which occurs together with *wꜣḏ n bꜣḫ*[13], and corresponds to no known source of malachite.

A general meaning is likewise more appropriate in some religious contexts where parts of figures and the like are said to be of *wꜣḏ* or *wꜣḏ šmꜥ*[14], and may also be suggested in the case of Horus lord of *wꜣḏ*[15], though a comparable mention of the goddess Neith as lady of *wꜣḏ* at Sinai[16] must refer to malachite. Significant too is the mention of *wꜣḏ* in Prisse[17], where it is evidently something rare, and for which malachite seems inappropriate. A translation 'emerald' is tempting, but cannot be justified, and it is possible that the meaning is simply 'green pebble'. Alternatively, this *wꜣḏ* may be the word for 'Röhrenperle' (Wb. I. 268.1–3), quite distinct from the precious stone, though it is often difficult to differentiate between the two. 'Green pebble' may also be the best rendering of *wꜣḏ n ꜥr*, found among the 24 *ꜥꜣt* at Dendera[18], and the somewhat similar *ṯꜣ nw wꜣḏ*[19], though both may equally refer to lumps of malachite.

---

[1] Dend. Mar. IV, 36, 49–50; 39, 140–2.
[2] Černý, Sinai 2 p. 3.
[3] Dend. Chass. IV, 159; ? Edfou VI, 202, 3.
[4] Wadi el Hudi 23, 4.
[5] Dend. Mar. I, 70; I, 71; II, 12 (*prst*); Düm. G. I. IV, 164 (*šdt*); ib. II, 63; IV, 165; IV, 174; II, 66 (*ỉspd*); Edfou VI, 203, 2 (*ỉmfn*); Edfou VI, 201, 10 (*nꜣpwtt, wkmtt*) etc.
[6] Dend. Mar. I, 71; II, 12; Düm. G. I. II, 69, 5; II, 75
[7] Düm. G. I. II, 67, 7.
[8] Dend. Chass. IV, 159; ? Edfou VI, 202, 3.
[9] Wb. I. 422.8–13; Sethe in Sahurēꜥ II p. 74.
[10] Lucas, Materials p. 446.
[11] Pyr. 567 b.
[12] Urk. IV, 875.
[13] Ann. Serv. 16, 223.
[14] Eg. Totb. Nav. 77, 3 (wings); Coff. Texts IV, 59 b (eyebrows); Luxor ⟨560⟩ (breast) etc.
[15] Pyr. 457 c.
[16] Sinai 121.
[17] Prisse 5, 10.
[18] Dend. Mar. IV, 39, 140–2.
[19] Edfou I, 590.

It seems therefore that *wꜣḏ*, although in most cases referring to malachite, also had a wider application consistent with its root meaning 'green', and perhaps included any green stone for which no more specific name existed. Thus, while *wꜣḏ* as a pigment and medicament is certainly malachite and the like, and *wꜣḏ mḥ* and *wꜣḏ šmꜥ* probably malachite from Sinai and the eastern desert, *wꜣḏ bꜣḫ* is almost certainly not, and *wꜣḏ mꜣnw* is suspect, as also are certain other examples of the word.

In this respect, *wꜣḏ* is closely parallel to the classical smaragdus, which included malachite, beryl, emerald and other green stones. Pliny[1] mentions ten varieties, of which the Egyptian holds third place and is found near Coptos, and also refers to Ethiopian smaragdi[2], which are probably beryls. On the other hand, Theophrastus[3] and Pliny[4] both speak of large objects such as obelisks and a colossus of Sarapis supposed to be smaragdus, which can only refer to a green stone like serpentine, breccia or 'schist'[5], and it is not impossible that in some instances these were called *wꜣḏ* by the Egyptians.

That *wꜣḏ* ever refers to emerald, however, seems highly unlikely, though smaragdus undoubtedly does in some cases. Epiphanius[6] speaks of smaragdi which sparkle, and are evidently emeralds, and mentions some from Ethiopia, some from an island near Berenike, and others mined near Kalabsheh (ⲦⲀⲖⲘⲈⲤ). Moreover, it is quite possible that a few beryls of sufficiently good quality to be called emeralds may from time to time have been obtained from the mines of Gebel Zebara and Gebel Sikait[7], though the latter can scarcely have been the principal classical source as Ball suggests[8]. However, fine specimens of this kind were probably rare, and it is improbable that the Egyptians ever recognised emeralds as such.

Some inconsistency in the green stones covered by *wꜣḏ* is thus to be expected, and it is significant that modern Egyptologists have frequently confused malachite, beryl, felspar, and other stones of similar colour[9]. The Egyptians themselves are equally likely to have done this, and except in those instances where *wꜣḏ* is evidently a pigment or a medicament, it is perhaps unwise to insist upon too precise a translation.

In conclusion, it must be noted that writings of *wḏ* (Wb. I. 399.9), accorded separate existence by the Wörterbuch, are merely graphic variants of *wꜣḏ*, the alternation of and  being 'too familiar to need comment'[10]. If any proof were required, it is provided by a passage in Harris where *wḏ* (sic.) and *msdmt* occur together[11], and by the writing *wḏ šmꜥ* in the Chester Beatty list[12]. Similarly, it is possible that *ḥḏt* (Wb. III. 211.20–21) is to be eliminated, since the Hearst reference[13] is regarded by Wreszinski as corrupt, and could well be *wꜣḏ* or *wḏ*. However, the other example[14] is less readily explained, and it is perhaps better to retain the distinction.

wꜣḏwꜣḏt (Wb. I. 270.8), regarded by the Wörterbuch as 'ein Mineral, wohl irrig statt *wꜣḏ*', is probably to be read *wꜣḏty*, and translated 'two pieces of *wꜣḏ*'. The passage in which it occurs[15] concerns a statue whose various parts are of precious materials, the eyes being *m*

---

[1] HN. 37, 16, 62. [2] HN. 37, 18, 69. [3] § 24. [4] HN. 37, 19, 74.

[5] Cf. Caley & Richards, Theophrastus p. 100f.

[6] Blake & De Vis, De Gemmis p. 105f.; p. 126; p. 241.

[7] The nature of the stones found in the Sikait-Zebara workings is discussed at length by Schneider, Zeitschrift für Ethnologie 24 (1892) p. 41f.

[8] Roman Book p. 260. [9] Lucas, Materials pp. 446–7; p. 451.

[10] Gardiner, Onom. 1 p. 22; cf. Grammar[3] p. 480, M. 13; Möller, Paläographie nos. 280, 474; Wb. I. 264.14; 394.11; 395.5; 396.11; 397.1; 397.11; 399.7 etc.

[11] Harr. 55a, 6. [12] P. Ch. B. IV vs. 7, 5. [13] Hearst 11, 8.

[14] LD. II, 119. [15] Fayum Pap. 3, 9.

. CHABAS[1] correctly interprets as 'les deux pierres *wꜣḏ*', but translates 'les deux cristaux', because the eyes of statues are usually of this material. This is so, but the figure in question is magical, and therefore not subject to normal conditions, and the eyes could well be two pieces of malachite or other green stone.

brgt (no Wb. ref.) is known from only one occurrence in the Sehel inscription[2], where BARGUET[3] reads *brḳt*, regarding as a mistake for [4]. In view, however, of the existence of a related verb *brg* (Wb. I. 466.13), 'leuchten, glänzen', *brgt* seems a preferable transcription, though the cognate words do in fact have *ḳ*. The suggested translation 'diamond' may be rejected, since the stone was not known to the Egyptians, and was certainly never a product of Elephantine. *brgt* is clearly derived from Hebrew *bârĕḳĕth*, Syriac *bâr'ḳa*, generally supposed to be emerald, from which Akkadian *barraḳtu* is also borrowed[5], and may be further connected with Sanskrit *marakata* and Greek σμάραγδος. A translation 'emerald' is therefore appropriate, though perhaps 'beryl' would be more suitable in this context, since it has been suggested above that the Egyptians are unlikely to have known the true precious emerald. Beryls are found in the Sikait-Zebara range of the Red Sea hills, and were certainly being exploited in Graeco-Roman times[6]. On the other hand, *brgt* might conceivably be peridot, Persian *zabargat*, Arabic *zebirget*, which was obtained from St. John's Island, Zeberged, and is mistakenly called 'topaz' by ancient authors, though this stone was scarcely used by the Egyptians[7].

prḏn (Wb. I. 533.5) is also confined to the Sehel inscription[8], where BRUGSCH[9] suggests that it may be an Egyptian rendering of the Greek πράσον, πράσινος, 'leek-green prase', an identification repeated by BARGUET[10] without comment. The equation seems quite convincing, though the meaning of the Greek is not altogether certain. LIDDELL & SCOTT think that the πρασῖτις λίθος is probably emerald, but PLINY's[11] description of prasius hardly supports this, and CALEY & RICHARDS[12] suggest that THEOPHRASTUS' πρασίτις was green jasper, though it may have been any dark green opaque stone. Chrysoprase is possible but unlikely, and either olivine or peridot may perhaps be meant, though none of these stones was well known to the Egyptians[13]. At all events, whatever the meaning of *prḏn*, it must be distinct from *brgt*, and both are evidently used to describe stones not clearly identified in the native language. A non-committal translation as 'prase' is perhaps to be preferred.

frtf (sic.) (no Wb. ref.) is known only from the Dendera table of amulets[14], where it is the material for 11 amulets. MARIETTE saw , but in the Supplement has , which he understands as 'Phrat de Ruten'[15]; PETRIE has 'Kartef of Rutennu'[16]. The true reading is not easily established, but may be *frtf rtnw*, though corruption seems likely,

[1] Études^2 p. 27. [2] Sehel 16. [3] Stèle de la Famine p. 24.
[4] Cf. Möller, Paläographie III nos. 468, 125; Breasted, Pap. Smith p. 196.
[5] Campbell Thompson, Dictionary p. 174; Quiring, Sudhoff's Archiv 38 p. 197.
[6] Lucas, Materials p. 445. [7] ib. p. 458; cf. Ball, Roman Book p. 283.
[8] Sehel 15. [9] Hungersnoth p. 122. [10] Stèle de la Famine p. 24.
[11] HN. 37, 34, 113, perhaps a green jasper spotted with red. [12] Theophrastus p. 138.
[13] Lucas, Materials p. 449; p. 458. [14] Dend. Mar. IV, 87.
[15] Dend. Supp. pl. I. [16] Amulets p. 6.

and it is possible as Prof. FAIRMAN suggests (in a letter) that collation might reveal *tfrr(t)*. In view of this it is perhaps significant that lapis lazuli is prominent among the stones from which the listed amulets are made. However, if *frtf* is the correct reading, no identification can be suggested, and the word may be regarded as an obscure Ptolemaic creation.

mꜣꜥ ḥḏ (no Wb. ref.) is found once in the Edfu mineral list[1], described as *mꜣꜥ ḥḏ m tꜣ imnty*. Both BRUGSCH[2] and WENDEL[3] suggest an identification with *ḥrst ḥḏ*, which came from the same region, and this is possible, though not altogether convincing. The most that can be said with certainty is that *mꜣꜥ ḥḏ* was a stone or mineral of white or light colour.

minw (Wb. II. 44.5) appears to be confined to two references in Harris, once as the material for a scarab[4], and once for a statuette of the Nile god[5], from which it would seem to be a species of semi-precious stone. BRUGSCH[6] compares Coptic ⲘⲒⲚⲈ, actually Ⲙ̄ⲘⲒⲚⲈ ⲘⲒⲚⲈ, ποικίλος, and renders 'der buntfarbige', while BIRCH[7] regards *minw* as glass, and PIEHL as 'le bariolé'. In fact, as WENDEL[8] rightly observes, no conclusion can be reached, though it is just possible that *minw* is a writing of *mnw*, which does not otherwise occur in Harris. Whether a similar word (Wb. II. 44.4) 'als Beischrift zu Beuteln unter Schmucksachen genannt'[9] is to be identified with the *minw* of Harris it is impossible to say.

mfkꜣt (Wb. II. 56.1–14)[10] has perhaps been subject to greater misunderstanding than any other Egyptian word, misunderstanding that is not yet entirely dispelled, in as much as there are still those who translate as malachite[11]. The word was originally taken by CHAMPOLLION[12] to mean green copper, and this was followed by GENSLER[13], but the first real progress was made by BRUGSCH[14], who saw turquoise at Wadi Maghara, and concluded, correctly, that this was what the Egyptians called *mfkꜣt*. He proposed this identification in his Wörterbuch[15], commenting that *mfkꜣt* 'bezeichnet nicht das Kupfer . . . sondern den Türkis, turquoise, einen himmelblauen Edelstein, der auf der Sinai Halbinsel in dem Wadi Maghara und in Sarabit-el-Khadem gefunden wird', and regarding the root meaning of the stem as 'blau, hell, heiter sein' from which the sense 'sich freuen' was derived.

But BRUGSCH's interpretation was attacked by CHABAS[16], who rejected both copper and turquoise as translations, though unable to offer any alternative suggestion[17], and was opposed in detail by LEPSIUS[18], with whom must lie the responsibility for subsequent confusion. He gave *mfkꜣt* a fairly wide range of meaning, including 'σμάραγδος, χρυσοκόλλα; smaragdus, molochites, chrysocolla; l'émeraude, la malachite, le vert de cuivre; le verre de couleur verte; les couleurs vertes'. The word was closely connected with *ḫsbḏ*, and had a Ptolemaic variant *ḥb*, which, however, he could not explain; the substance to which it referred was of a green colour, as was shown by painted representations and by the description

[1] Edfou VI, 202, 5. [2] Dict. géog. p. 35. [3] Bau- und Edelsteine p. 113.
[4] Harr. 15b, 3. [5] Harr. 41a, 1. [6] Wb. Erg. p. 549.
[7] Hier. Pap. p. 9. [8] Bau- und Edesteine p. 116. [9] B. M. 6655.
[10] Cf. Hebrew *nōpek*, JAOS. 73 p. 152; Sudhoff's Archiv 38 p. 199.
[11] Eg. Forbes, Metallurgy p. 341, says *mfkꜣt* means 'malachite or turquoise . . . which the Egyptians do not seem to have distinguished properly'.
[12] Not. Descr. I p. 509. [13] ZÄS. 8 p. 144. [14] Wanderung p. 66f.
[15] Wb. p. 603. [16] Études[2] p. 21. [17] In the list of contents, p. 591, he has malachite.
[18] Métaux p. 35f.

of a tree and a pool of water as of *mfkꜣt*[1]. The ▭ determinative and the connection with *ḫsbḏ* proved that Champollion's translation 'copper' was incorrect, and Brugsch's 'turquoise' was also wrong because Sinai turquoise was of poor quality, few objects of turquoise were known, and the stone was not green but blue. The parallelism between *ḫsbḏ* and *mfkꜣt* was similar to that between κύανος and χρυσοκόλλα, and *mfkꜣt* was therefore malachite or verdigris. Moreover, Theophrastus seemed to identify χρυσοκόλλα and ψευδὴς σμάραγδος so that *mfkꜣt mꜣᶜ* might be true smaragdus, ie. emerald, though it had to be admitted that the classical term had a variety of meanings, and it was difficult to believe that emerald was made into amulets.

Lepsius' views were accepted by Dümichen[2], who quoted the connection with *wꜣḏ n bꜣḫ* to prove that *mfkꜣt* was a green stone, and also by Brugsch[3] who abandoned his original idea, and accepted that *mfkꜣt* meant '1. Smaragd oder grünen Beryll, 2. den nachgeahmten (*ỉryt*)[4] Smaragd (einen grünen Glassfluss), 3. Malachit, Kupfergrün, Berggrün'. The translation malachite was likewise adopted by Wendel[5], who, however, admitted that objects of the stone were rare, by Weill[6], and by the Wörterbuch, whence it has gained general acceptance.

However, Birch[7] took *mfkꜣt* as turquoise, and thought that artificial *mfkꜣt* might be 'blue glazed steatite or steaschist', and both Jéquier[8] and Thomas[9] maintained that turquoise must have been among the minerals represented by *mfkꜣt* at Sinai. Similarly Vernier[10] denied that emerald was included under *mfkꜣt*, and pointed out that turquoise was in fact much used in jewellery[11].

The reinstatement of the true meaning turquoise was, however, due to Loret[12], who surveyed the previous history of the controversy, and concluded that the choice lay between the malachite of Lepsius and the Wörterbuch, and the turquoise of Brugsch and Gardiner[13]. He specifically rejected Lepsius' assumption that the Egyptians lumped together all precious stones of one colour, regardless of transparency, brilliance, depth of colour or hardness, and thought it far more likely that they should have given separate names to scarcely distinct varieties of the same stone. Concerning the objections to turquoise, he observed that the green colour by which *mfkꜣt* was represented was easily explained, since turquoise, although ideally blue, was generally green when found, or rapidly became so on exposure to light. That turquoise was not found among the stones used in Egyptian jewellery was not true; on the contrary, it was frequently used, whereas malachite hardly ever appeared. Moreover, though malachite was found lying about at Sinai, the only stone actually occurring in the galleries was turquoise. Loret further pointed out that the expression *mfkꜣt mꜣt*, 'new *mfkꜣt*'[14], would be quite inexplicable were it not known that turquoise soon faded, and that new turquoise was for that reason particularly desirable. Proof of this was afforded by a Sinai inscription[15] where the colour of *mfkꜣt* and its vulnerability to heat and bright light were clearly understood. This could only apply to turquoise, and certainly not to malachite, which was highly durable. The 'right colour' may have been either blue or

[1] Totb. Leps. 109, 3; 39, 13. [2] ZÄS. 10 p. 98. [3] Wb. Erg. p. 560.

[4] From Lepsius, Métaux p. 62. No example of *mfkꜣt ỉryt* can be quoted, unless the of Urk. IV. 722.9 is to be interpreted as such.

[5] Bau- und Edelsteine p. 95. [6] Sinai p. 23–5.

[7] ZÄS. 11 p. 153.

[8] in De Morgan, Recherches p. 217 n. 1.

[9] Cairo Sc. Journ. 6 p. 56. [10] Bijouterie p. 16–23.

[11] Cf. also Lucas, Materials (1) p. 168. [12] Kemi 1 p. 99.

[13] Grammar[1] p. 543. [14] Urk. IV, 932; 1213; Westc. 6, 3. [15] Sinai 90.

green, the latter being the colour shown on the monuments; what were despised were stones of a very pale, yellowish, or brown hue.

Since LORET's article, *mfkꜣt* has generally been regarded as turquoise, and LUCAS[1] has reiterated the point. Recently, the question has been re-examined in detail by ČERNÝ[2], who briefly summarises the previous discussion, citing LUCAS and LORET for the definitive identification of *mfkꜣt* as turquoise, He then examines two questions, 1. did the Egyptians include under the term *mfkꜣt* any other green or blue mineral substances besides turquoise?, and 2. was the turquoise itself prized as a gem, or only for the sake of the green pigments and glazes made from it? Sinai produces only malachite, silicate of copper, and turquoise, the latter of unequal merit. That turquoise was used as a gem is well known, but there is no evidence for its use as a constituent of pigments, and the source of the colour in green glaze was a copper compound. LUCAS has shown doubts about the use of turquoise in jewellery to be incorrect. *mfkꜣt* is found very frequently in texts, usually in conjunction with *ḫsbḏ*, and was evidently a green stone obtained in round or oval lumps, and used for a variety of objects. These uses correspond to turquoise, whereas few instances of the employment of malachite can be quoted. References to 'real *mfkꜣt*' imply that an artificial kind existed, and this was turquoise-coloured glass. The few mentions of *mfkꜣt mꜣt* must refer to turquoise, and Sinai 90 is explicit concerning the fading of the stone. In short, it may be concluded that the Egyptians obtained most of their turquoises from Sinai, and used them as gems in their natural state. There is no evidence for powdered turquoise as a pigment, or for thinking that any blue or green minerals other than turquoise were included under *mfkꜣt*.

Of the validity of the arguments put forward by LUCAS, LORET and ČERNÝ there can be no doubt whatever, and the meaning of *mfkꜣt* is clearly established as turquoise and nothing else. It is, not, however, always possible to determine whether the word refers to the precious stone, or to the faience or glass imitation which bore the same name, though there are clear indications of both.

Various sources of the precious stone are mentioned in the texts, the most important being *ršꜣwt* or *ršt*[3], which ČERNÝ[4] thinks may possibly refer to Sinai or a part of it, either Wâdy Maghâra or Serâbiṭ el-Khâdim. Sinai or some specific region therein may also be indicated by *mfgt*[5] and *ḫrwtt*[6], and by the phrase *mfkꜣt mꜣʿ n bꜣẖ*[7], though the latter is a little puzzling[8]. Other sources of *mfkꜣt* include *tfrrt*[9], well known as the chief locality from which *ḫsbḏ* was obtained, and probably a trading station on the route from Badakhshan, *ẖꜣrw*[10], possibly a reference to Persian turquoise, and 'the west'[11], the significance of which is obscure, since no source of turquoise is known in Libya. In all these instances, *mfkꜣt* can only refer to the genuine stone, since the imitations were produced in Egypt, and are unlikely to have been sought elsewhere.

As has already been observed, *mfkꜣt mꜣʿt*, genuine turquoise[12], and *mfkꜣt mꜣt*, new turquoise[13], must indicate the real mineral, and to these may be added cases where *mfkꜣt* is depicted as lumps or in baskets[14], all mentions of *mfkꜣt* in inscriptions at the mines them-

[1] Materials p. 458; p. 461. [2] Sinai 2 p. 7f.

[3] Dend. Mar. I, 70; I, 71; II, 12; III, 32; III, 67a; Düm. G. I. II, 63; II, 69; IV, 102; IV, 170; Urk. IV, 373 etc.; cf. Gauthier, Dict. géog. III p. 127. [4] Sinai 2 p. 3.

[5] Mem. Inst. 24, 108 = Coff. Texts VI, 212–3; cf. also *ḥtyw mfkꜣt*, *ẖꜣst mfkꜣt*, *ḏw mfkꜣt*.

[6] JEA. 4 pl. 9. [7] Düm. G. I. II, 67, 7. [8] Cf. above p. 103.

[9] Dend. Mar. III, 32; 67a; Abydos, Sethos Saal D ⟨38⟩.

[10] Dend. Mar. IV, 39, 140; IV, 87. [11] Dend. Mar. IV, 87.

[12] Urk. IV, 875; Edfou II 215, 4–5; VI, 165, 15–16; Harr. 40b, 10; 62b, 1; 70a, 10 etc.

[13] Urk. IV, 932; 1213; Dend. Mar. I, 71; Westc. 6, 3; Cairo 20539 IIb, 8; Nitocris 7.

[14] Urk. IV, 638; 1101; 1146.

selves[1], and the expression *mfkꜣt n ꜥr*[2], which obviously refers to pebbles of turquoise found at random. Moreover, it is probable that whenever *mfkꜣt* follows *ḫsbḏ* in the enumeration of precious minerals, the true turquoise is meant, though *nbw, ḥḏ, ḫsbḏ, mfkꜣt* has very much the appearance of a formula, and this may perhaps be doubted.

Less certain, however, are the numerous citations of *mfkꜣt* as the material for specific objects, such as scarabs[3], beads[4], necklaces and collars[5], rings[6], statues of the Nile god[7] etc.[8], for although turquoise was undoubtedly used for many of these, green glazed faience was also frequently employed. Further, there is definite evidence that in some cases *mfkꜣt* must refer to faience or the like, as, for example, the passage noted by Newberry[9] where Pi-Ramesse is said to be 'dazzling with halls of *ḫsbḏ* and *mfkꜣt*', almost certainly in reference to glazed faience tiles[10].

However, it must not be thought that the use of ersatz turquoise exceeded that of the real material, or that the imitations were composed of turquoise in any form. Both these misapprehensions can be traced to statements by Peet[11], to the effect that turquoise is rare in jewellery, and that 'the Egyptians used it chiefly as a colouring matter for their glazes and paints, after crushing it down'. Nothing could be further from the truth, as has been shown by Lucas[12] and Černý[13], and Iversen[14] rightly points out that the green glazed faience or glass, which owed its colour to a copper compound, was called *mfkꜣt* in reference to the stone it imitated, and not to the substance from which it was made. The same is also true of artificial *ḫsbḏ*.

Concerning the green colour of *mfkꜣt* on the monuments, it has already been observed that the majority of Egyptian turquoise was undoubtedly green. It may, however, be added that since the green colour is known to have been preferred by eg. the Persians[15], it is possible that the Egyptians had the same preference, and actually regarded the bluer turquoise as inferior[16]. Moreover, in depicting stones, blue was generally confined to lapis lazuli, the only true blue stone, and the green for turquoise may have been a fixed convention. It is therefore possible that the 'good colour' desired at Sinai[17] was green, and not blue, which may, like the pale yellowish-brown shades, have been less valued.

To sum up, it is now beyond doubt that *mfkꜣt* refers only to turquoise or to its imitations in faience or glass, and does not embrace any other green stone such as malachite or felspar. The majority of examples are probably to be taken as the genuine mineral, which was found largely in Sinai, and used in its natural form, and was not crushed into a paste as has been suggested. The remainder indicate faience or glass coloured green or greenish-blue in imitation, the colour being due to a copper compound, and again not to the presence of turquoise

---

1 Sinai 90; 182; 200; 275.

2 Dend. Mar. IV, 36, 49–50; 39, 140–2.

3 Harr. 52b, 8; 15b, 2; Dend. Mar. IV, 87 etc.

4 M. u. K. 1, 1; 1, 3.

5 Urk. IV, 22; Cairo 28023; 28091; 20593 IIb, 8.

6 Harr. 64a, 15.

7 Harr. 40b, 10; 73, 13.

8 Dend. Mar. IV, 87; Edfou VI, 299; P. B. M. 10052, 6, 12.

9 Anast. III, 7, 5; JEA. 25 p. 120.

10 Cf. also a definite reference to the making of artificial *mfkꜣt*, Naville, XIth Dyn. Temple at Deir el-Bahari III p. 18.

11 Journ. Man. E & O. Soc. 1918–9 p. 21–2; cf. Vernier, Bijoux pp. 88; 298; 299; 307; 336.

12 Materials p. 212; p. 460–1; cf. also Ransom-Williams, Jewelry p. 22.

13 Sinai 2 p. 9.

14 Paints and Pigments p. 18.

15 Ball, Roman Book p. 285. Pliny, HN. 37, 33, 112, states that the best callaina, almost certainly turquoise, is that which is nearest the colour of emerald, and that the better the colour is, the more likely it is to fade.

16 Lucas, Materials p. 216, thinks that light blue glass imitates turquoise, which may be doubted in view of the greenness of the actual material and the colour on the monuments; cf., further, such expressions as *wšbyt wꜣḏ nt mfkꜣt*, M. u. K. 1, 1.

17 Sinai 90.

in any form. Most Egyptian turquoise was green, though some was blue, and less desirable specimens were faded pale or yellowish. The tendency of turquoise to fade and discolour in bright light or heat was undoubtedly recognised.

mfkȝt wḏḥ (Wb. II. 56.7) is known from only one passage[1], where it is coupled with *ḫsbḏ wḏḥ*, and the Wörterbuch rendering 'grüner Glasfluss' is almost certainly correct. Turquoise was imitated by a greenish-blue glass[2], owing its colour to a copper compound, and this would suit *mfkȝt wḏḥ*, though it is not impossible that the expression refers to some specific ersatz, such as one of Lucas' faience variants, eg. variant D[3]. A translation 'artificial turquoise' may therefore be more appropriate.

mnw (Wb. II. 72.4–7; V. 123.5) is a semi-precious stone of two distinct colours, light and dark, occurring frequently as the material for *ḫnt* and *ḥȝts* vessels. Opinions have differed as to the identification of the stone, though it is generally accepted as quartz. Brugsch[4] recognises *mnw*, but is unable to offer any suggestion as to the meaning, while Wendel[5] concludes that it represents quartz, since it has apparently both white and red varieties[6]. Quartz is also accepted by Bissing[7], on the grounds of the white colour, but Balcz[8] does not take *ḥḏ* and *km* literally, interpreting them as 'hell' and 'dunkelgrau', and translating *mnw*, which he identifies with *mntt*, as 'schwarzer und schwarzweiß gefleckter Granit (?)'. He notes that specimens of the *ḫnt* and *ḥȝts* vessels are made of basalt, granite, diorite, breccia and other stones. Lacau[9], on the other hand, cites *ḫnt* and *ḥȝts* vessels of rock crystal and obsidian, and identifies these two materials as *mnw ḥḏ* and *mnw km*, the Egyptians having considered them as different coloured varieties of the same substance. This interpretation is accepted by Barguet[10], who regards *mnw* as 'sans doute le cristal de roche', and quotes Lucas for its use and colour.

The majority of references to *mnw* come from the Pyramid texts[11], and from offering-lists[12], where, as already mentioned, *mnw ḥḏ* and *mnw km* occur as the materials for *ḫnt* or *ḥȝts* vessels. But *mnw* is also listed three times among tribute of Retenu[13], both as material for vessels and in *dbn*, and both *mnw km* and *mnw ḥḏ* are found in a list of precious stones from the eastern desert[14]. Other references include *mnw* among the stones of the Elephantine region[15], and with *ḫrst*[16], and as the material of a decan[17] etc.[18] A further example quoted by the Wörterbuch[19] is not in fact *mnw*, and must be regarded either as *mn*, or as a corruption of *mḫn(m)t*.

[1] Harr. 64 b, 9. [2] Cf. p. 109 n. 16. [3] Lucas, Materials p. 187.

[4] Wb. Erg. p. 604. [5] Bau- und Edelsteine p. 112.

[6] Wendel deduced the red colour from Düm. H. I. II, 57 c a 19, where a figure of the setting sun is of *mnw*, but this seems somewhat far-fetched.

[7] Stat. Taf. p. 44. [8] Mitt. Kairo 3 p. 68.

[9] Ann. Coll. de France 1947–8 p. 143. [10] Stèle de la Famine p. 24.

[11] Pyr. 33 b; 36 b; 36 c; 37 a; 39 b; 39 c etc.

[12] Dendereh pl. 4; Beni Has. I, 17 list 16 A 30; Antefoker 32; Theb. Grab. 107 ⟨1633⟩; P. Abusir rt. 21 unpublished [18].

[13] Urk. IV, 706; 718; 722 (as *mnw ḥḏ* and  ? *mnw km*).

[14] Wadi el Hudi 23, 4, written  and 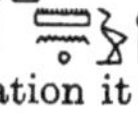. Whether the apparent corruption of the former is original or due to the untidy publication it is impossible to say.

[15] Sehel 16. [16] Lit. Frag. 9, 2, 14. [17] Thes. 19.

[18] Wendel's reference to eyes and lips of *mnw* is based on a misunderstanding of Pleyte, Rouleau Magique p. 21–2. [19] Karn. R. III pl. 90, 28.

Now it is clear from these examples that *mnw* is a stone of some value, so that BALCZ's translation can hardly be accepted; nor does any other of the monumental stones that he mentions seem appropriate either. But quartz is also cited by him as material for the particular vessels[1], and this would suit *mnw*, since it is both white or transparent, (milky quartz and rock crystal), and has various darker shades (smoky and amethystine quartz)[2]. Moreover, quartz is found both in the eastern desert and at Aswan, and rock crystal seems to have been imported[3], which coincides with what is known of *mnw*. That there are relatively few references to *mnw* would not be inconsistent with the identification, since quartz was not greatly used, probably because of its extreme hardness.

Nor is it necessary to deny LACAU's contention that *mnw ḥḏ* and *mnw km* refer to rock crystal and obsidian respectively, since the former is, as has been stated, a form of transparent quartz[4], and the latter, for which no other term is known in Egyptian[5], may well have been mistaken for smoky quartz. It is, however, impossible that *mnw* should refer solely to these materials, since neither is found in the eastern desert, and obsidian is not native to Egypt at all.

It therefore seems reasonable to suppose that *mnw* is primarily quartz, and that *mnw ḥḏ* refers to both transparent variety, rock crystal, and the translucent and opaque varieties known as milky or cloudy quartz. *mnw km*, however, includes not only the brown and black species of smoky quartz, but also obsidian, which was probably confused with them; that it likewise embraces amethystine quartz, which has a somewhat distinctive colouring, is extremely improbable. But it is evident that in making the *ḥnt* and *ḥ3ts* vessels, other stones were also used, probably because of the intractability and relative rarity of quartz, so that in the Pyramid texts and the offering lists, *mnw* is not perhaps to be taken literally.

mḫn (Wb. II. 132.2) seems to be known only from the Berlin amulet board[6], where *nbw mḫn* is the description of an amulet of gold and lapis lazuli. The WÖRTERBUCH regards it as 'Art Halbedelstein oder Glasfluss von blauer Farbe', and it is difficult to see how any other conclusion is possible. Had the word occurred in another context, it might have been interpreted as a writing of *mḫn(m)t*, which seems to be jasper, but to do so here would be to assume a very considerable mistake in the material inserted into the board, which is perhaps unlikely. In default of further evidence, no conclusion is possible.

mḫn(m)t (Wb. II. 132.4; cf. III. 301.7–8) is the subject of much confusion in the WÖRTERBUCH, for although the word is recognised, only three examples are quoted[7], while four others are to be found among the references to *ḫnmt* (Wb. III. 294.4–8)[8], and a further four are regarded as a distinct word *ḫnt* (sic.) (Wb. III. 301.7–8), which does not in fact exist[9]. But this inconsistency is due largely to the incorrect copying of the word both in published texts and on the Zetteln, and could scarcely have been avoided in the circumstances.

---

[1] Q. Saqq. VI pl. 10; 12. [2] Lucas, Materials p. 459. [3] ib.

[4] Cf. below p. 137. [5] Cf. Appendix VI.

[6] Berl. 20600.

[7] Anast. III, 5, 1; Ostr. Petrie 36 ⟨67⟩ = Hier. Ostr. 23, 1 rt. 6; P. Boul. 19 (nine times).

[8] Karn. ⟨719⟩ = Karn. I, 27 = Urk. IV, 1668; LD. III, 258a b, 8 = Bub. Port. 22; Cairo Wb. 375 ⟨636⟩ = Müller Res. I, 14; Düm. G. I. II, 97, 13–14 = Edfou II, 215, 4–5.

[9] Urk. IV, 638 (twice); 1096; Karn. Mar. 34, 30 = Urk. IV, 1729.

Of the four instances of *ḫnt* (sic.), three are from Urkunden, where SETHE shows the first sign as ●, and one is from MARIETTE's Karnak, where, however, the writing is clearly ◎, which the WÖRTERBUCH editors were presumably at a loss to understand. But the latter is certainly the correct reading, as is shown by other examples which can now be quoted[1], and examination of other copies of the Urkunden text from the tomb of Rekhmirēʿ suggests that it was in fact the reading there also. SETHE's version undoubtedly shows [2], with the first sign interpreted as *ḫ*, but HAY has [3], and WILKINSON [4], from which it appears that the sign was not familiar, and CHAMPOLLION actually [5], which must surely be the correct reading. That SETHE could misinterpret ◎ as ● is proved by a supposed example of *ḫnmt*[6], copied as in his notebook, whereas the correct reading is [7], the first sign not even being round. In view of this it seems reasonable to suppose that the other two writings of *ḫnt* (sic.) in Urkunden are in fact incorrectly copied, though neither can be checked.

There has been some question as to the transliteration of the word , and ROWE[8] notes the tentative reading *nḫnwt* of the Cairo Museum authorities, while himself proposing *tḫ* from a rather over-ingenious comparison of Hathor lady of with Hathor lady of *tḫ*, 'drunkenness'. But GARDINER[9] is of the opinion that the word is to be read *mḫnmt*, and identified with the *mḫnmt* written more fully elsewhere, and further suggests that *mḫnmt* and *ḫnmt* are in fact the same and that possibly there has been a metathesis. The appearance of ◎ in the writing is accounted for by the existence of *mḫn* as a variant for the place-name *nḫn*[10].

The full range of writings of *mḫn(m)t* is thus considerable, but examples fall roughly into four groups, 1. those written or the like[11], 2. those written with ◎ and a more or less full phonetic complement[12], 3. those written with ◎ to which only *nt* is added[13], and 4. those written with ◎ to which only *t* is added[14]. That these form a unity there can be little doubt, particularly since a strikingly similar variety of writings is exhibited by the place-name *mḫnt*[15], and the sign ◎ must have the value *mḫn*[16]. It therefore seems likely that the reading is either *mḫnmt* or *mḫnt* according to the writing, since in examples where ◎, *mḫn*, has only *nt* or *t* as phonetic complement, the second *m* is certainly not present. This uncertainty might support GARDINER's suggested identification with *ḫnmt*, the three forms *mḫnmt*, *mḫnt* and *ḫnmt* being due to metathesis, and the apparent existence of *mḫnt* as a variant for *ḫnmt* in a passage in the 'Book of the Dead'[17] would tend to confirm this.

---

[1] Eg. , JEA. 38 pl. 5, 42; , Ann. Serv. 33 p. 71; pl. 2, 4; 3, 3; 4.

[2] Urk. IV, 1096; cf. also Brugsch, Thes. 1111.

[3] Quoted by Davies, Rekhmirēʿ p. 19, where, however, Champollion is wrongly cited. [4] ib.

[5] Not. Descr. I p. 506; cf. Grammaire p. 90, translated 'grenat'.

[6] Karn. ⟨719⟩. [7] Karn. I, 27 = Urk. IV, 1668.

[8] Ann. Serv. 38 p. 683. [9] JEA. 38 p. 13.

[10] Cf. Wb. II. 132.3; Onom. 2 p. 7; Gauthier, Dict. géog. III p. 18.

[11] P. Boul. 19; Anast. III, 5, 1; Ostr. Petrie 36 = Hier. Ostr. 23, 1 rt. 6; Ostr. Nash 6 = Hier. Ostr. 66, 1, 8; 1, 9; JEA. 28 pl. 5; LD. III, 258a b, 8 = Bub. Port. pl. 22; Ostr. IFAO/AG. 85 rt. 5 [19].

[12] Cairo Wb. 375 ⟨636⟩ = Müller Res. I, 14 (;) Karn. I, 27 = Urk. IV, 1668 ( ); LD. III, 127b ( ).

[13] Urk. IV, 638 (twice); 1096; 1259 = JEA. 38 pl. 5, 42; Karn. Mar. 34, 30 = Urk. IV, 1729; Menkhep. pl. 9; Totb. Nav. 156, 4 Pb; M. u. K. A 1, 1; A 1, 3 – hieratic. Erman noted that the sign is nearer ⊙ than ●, but ◎ is more likely.

[14] Ann. Serv. 33 p. 71; pl. 2, 4; 3, 3; 4 (twice). [15] Gauthier, Dict. géog. III p. 18.

[16] Note also a Ptolemaic writing , Edfou II, 32, 13–14, but cf. Edfou II, 215, 4–5.

[17] Totb. Nav. 156, 4. Pb. has where other versions have *ḫnmt*; cf. also Edfou II, 215, 4–5 (*mḫnmt*) and Edfou VI, 165, 15–16 (*ḫnmt*).

Moreover, the meaning of *mḫn(m)t* appears to correspond to that of *ḫnmt*, which is in itself significant. From the references collected above, it is evident that *mḫn(m)t* is a semi-precious stone[1] brought from Pwenet[2], Coptos[3] and Nubia[4], and mined in the region of the diorite quarries at Tushkah in the western desert[5], where various stones including some carnelian, poor quality amethyst, and 'large quantities of strange multi-coloured quartz'[6] are found, but not apparently red jasper. However, it is always possible that a deposit may exist at a lower level than any yet explored, or that the source is now completely worked out. Like other precious stones, *mḫn(m)t* was obtained in large pieces[7], and in lumps of varying size[8]; it was primarily red[9], but also yellow[10] in colour, and was used for the making of beads[11], *wḏ3t* eyes[12], and the *tit* amulet[13], of which the usual material is *ḫnmt*[14], red jasper. The latter was the most typically red of all Egyptian stones, and would suit the colour and uses of *mḫn(m)t* admirably. Furthermore, jasper is perhaps the only stone which would also explain the yellow lumps, since a yellow variety of jasper occurs in Egypt, and both red and yellow are found together[15], the one streaked with the other. It therefore seems probable that in spite of the negative evidence of the Tushkah source, *mḫn(m)t* is to be regarded as red and yellow jasper, and that since *ḫnmt* also refers to jasper, the two are to be identified.

At all events, it is quite clear that *mḫn(m)t* is a rather more common word than the WÖRTERBUCH would suggest, being known from some two dozen examples, and in a number of different forms. Whether it is to be regarded as one with *ḫnmt* must remain an open question, though identification seems to be suggested both by the writing and by the apparently similar meaning.

nmḥf (Wb. II. 268.17) occurs in a number of different writings, of which *nmḥf*[16] appears to be the oldest and the most correct, the others, including *nmḥ*[17], *mḥf*[18] and *mḥ*[19], being doubtless corruptions, due to omission of the final *f*, or misinterpretation of the initial *n* as the genitival particle. That all are the same word is quite evident, since, apart from later variants with *inr*, all are found in similar contexts in versions of the 'Book of the Dead'.

Various suggestions as to the nature of the stone *nmḥf* have been advanced, though none is altogether satisfactory. BRUGSCH[20] translates 'agate', BIRCH[21] 'steatite', and NAVILLE[22] 'jasper', while EBERS[23] favours 'black serpentine', on the grounds of a scaraboid of that material inscribed *r n ib n nmḥw*. This latter identification is accepted by WENDEL[24], who regards the true name of the stone as *mḥ*, and states it was found near Elephantine, and

---

[1] Karn. Mar. 34, 30 = Urk. IV, 1729; Karn. I, 27 = Urk. IV, 1668; LD. III, 127b etc.
[2] Urk. IV, 1096. [3] Menkhep. pl. 9. [4] JEA. 28 pl. 5.
[5] Ann. Serv. 33 p. 71 etc. [6] Ann. Serv. 38 p. 372.
[7] LD. III, 258ab, 8 = Bub. Port. pl. 22. [8] Urk. IV, 638; 1096; JEA. 28 pl. 5.
[9] Urk. IV, 1096 ('rouge très vif et très éclatante' – Not. Descr. I p. 506); JEA. 28 pl. 5 (light red); M. u. K. A 1, 1; P. Boul. 19. [10] Urk. IV, 638.
[11] P. Boul. 19, at least two in the shape of (red) fruits; M. u. K. A 1, 1 (a red bead); A 1, 3.
[12] Ostr. Nash 6, 9 = Hier. Ostr. 66, 1, 9. [13] Totb. Nav. 156, 4 Pb.
[14] ib. and eg. Berl. 20600. [15] Lucas, Materials p. 455.
[16] Totb. Nav. 30B Rub.; Totb. Budge 30B Rub. (Nu); Yuya 30B (pl. 16); Neferrenpet 30B (pl. 11).
[17] Totb. Nav. 30B Rub.; ZÄS. 18, 56; Berl. 20600; cf. *inr nmḥ* P. Ch. B. IV vs. 7, 12.
[18] Totb. Nav. 30B Rub.
[19] Totb. Leps. 64, 33; R. Fun. Guieysse 1.33; cf. *inr mḥ* Dend. Mar. IV, 39, 140–2, and *inr n mḥ* Dend. Mar. IV, 36, 52–3, though the latter could be 'inlay-stone'.
[20] Wb. p. 692. [21] ZÄS. 5 p. 54. [22] Totb. Einl. p. 128.
[23] ZÄS. 18 p. 56. [24] Bau- und Edelsteine p. 117.

came as tribute from the first nome of Upper Egypt[1]. GARDINER[2], on the other hand, gives *nmḥf* the literal meaning 'it floats not', and translates 'green basalt', while noting that Yuya's heart scarab was green felspar and that the scarab on the Kennard board (now in Berlin) is nephrite. The WÖRTERBUCH takes this last example literally, and renders 'Nephrit'.

The one significant fact in determining the meaning of *nmḥf* is that it was the material par excellence for the heart scarab, and it is clear that the majority of these scarabs were made of green stones such as jasper, felspar, serpentine, 'basalt' etc. The nephrite scarab of the Berlin amulet board[3] is unique, and was obviously a substitute, perhaps accidental, for the correct stone, since neither nephrite nor jadeite is at all common in connection with ancient Egypt[4].

In considering which of the various green stones used is the one to which *nmḥf* really refers, it is important to discover the material of the earliest heart scarabs, and of those made for royal or noble persons, since these are more likely to be of the true *nmḥf* and not imitations. Of the early examples, that of Sebekemsaf in the British Museum is green jasper[5], and that of Tetimes of 'durite'[6], while among other more notable specimens are those of Yuya in green felspar[7], of Amenardas in green porphyry[8], and of Shabaka and Tanutamun in green serpentine and green slate respectively[9]. Possible heart scarabs of late date are of 'green stone' (pierre verte)[10], green jasper[11] and green felspar[12], while of the many cited by PETRIE[13], 35 are 'basalt', 20 serpentine, 13 steatite, 10 green jasper, and 10 porphyry.

To which of these stones *nmḥf* properly refers it is not easy to decide, though green felspar may be eliminated, since it is certainly *nšmt*. The most characteristically green is perhaps green jasper, of which the earliest royal scarab is made, and this may well be the correct stone. The nephrite scarab of the Berlin board indirectly confirms this, since jade and jasper are similar in appearance, and were apparently confused by PLINY, BOETIUS (1609) and other lapidaries[14], so that the maker of the board may have inserted the nephrite scarab thinking it was jasper. Moreover, green breccia, serpentine, green 'basalt' and certain other stones are all sufficiently similar in appearance to have been substituted for jasper, which was relatively rare[15]. On the other hand, *nmḥf* may equally well be either serpentine or 'basalt', both of which occur with greater frequency than jasper as material for the heart scarab, and no satisfactory conclusion seems possible.

The significance of the name *nmḥf* is not altogether clear, and it may be translated either 'it floats not', as GARDINER, or 'may it not forget', neither of which can have any reference to the actual material, but must be connected with the use of the stone for heart scarabs. The latter is perhaps the less likely interpretation, and would only make sense were it taken to mean 'may it not forget (to be favourable)' or the like, expressing a pious hope that the scarab will protect its owner in the moment of judgement. 'It floats not' is at first sight even more obscure, but may in fact be understood as referring to the actual weighing of

---

[1] I find no support for these statements. [2] Amenemhet p. 112.

[3] Berl. 20600, described as nephrite by Möller, Amtliche Berichte 34, no. 2, col. 24, and Roeder, Äg. Inschr. Berl. 2 p. 312, but tentatively identified as green jasper in Cat. Kennard Coll. p. 54, no. 530.

[4] Lucas, Materials p. 452. [5] Hall, Scarabs p. 22 no. 211.

[6] Petrie, Scarabs p. xlvii. [7] Gardiner, Amenemhet p. 112.

[8] MMA. Bull. X, 1915, p. 116. [9] Dunham, El Kurru pl. 48; p. 62.

[10] Ann. Serv. 1 p. 234. [11] Ann. Serv. 3 p. 248.

[12] Ann. Serv. 4 p. 82. [13] Amulets p. 23–4, nos. 89, 90.

[14] Ball, Roman Book p. 289; Caley & Richards, Theophrastus p. 108; Moore, Ancient Mineralogy p. 219 etc.

[15] It seems probable that classical iaspis included several similar green stones, cf. the refs. cited in the previous note.

the heart. In this ordeal, equilibrium seems to have been sufficient, and perhaps normal, and no representations are known with the scale pans other than level. It is therefore uncertain whether a particularly good heart went up in the scale because it was free from vice, or down because it was full of virtue. But the provision of large and heavy scarabs in place of the heart would suggest the latter[1], and this would also explain the name 'it floats not', as meaning 'it does not rise in the balance' or something similar, which seems singularly appropriate. That the literal meaning of *nmḥf* was frequently overlooked by the Egyptians themselves is clear from the abbreviated and unintelligible forms in which the word occurs.

nšmt (Wb. II. 339.19f.), one of the half-dozen more important semi-precious stones, has frequently been misinterpreted in the past, and is wrongly translated in the WÖRTERBUCH. CHAMPOLLION[2] identifies *nšmt* correctly as 'spath vert', and this is accepted by BIRCH[3], but BRUGSCH[4] rejects it in favour of a comparison with the Hebrew *lešem*, apparently opal, the Greek λιγύριον[5]. Similarly, LEPSIUS[6], mistakenly thinking the stone to be light blue as opposed to the dark blue of *ḫsbḏ*, takes it to be turquoise, which is criticised by WENDEL[7], who, however, also abandons both green felspar and opal. He discusses *nšmt* in some detail, noting its provenance and uses, and takes it to be of a brown colour, concluding that *nšmt* is brown glass, while *nšmt mꜣꜥ* is a brown stone, probably agate. More recently, however, identification of *nšmt* has been based on the Berlin amulet board, from which GARDINER[8] deduces that it is green felspar, while the WÖRTERBUCH takes it too literally and renders 'weißblauer Feldspat', which is accepted by SÄVE-SÖDERBERGH[9], though not apparently by BARGUET[10], who seems to prefer comparison with Hebrew.

The evidence of the board, if correctly interpreted, is in fact conclusive, and can be supported by indications from other sources. Let into the board is a *wꜣḏ* sceptre of bluish felspar, beside which is the description *nšmt*[11], indicating that such was the correct material for the object. But it does not necessarily imply that the sceptre in the board is the ideal *nšmt*, as is shown by the scarab which is nephrite, and the *tit* amulet of *ḥnmt*, which is actually carnelian, and it is therefore wrong to base a translation on the single example. In fact, the common material for the *wꜣḏ* sceptre, apart from glazed faience, is green felspar, and 16 of those listed by PETRIE[12] are made from it, while of the *wꜣḏ* sceptres on plaques[13], no less than 21 out of 26 are green felspar, which PETRIE rightly recognises as *nšmt*. It is therefore clear that the bluish hue of the specimen in the Berlin board is purely accidental, and is not to be taken literally as is done by the WÖRTERBUCH. Indeed, it seems unlikely that the Egyptians would have distinguished between the green, bluish-green and blue varieties of felspar[14], and since the green is the most usual and the most widely used, it is to this which *nšmt* must primarily refer.

---

1 That the heart was heavy rather than light is also suggested by representations showing the heart being weighed against the deceased, cf. Budge, Ani p. 261. However, for a full discussion of the evidence pointing to the heart being light, cf. Clère, BIFAO. 30 p. 430f.

2 Grammaire p. 90.

3 Hier. Pap. p. 15.

4 Wb. p. 810; Erg. p. 699.

5 Cf. JAOS. 73 p. 152; Sudhoff's Archiv 38 p. 202.

6 Métaux p. 38.

7 Bau- und Edelsteine p. 103.

8 Hier. Pap. B. M. p. 113.

9 Äg. und Nub. p. 217.

10 Stèle de la Famine p. 24.

11 Berl. 20600. Written , where  is probably for 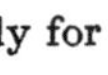 ie. .

12 Amulets p. 12 no. 20. 4 more are called beryl, and may well be felspar.

13 ib. p. 13 no. 21. 3 more are called beryl.

14 Cf. Lucas, Materials p. 450–1.

nšmt is included among the six most valuable stones[1], and is frequently mentioned together with *ḫsbḏ* and *mfkꜣt*[2], though for some obscure reason it is omitted from the lists of the 24 *ꜥꜣt* at Dendera[3]. It occurs with *mḫn(m)t* in a list of tribute[4], is found among the products of Sinai[5], and the Elephantine region[6], and seems also to appear in an enumeration of stones from the eastern desert[7]. *nšmt* is the material of the papyrus sceptre, not only on the Berlin board[8], but in the 'Book of the Dead'[9], where also Thoth is given the name *nšmt*[10], which may perhaps refer to the green colour commonly ascribed to the sacred baboon. Moreover, the phrase *r n ỉb n nšmt*[11] would indicate that it was an alternative to *nmḥf* as material for the heart scarab, and this would correspond to the occasional appearance of scarabs of green felspar[12]. A scarab of what must also be *nšmt* is mentioned in the Dendera amulet table[13], and *nšmt* occurs among the minerals from which statuettes of the Nile god are made[14]. The frequent citation of *nšmt mꜣꜥ*[15] shows that the stone was imitated in faience or glass, and it is interesting that 38 of the papyrus sceptres known to Petrie[16] are of this description.

Lastly, a word must be said concerning the passage in the Turin love-song[17] where the *ḫt* of a sycomore tree is compared to the colour of *nšmt*, and as a result of which Wendel took *nšmt* to be brown. To interpret this quasi-poetic comparison literally is altogether unreasonable, and it may be pointed out that elsewhere a moringa is likened to the colour of *ḫsbḏ*[18], another sycomore is said to be of *mfkꜣt*[19], and a type of wood is described as the colour of *ṯḥnt*[20]. It is therefore apparent that the reference is not to an exact colour comparison, but to the generally fresh and pleasing appearance of the tree[21]. Similarly, it is misleading to interpret the description of *nšmt* as *snn*[22] as an indication of either resinous appearance or golden colour, since there is reason to suppose that *snn* has another meaning in this context, and *nšmt* is clearly grouped with two other stones, *wꜣḏ m bꜣḫ*, and *ḥb*, which are undoubtedly green.

In short, there is nothing to suggest that *nšmt* was other than green in colour, and its use for the green papyrus sceptre shows that it was undoubtedly green felspar. That it covered bluish varieties of the same stone seems likely, and it is also possible that other green stones were on occasion confused with it[23].

rpy (no Wb. ref.) is altogether problematic, and may not in fact exist, since both the possible examples might equally be writings of *ry(t)*, 'ink', as Prof. Fairman suggests in a letter. The word seems to occur in one of the Dendera tribute tables[24] as a variant for *ḫsbḏ*

[1] Edfou II, 215, 4–5; VI, 165, 15–16; Dend. Mar. I, 22.

[2] Eg. Harr. 34a, 1; 52b, 5; 70a, 9; LD. III, 258a b; Ann. Serv. 16, 223; P. Ch. B. IV vs. 7, 5; P. Ch. B. IX vs. B 18, 7 etc.

[3] Dend. Mar. IV, 36, 49–50; 39, 140–2.

[4] Cairo Wb. 375 ⟨636⟩ = Müller Res. I, 14.

[5] Sinai 182; 200.

[6] Sehel 16.

[7] Wadi el Hudi 23, 4, written , where  may well be  or  cf. the writing of the Berlin board; cf. also Lucas, Materials p. 451.

[8] Berl. 20600.

[9] Totb. Leps. 159; Totb. Nav. 160.

[10] Totb. Nav. 160, 5 Ik.

[11] Totb. Nav. Ch. 27 Ba.

[12] Cf. Amenemhet p. 112; Ann. Serv. 4 p. 82; Petrie, Amulets p. 23–4, nos. 89–90.

[13] Dend. Mar. IV, 87, written 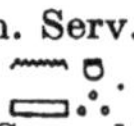.

[14] Harr. 41a, 3.

[15] Harr. 52b, 5; 70a, 9; Ann. Serv. 16, 223; Düm. G. I. II, 97 etc.

[16] Amulets p. 12, no. 20.

[17] Love Tur. 2, 4.

[18] Nauri 11.

[19] Totb. Leps. 109, 3.

[20] Düm. G. I. II, 88, 28.

[21] Cf. Appendix IV.

[22] Edfou VI, 202, 4.

[23] Cf. Lucas, Materials p. 451.

[24] Dend. Mar. I, 71, noted by Lepsius, Métaux p. 20, where it is transliterated *repi*.

brought from *tfrrt*, clearly written *rpy*, though the □ may possibly be a badly made ▯ This would suggest an artificial variant parallel to *ḥb* and *ḥkn*, and it is significant that in the Edfu mineral list[1], the only context outside the Dendera tables where these two are found, there also occurs, together with *ḫsbḏ*, a term *ry* or *rpy m pwnt*[2]. That the two examples are related seems likely, and although interpretation as *ry*(*t*) is not improbable, it is difficult to see how *ḫsbḏ* could be described as 'ink' in the Dendera example, since there is no apparent connection between the two. In view of this, it may be reasonable to regard *rpy* or *ry*, if it is so to be read, as a quite distinct word of the same type as *ḥb* and *ḥkn*, though whereas their genesis may plausibly be explained, it is not easy to suggest a pun or word play which would account for this variant of *ḫsbḏ*.

ḥb (Wb. III. 62.1) seems to be known from only three examples, all of late date, in two of which[3] it occurs as a variant of *mfkꜣt*, while in the third[4] it accompanies *wꜣḏ m bꜣḫ* and *nšmt*, and is described as *ḥb m ḏt.f*. Now whereas some of the variants in the Dendera tables are real, in as much as they represent closely allied but yet distinct substances[5], others are quite clearly artificial[6], and seem merely to be imaginative by-words for materials more usually known under other names. It is to this latter category that *ḥb* appears to belong and it is, moreover, possible to suggest approximately how the word was created. It has long been observed that *mfkꜣt* frequently occurs parallel to *ḥb* in such phrases as *pt m ḥb*, *tꜣ m mfkꜣt*, and that both have a similar meaning 'joy' or 'festivity', *mfkꜣt* acquiring the same sense as *ḥb* because of its pleasing appearance. But it seems probable that by an ingenious reversal of the process by which *mfkꜣt* came to mean 'joy', *ḥb* was made to refer to turquoise and used as a variant name for the stone in these few instances. However, it must be admitted that the expression *ḥb m ḏt.f* does rather suggest a specific mineral, and it may be that in this instance *ḥb* refers to some other green stone closely connected with turquoise, though what this would be it is difficult to imagine.

ḥmt (Wb. III. 86.15f.) is etymologically connected with *ḥmt*, 'craft', and *ḥmwty*, 'craftsman', which in itself suggests an artificial product. Ebbell[7] translates 'frit' in the Ebers examples, and Rowe[8] takes *ḥmt* to be 'a kind of vitreous paste (sometimes green) used as material for seals ..... also ..... pulverised as pigment for ink'. Forbes[9] refers to *ḥmt* as a material for seals and green frit in particular. From the textual evidence it appears that *ḥmt* was a substance obtained either as lumps[10] or in granular or crystalline form[11], and used for making seals[12], as a pigment[13], and in medicine[14]. It was apparently a product of *Kftyw*[15], and was certainly green in colour, as is shown by the parallelism between *dḳw nw ḥmt wꜣḏt* and *dḳw nw wꜣḏ*[16], though simple references to *ḥmt wꜣḏt* in Ebers[17] might equally well be fresh *ḥmt*.

---

[1] Edfou VI, 202, 2.

[2] The second sign is definitely ▯, but *ry* may have been written because *rpy* was not understood.

[3] Dend. Mar. I, 70; I, 71.

[4] Edfou VI, 202, 4.

[5] Eg. *msdmt*, *ḥtm* and *gsfn*; *bd* and *ḥsmn*.

[6] Eg. *ḥkn*, *inf*.

[7] Pap. Ebers p. 84.

[8] Ann. Serv. 38 p. 683.

[9] Studies V p. 121.

[10] P. Kah. 28, 26; 28, 41.

[11] *dḳw nw ḥmt*: Eb. 70, 16; 70, 21; P. Med. Berl. 5, 5; Totb. Nav. 100, 8.

[12] Harr. 33b, 13; 34a, 4; 64a, 16; P. Kah. 19, 28; 20, 30.

[13] Totb. Nav. 100, 8.

[14] Eb. 70, 16; 70, 21; 87, 14; P. Med. Berl. 5, 5.

[15] Edfou VI, 203, 3.

[16] Totb. Nav. 100, 8.

[17] Eb. 70, 16; 70, 21; 87, 14; cf. also Edfou VI, 203, 3.

That *ḥmt* was a mineral substance is shown by two occurrences in the Chester Beatty mineral list[1], where, however, nothing can be deduced as to its precise meaning. But the parallelism with *wꜣḏ* and its use as a pigment suggest a green frit, and this would be appropriate in every other instance, though where the colour is not specified, frit of any description may be meant. Furthermore, the fact that *ḥmt* was obtained from Keftiu might indicate that it was not a common frit like *šsyt*, but a specific variety, perhaps akin to faience.

No very satisfactory conclusion is, however, possible, and *ḥmt* can only be regarded as an artificial substance analogous to frit, and probably, though not necessarily, of a green colour.

ḥmꜣgt (Wb. III. 95.1; cf. III. 99.10; 99.12; 174.21) like *mḫn(m)t* discussed above, is written in a variety of different ways, the majority of which are recognised as one by the Wörterbuch, though three are regarded as separate words, without perhaps sufficient justification. These are *ḥmgt* (Wb. III. 99.12), *ḥmḳ* (Wb. III. 99.10), and *ḥḳm* (Wb. III. 174.21), to which may be added *mꜣky* (sic.) (Wb. II. 33.9), where the *ḥ* has in all probability been lost.

Of *ḥmgt*[2] there can be little doubt, since the form is close to several writings of *ḥmꜣgt*[3], and the colour of the substance referred to is apposite. Similarly *ḥmḳ*[4], identified with *ḥmꜣgt* by Caminos[5], is akin to known writings of the latter[6], the interchange of *g*, *k*, and *ḳ* being parallelled in the case of *ḳʿ* discussed below (p. 133). *ḥḳm*[7] is somewhat more doubtful, though it is probable that metathesis has taken place, and that the word is to be read *ḥm(ꜣ)ḳ*. Concerning *mꜣky* (sic.)[8] a little more must be said, since identification with *ḥmꜣgt* requires the assumption that *ḥ* has been lost. However, both the commentators on the Sehel text are of the opinion that this is the case, Barguet[9] thinking that a break in the stone has resulted in the loss of 𓎛, while Brugsch[10] actually saw ·· and prefers to restore —. A reading *ḥmꜣky* may therefore be accepted, and compared with similar writings of *ḥmꜣgt*[11].

Numerous suggestions have been made as to the meaning of *ḥmꜣgt*, all of which may, for one reason and another, be rejected. Lepsius[12] tentatively suggests red japser, while Brugsch[13] does not commit himself, and Wendel[14] proposes amethyst, abandoning jasper as being too common for the small number of references to *ḥmꜣgt*, and also rejecting identification with ἄνθραξ. The uses of amethyst are, he claims, consistent with those of *ḥmꜣgt*, and the apparent red colour can be explained by the wide range of meaning implicit in *dšr*. A similar view is put forward by Jéquier[15], who sees *ḥmꜣgt* as the only possible word for amethyst, and explains the dark red colour given to it as due to the limitations of the Egyptian palette, and the absence of any violet pigment. A translation as amethyst is also accepted by Barguet[16], and by Gardiner[17], who, however, thinks that *ḥmꜣgt* might equally well be either carnelian or red jasper. Similarly, Rowe[18] takes carnelian to be either *ḥmꜣgt* or *ḥnmt*, and Säve-Söderbergh[19], wrongly identifying the latter as carnelian, gives granite, haematite, amethyst and red jasper as possible alternatives for *ḥmꜣgt*.

---

[1] P. Ch. B. IV vs. 7, 9; 7, 10 = Ostr. Tor. A. 4.
[2] Urk. IV, 1099.
[3] Eg. Wadi el Hudi 23, 4; P. Ch. B. IV vs. 7, 6 etc.
[4] Koller 4, 2.
[5] LEM. p. 442; cf. Burchardt, Fremdworte II p. 35 no. 677.
[6] Eg. P. Ch. B. IV vs. 7, 6; Dend. Mar. IV, 36, 49–50 etc.
[7] Harr. 64c, 1; cf. Burchardt, Fremdworte II p. 36 no. 693.
[8] Sehel 16.
[9] Stèle de la Famine p. 25.
[10] Hungersnoth p. 129.
[11] Eg. Düm. G. I. II, 70; II, 63; Thes. 20.
[12] Métaux p. 21 n. 6.
[13] Hungersnoth p. 129; Wb. p. 959; cf. Erg. p. 823, where Lepsius' 'jasper' is quoted.
[14] Bau- und Edelsteine p. 107.
[15] BIFAO. 19 p. 121.
[16] Stèle de la Famine p. 25.
[17] Hier. Texts I p. 41.
[18] Ann. Serv. 38 p. 682.
[19] Äg. und Nub. p. 216.

Before attempting a more satisfactory identification of *ḥm3gt*, it must first be made quite clear that red jasper, carnelian, and amethyst are all impossible as translations. Red jasper and carnelian, quite apart from being far more common than the relatively few references to *ḥm3gt* would suggest, are certainly to be identified as *ḫnmt* and *ḥrst*, and amethyst, though for a long time unknown, can now be shown to be *ḥsmn*. Thus, previous suggestions that *ḥm3gt* refers to one or other of these stones may be disregarded entirely.

The evidence for the meaning of *ḥm3gt* is largely confined to its colour and provenance, though a number of specific uses are also mentioned. That it was a precious stone is clear from its appearance among the 24 *ˁ3t* at Dendera[1], and its occurrence together with other known stones[2], as well as from the form in which it is depicted[3]. The chief sources were undoubtedly in Nubia[4], though the eastern desert[5], and the region round Elephantine[6] are also noted. The stone was of a red colour, perhaps particularly dark[7], and was used for *wḏ3t* eyes[8], for beads[9], for a (?) ring[10], and as the material of a decan[11]. It is also one of the stones mentioned in the text of filling the sound eye[12].

The only information of assistance in establishing the meaning of *ḥm3gt* is that it was a red stone, primarily of Nubian provenance, but also coming from the eastern desert and near Aswan. In itself this is insufficient for an identification, but given that red jasper and carnelian are otherwise known, the possibilities seem to be reduced to three, namely garnet, sard, and haematite[13], with perhaps the addition of resin, which was frequently used as if it were a semi-precious stone[14].

Of these, the least likely is haematite, since although it occurs in a form suitable for the making of amulets, beads and other small objects, it was not commonly so used, and would scarcely have been reckoned among the precious stones as was *ḥm3gt*. Similarly, sard, the dark red variety of carnelian, though it may have been regarded as a distinct stone[15] was in all probability recognised as being a species of *ḥrst*, and may be specifically represented by the term *ḥrst dšr*. This seems all the more likely in that there is no very clear line of demarcation between carnelian and sard, but it is always possible that *ḥrst dšr* may refer simply to red carnelian as opposed to the white variety which we now call chalcedony.

Garnet, on the other hand, is more appropriate, since it is widely distributed in a number of rocks[16], and occurs both in the eastern desert and near Aswan[17], and also, according to PLINY[18], in Ethiopia. Moreover, it is dark red or reddish-brown in colour, and was used for beads and other objects[19], which would accord well with the little that is known of *ḥm3gt*.

Concerning the possibility that *ḥm3gt* is resin, it can only be observed that the latter was obtained largely from Nubia, and was used for a number of comparatively small objects,

---

[1] Dend. Mar. IV, 36, 49–50; 39, 140–2.

[2] Wadi el Hudi 23, 4; Admon. 3, 2; P. Ch. B. IV vs. 7, 6; Koller 4, 2; Sehel 16.

[3] Urk. IV, 1099.

[4] Dend. Mar. IV, 39, 140–2; Koller 4, 2; Urk. IV, 1099; Düm. G. I. II, 63; II, 70 etc.

[5] Wadi el Hudi 23, 4. [6] Sehel 16. [7] Urk. IV, 1099.

[8] Totb. Leps. 140, 11, written with *ḥ* omitted. [9] Harr. 64c, 1.

[10] M. u. K. vs. 2, 4; 2, 6. [11] Thes. 18; 20.

[12] Edfou VIII, 136–7; Thes. 41–2; Philae Phot. 524–19.

[13] Ie. the less common red variety, cf. Lucas, Materials p. 452. [14] Lucas, Materials p. 444.

[15] Theophrastus, § 30, distinguishes two varieties of σάρδιον, one female (carnelian), one male (sard).

[16] Campbell Thompson, Dictionary p. 164. [17] Lucas, Materials p. 451.

[18] HN. 37, 25, 92.

[19] Lucas, Materials p. 451; Ransom-Williams, Jewelry pp. 237; 238; 242 etc. Garnet was probably not as rare as Lucas seems to imply, and Ransom-Williams, op. cit. p. 6 states that it was even imitated in glass.

specimens mentioned by LUCAS[1] being dark red by transmitted light and almost black by reflected light. This would be applicable to *ḥmꜣgt*, but it must be admitted that although the resin was used in the same way as a semi-precious stone, it seems unlikely that the Egyptians were ignorant of its true nature.

As far, therefore, as can be ascertained, garnet appears to be the most suitable translation of *ḥmꜣgt*, though identification with sard or even resin is not altogether impossible, and the problem cannot finally be solved.

ḥrst (Wb. III. 150.9–15) and its two sub-species *ḥrst ḥḏ* and *ḥrst dšr* have been somewhat misunderstood in the past, though there cannot in fact be any doubt that the WÖRTERBUCH's translation 'carnelian' is correct, the significance of the two varieties being perhaps less readily determined.

LEPSIUS[2] notes that there are two kinds of *ḥrst*, the one white, the other red, maintaining that the former was the more esteemed, and translating therefore as milky quartz. BRUGSCH[3] also regards *ḥrst* as quartz, while BIRCH wavers between alabaster and arragonite[4], and red and white carnelian[5]. WENDEL[6] follows LEPSIUS and BRUGSCH, noting that *ḥrst* came from Nubia and was often of two colours, *ḥrst ḥḏ* being white quartz, and *ḥrst dšr* red quartz and perhaps carnelian; in older inscriptions, where *ḥrst* is mentioned alone, it is to be regarded as quartz, and where an artificial material is meant, white glass. JÉQUIER[7], taking *ḥnmt* as carnelian[8], thinks that *ḥrst* may also be the name of a sort of carnelian or jasper, while BARGUET[9] seems to regard *ḥrst* as a word for red ochre. But SETHE[10], arguing from objects stated to be of *ḥrst*, shows that it is in fact carnelian, and that red was the more important of the two colours.

The textual evidence for *ḥrst* is considerable, and it was evidently regarded by the Egyptians as one of the more important semi-precious stones[11]. It occurs among the 24 *ꜥꜣt* at Dendera[12], and in many other lists of minerals and miscellaneous tribute[13], and was obtained from[14] *kš* and other localities in Nubia[15], as well as from *tꜣ ỉmnty*[16], the region round Elephantine[17], and an unknown district *šꜣyt*[18]. Although in the Graeco-Roman period both white[19] and red[20] varieties of *ḥrst* were recognised, the typical colour was undoubtedly red, as is shown by the objects in the Berlin board[21], the colour of some beads of *ḥrst* painted on a coffin[22], and the description of a baboon's phallus as being of *ḥrst*[23]. The same colour is perhaps also implied by the frequent opposition of *mfkꜣt* and *ḥrst* as words for joy and sorrow, though it seems possible that there may be some other implication[24].

---

[1] Materials p. 444. [2] Métaux p. 39. [3] Hungersnoth p. 133; Wb. Erg. p. 835. [4] ZÄS. 11 p. 153. [5] Hier. Pap. p. 15. [6] Bau- und Edelsteine p. 109. [7] BIFAO. 19 p. 121. [8] Cf. below p. 123 for other incorrect identifications of carnelian with *ḥnmt*. [9] Stèle de la Famine p. 25. [10] Dram. Texte p. 181.

[11] Edfou II, 215, 4–5; VI, 165, 15–16; Dend. Mar. I, 22.

[12] Dend. Mar. IV, 36, 49–50; 39, 140–2.

[13] P. Ch. B. IV vs. 8, 4; IX vs. B 18, 8; Edfou VI, 202, 6; Sehel 17; 26; Antefoker 14; P. Tur. PR. 67, 11; Dend. Mar. I, 70; I, 71; II, 12; Harr. 33b, 14 etc.

[14] Dend. Mar. II, 12; Edfou VI, 202, 6.

[15] Dend. Mar. I, 71; Düm. G. I. II, 63, 7; II, 70, 8; IV, 164; P. Tur. PR. 67, 11.

[16] Dend. Mar. I, 70. [17] Sehel 17; 26.

[18] Mem. Inst. 24, 108 = Coff. Texts VI, 212–3.

[19] Dend. Mar. IV, 36, 49–50; 39, 140–2; I, 71; Edfou VI, 202, 6.

[20] Dend. Mar. IV, 36, 49–50; 39, 140–2.

[21] Berl. 20600. [22] Cairo 28092. [23] Anast. III, 5, 2. [24] Cf. Appendix IV.

ḥrst is shown being bored and polished[1], and was used for a variety of small objects including *wḏ3t* eyes[2] and other amulets[3], various kinds of beads and similar ornaments[4], figures of the four sons of Horus[5], and statuettes of the Nile god[6]. Further, in religious contexts it appears as the material for decans[7], the fore-part of a lion[8], and a temple[9], but is not included among the stones used in filling the sound eye, perhaps because of its untoward significance in Ptolemaic times.

Direct evidence for the identification of *ḥrst* as carnelian is provided by the Berlin amulet board[10], where *ḥrst* is the label beside two objects, a *wḏ3t* eye and a bead, which are in fact carnelian. Moreover, of the *wḏ3t* eyes cited by Petrie[11], no fewer than 32 are either carnelian or sard, which confirms that in this instance the material inserted into the board is the correct one. The other examples of *ḥrst* are all consistent with such a translation, which is also appropriate to the sub-species *ḥrst ḥḏ* and *ḥrst dšr*, as may now be pointed out.

Carnelian is a translucent red chalcedony, which owes its colour to the presence of small amounts of iron oxide, and although the ideal tint is an orange-red, all shades from pale to dark are known, and specimens occur blotched and mottled with light and dark patches. At first the Egyptians seem to have regarded all these varieties as one, and to have referred to them indiscriminately as *ḥrst*, but later, in the Ptolemaic period, they clearly distinguished three main species, *ḥrst*, *ḥrst ḥḏ*, and *ḥrst dšr*[12]. Of these, *ḥrst* no doubt refers to carnelian of the ideal flame-red hue, while it seems reasonable to suppose that the others represent the lighter and darker nuances respectively, that is, what would now be called chalcedony and sard. Both these stones were employed to a small extent in Egypt[13], and unless sard is *ḥm3gt*, no name for either is known. Modern science regards both carnelian and sard as forms of chalcedony, realising that the colour is incidental, but there is no reason why the Egyptians, to whom the red variety was more familiar, should not have taken chalcedony and sard to be species of carnelian.

In the majority of cases, therefore, *ḥrst* is to be identified as carnelian, or the glass and faience imitations which were sometimes substituted for it, though it may also refer to sard or chalcedony, which appear to have been differentiated only at a late date.

ḥsmn (Wb. III. 163.25)[14] has frequently been confused with the similarly written terms for bronze and natron, though Brugsch[15] recognised a separate word referring to a particular kind of stone, which he compared to the Greek ἤλεκτρον. Erman[16] further stressed the distinction, and noted Brugsch's comparison, concluding that both the stone and the metal must have been bright and shining. But no precise identification of the stone *ḥsmn* was possible until comparatively recently, when Fakhry[17] noted the word on a number of Middle Kingdom stelae from Wadi el Hudi.

---

1 D. el Geb. I, 13; I, 14.
2 Berl. 20600; Cairo 28083.
3 Harr. 15a, 15.
4 Harr. 64b, 16; Cairo 28083; 28091; 28092; Berl. 20600; P. Med. Lond. 13, 2; Dendereh pl. 3 etc.
5 Dend. Mar. IV, 36, 52.
6 Harr. 41a, 6.
7 Thes. 19; 20.
8 Coff. Texts II, 46a.
9 Coff. Texts II, 377b.
10 Berl. 20600.
11 Amulets p. 32 n. 138.
12 Dend. Mar. IV, 36, 49–50; cf. also ib. 39, 140–2; Edfou VI, 202, 6; Dend. Mar. I, 71 etc.
13 Lucas, Materials p. 448–9.
14 Cf. Akkadian *ḫašmanu*, a stone or metal from which beads are made, Bottéro, Archives Royales de Mari VII p. 296f.
15 Wb. Erg. p. 853.
16 ZÄS. 30 p. 31.
17 Ann. Serv. 46 p. 51; cf. also the later, inadequate, publication Wadi el Hudi.

Now since *ḥsmn* is not only the most important, but practically the only material mentioned on these stelae, it must inevitably be regarded as the name of the mineral obtained from the Wadi. But Murray has confirmed that the workings there produced only amethyst, so that there can be no doubt that it is to this that *ḥsmn* refers.

Concerning the actual references to *ḥsmn*, there is little to be said. The majority occur on the stelae from Wadi el Hudi itself, and record expeditions to fetch *ḥsmn*[1] from the region, which is called *ḫꜣst nt ḥsmn*[2], and whose patron deity is Hathor lady of *ḥsmn*[3]. In one instance[4], *ḥsmn* heads a list of semi-precious stones described as products of the area. Another stela, of the reign of Amenemhet III, also mentions an expedition to fetch *ḥsmn*[5], and although the monument is said to come from Dehmît, it seems highly probable that it was in fact brought from the mines at Wadi el Hudi. Other examples of *ḥsmn* are few, though it is once listed among tribute of Retenu[6], twice included among stones for jewellery or the like[7], and mentioned as the material for a scarab[8] and a seal[9], in the last instance qualified as *ḥsmn wꜥb*, perhaps through confusion with *ḥsmn*, natron.

Although the identification with amethyst is certain, two confirmatory points are worth noting, namely that *ḥsmn* occurs chiefly in texts of the Middle Kingdom, when the use of amethyst was relatively popular, and that its absence from the normal enumerations of stones corresponds to the absence of amethyst from composite jewellery, to which its distinctive colour was unsuited[10]. Lastly it must be observed that the equation of amethyst with *ḥsmn* is of particular importance in refuting the suggestion that either *ḥmꜣgt* or *ỉbhty* refers to this stone, since all three terms occur together in one context[11].

ḥkn (Wb. III. 180.14) is apparently known from only three examples, all of late date, in two of which[12] it occurs as a variant for *ẖnmt*, while in the third[13] it is referred to as *ḥkn pr m nst tꜣwy*, and seems to describe some stone or stones that precede it, though a lacuna at the crucial point makes interpretation difficult. As has been pointed out above[14], some of the variants in the Dendera tables in fact represent distinct substances sufficiently akin to be identified, while others have no real meaning and are simply artificial synonyms for the more common terms. *ḥkn*, like *ḥb* and *ỉnf*, is one of the latter type, and seems to be the result of a play on words. *ẖnmt*, red jasper, is connected with, and probably derived from, the verb *ẖnm* (Wb. III. 292.10), to delight or gladden, an approximate equivalent to which is the verb *ḥkn* (Wb. III. 178.2) of similar meaning. When a synonym for *ẖnmt* was required, it seems that this parallelism was made use of, and that *ḥkn* was created from the latter verb. Thus the word has no precise meaning in the Dendera tables, nor, possibly, in the Edfu mineral list, where a translation 'red stone' or the like would perhaps be appropriate, though *ḥkn* may again simply replace *ẖnmt*, the red stone par excellence.

ḥts (Wb. III. 203.11) seems to occur only twice, at Philae, once written 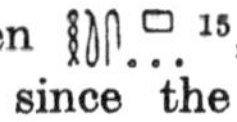[15], once [16]. Brugsch[17] identifies *ḥts* with *ḥrst*, and this appears likely since the latter

[1] Wadi el Hudi 4, 4; 8, 6; 14, 9; 14, 17. [2] ib. 21, 5; 23, 3.
[3] ib. 16, 3; 17, 3. [4] ib. 23, 3. [5] Ann. Serv. 9 p. 109 fig. 4.
[6] Urk. IV, 722. [7] Admon. 3, 2; Cairo 20539 IIb 8.
[8] Harr. 15b, 3. [9] B. M. 586.
[10] Cf. Ransom-Williams, Jewelry p. 6–7; Hayes, Scepter (1) p. 229. [11] Admon. 3, 2.
[12] Dend. Mar. I, 70; II, 12. [13] Edfou VI, 202, 6. [14] p. 117.
[15] Philae ⟨1991⟩ Phot. 241. [16] Philae ⟨2004⟩ Phot. 239.
[17] Wb. Erg. p. 873; cf. Dict. géog. p. 1033.

displays a wide variety of writings. No exact parallel can be quoted, but *ḥrts* is found on at least one occasion[1], and *ḥrds* is fairly common[2]. The appearance of *ḥrst* in these passages would be quite appropriate, and the Nubian regions *ḳns(t)* and *ỉtfty(t)* cited as provenance for *ḥts* would suit carnelian. In default therefore of further evidence, it seems reasonable to regard *ḥts* as a graphic variant of *ḥrst*, perhaps peculiar to Philae, and in this case, identification with 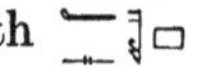 of Harris[3] is impossible.

ḫnmt (Wb. III. 294.4–8) has been variously identified in the past, most authorities being in favour of red jasper, though this translation has not gained complete acceptance, and does not appear in the WÖRTERBUCH. Red jasper was first proposed by CHAMPOLLION[4], and noted by BRUGSCH in his Wörterbuch[5], where, however, he prefers to compare *ḫnmt* with Hebrew *'aḥlāmāh*[6] rendered by ἀμέθυστος in the Septuagint. But LEPSIUS[7] notes that the stone was red, and suggests ruby, though admitting that amulets of *ḫnmt* such as the *ḏd*, *tỉt*, and *wḏꜣt* are normally 'en pierre rouge, ou en verre', and BRUGSCH in the supplement to his Wörterbuch[8] accepts this, taking *ḫnmt* as either ruby or red jasper. WENDEL[9] also notes the red colour of the stone, and the existence of a late variant *ḫkn*, and rejects LEPSIUS' ruby because the stone was not found in Egypt. He further discounts MASPERO's suggestion carnelian[10] on the strange ground that carnelian was rare (!), and inclines towards red jasper, red felspar, and red glaze as the materials to which *ḫnmt* referred. JÉQUIER[11], on the other hand, favours identification with carnelian because the *tỉt* amulet is made of that material, and ROWE[12] thinks carnelian was known as *ḥmꜣgt* or *ḫnmt*. SÄVE-SÖDERBERGH[13] also regards *ḫnmt* as carnelian, and GARDINER[14] wavers between carnelian, and red jasper perhaps in favour of the latter[15]. BARGUET[16], however, returns to the translation red jasper, and points out, against JÉQUIER, that the *tỉt* amulet occurs equally frequently in red jasper. FAIRMAN[17], too, discussing Seth as a hippopotamus of *ḫnmt*, adopts red jasper, though KEES[18] takes it to be 'in bewußter Polemik gegen die beliebten Thoërisamulette aus Karneol'.

One source of uncertainty has undoubtedly been the Berlin amulet board[19], where the description *ḫnmt* accompanies a *tỉt* amulet which is in fact dark carnelian. But it has been shown in discussing *nmḥf* and *nšmt* that the evidence of the board is not to be interpreted too literally, and it is quite clear that in this case too the material inserted was a substitute, deliberate or accidental, for the correct one. Of a total of 110 similar amulets cited by PETRIE[20], 57 are red or brown in colour, and of these no fewer than 34 are jasper, 21 red and 13 brown, while 18 are red glass; only 3 are carnelian. From this it must surely be apparent that *ḫnmt* refers not to carnelian, but to red and reddish-brown jasper.

The textual evidence for *ḫnmt* is considerable, and it is evident that it was regarded by the Egyptians as one of the more important semi-precious stones[21], and as the red stone par excellence. It occurs in the lists of the 24 *ꜥꜣt* at Dendera[22], and is frequently found

---

[1] Dend. Mar. I, 70.
[2] Dend. Mar. I, 71; Berl. 20600; Dum. G. I. II, 97 etc.
[3] Harr. 4, 8; cf. above p. 85.
[4] Grammaire p. 90.
[5] Wb. p. 1100.
[6] Cf. JAOS. 73 p. 147; Sudhoff's Archiv 38 p.205.
[7] Métaux p. 37.
[8] Wb. Erg. p. 941; cf. Hungersnoth p. 125–7.
[9] Bau- und Edelsteine p. 101.
[10] Ét. Ég. I p. 226.
[11] BIFAO. 19 p. 121.
[12] Ann. Serv. 38 p. 682.
[13] Ag. und Nub. p. 216.
[14] Hier. Texts I p. 41; Huy p. 22; p. 24.
[15] JEA. 38 p. 13.
[16] Stèle de la Famine p. 24.
[17] JEA. 21 p. 33.
[18] Farbensymbolik p. 461.
[19] Berl. 20600.
[20] Amulets p. 23 n. 88.
[21] Edfou VI, 165, 15–16; Dend. Mar. I, 22.
[22] Dend. Mar. IV, 36, 49–50; 39, 140–2

together with other stones and minerals, and among miscellaneous tribute[1]. *ḫnmt* was obtained from various districts in Nubia[2], and from the region round Elephantine[3], and is shown as lumps of red material[4]. The red colour of the stone is also exemplified by a passage in the Turin love-song[5], where the fruit of the sycomore is said to be redder than *ḫnmt*, and by a reference to the god Seth as a hippopotamus of *ḫnmt*[6], and colour is also in question in two other passages[7].

ḫnmt was the material for a number of objects, including the *tit* amulet[8], a *wḏ3t* eye[9], scarabs[10], beads[11], statuettes of the Nile god[12] etc.[13], and is also mentioned as the substance of a decan[14], and as one of the stones used for filling the sound eye[15]. Of particular interest is an account[16] in which *ḫnmt m3ʿ* is listed at least twelve times as a form of payment, both in *dbn*, and in terms of statues and parts thereof, the total quantity being 1499 *dbn*, ie. some 300 lbs. The weight of the statues is unfortunately lost, but the head or face of one is given as weighing 85 *dbn*, about 17 lbs., and 12 limbs together total 55 *dbn*, about 11 lbs. This represents a considerable quantity of jasper, some of it in large pieces, but the use of jasper for statues can be confirmed, and Lucas[17] mentions a large hand of red jasper and another of yellow.

From the evidence as a whole, it is quite apparent that *ḫnmt* is to be interpreted as red jasper and the glass and faience imitations frequently substituted for it[18]. Moreover, it seems likely that the reddish-brown and perhaps even wholly brown varieties of jasper were also referred to as *ḫnmt*, though the red colour was undoubtedly typical. Whether yellow shades were included too it is impossible to tell, though they appear to have been covered by *mḫn(m)t*, which may prove to be the same word, and both red and yellow occur together in nature[19].

ḫsbḏ (Wb. III. 334.1–13) has been universally accepted as lapis lazuli, at least since the publication of Lepsius' Metalle, though not until comparatively recently has any direct proof of the identification been forthcoming.

Lepsius[20] gives *ḫsbḏ* a wide range of meaning, including 'σάπφειρος, κύανος; sapphirus (cyanus), caeruleum; le lapis lazuli et le bleu d'outre-mer; l'oxyde bleu de cuivre; les émaux bleus et la couleur qu'on en fabrique; l'émail à base de cobalt, le bleu de cobalt; l'émail à base de cuivre, et le sulfate de cuivre bleu'. He notes different writings of the word, and also the variant *rpy*, and concludes that *ḫsbḏ* is feminine. It was blue in colour, as is shown by heaps of blue material labelled *ḫsbḏ*, and was used as a pigment, and for various amulets, such as the heart, *ḏd* pillar and *wḏ3t*, as well as for other objects. *ḫsbḏ* was therefore a blue stone[21], and lapis lazuli is the most obvious equivalent, since other blue stones would not

[1] Dend. Mar. I, 70; I, 71; II, 12; III, 37i; Harr. 34a, 2; Sehel 16; Koller 4, 2; P. Tur. PR. 67, 11; Kenam. 44 = Urk. IV, 1406; Ombos I, 168, 216b 4; Philae ⟨1991⟩ Phot. 241 etc.

[2] Koller 4, 2; Urk. IV, 1375; Huy 27; Dend. Mar. I, 70; I, 71; II, 12; P. Tur. PR. 67, 11 etc.

[3] Sehel 16. [4] Huy 27. [5] Love Tur. 2, 3. [6] Edfou VI, 123.

[7] P. Tur. PR. 32, 5; Lansing 14, 6. [8] Totb. Nav. 156, 4 Rub.; Berl. 20600.

[9] Totb. Leps. 140, 12. [10] Harr. 15b, 1; 64b, 11; cf. Lucas, Materials p. 454.

[11] Harr. 64b, 17; P. Tur. PR. 39, 14 = RAD. 48.

[12] Harr. 41a, 5. [13] Harr. 64b, 10. [14] Thes. 21.

[15] Philae ⟨1672⟩ Phot. 520. [16] P. Tur. 1900 rt. col. II [20].

[17] Materials p. 454. For other fragments of jasper statues cf. Anc. Eg. Art (Burlington Fine Arts Club 1922) p. 80.

[18] Eg. the 18 glass *tit* amulets cited by Petrie.

[19] Lucas, Materials p. 455. [20] Métaux p. 20.

[21] Not a metal, as Gensler, ZÄS. 8 p. 148, who took *ḫsbḏ* (= κύανος) as a compound material including copper, in order to account for its position in the lists.

serve as pigments, and none is found in such profusion. The term *ḫsbḏ m3ᶜ* implies that imitation lapis was also known, and an expression *ḫsbḏ iryt* does in fact occur, but the *ḫsbḏ* shown in baskets was probably genuine, and so too the species called *ḫsbḏ nfr n bbr*. Blue glass and faience are found in considerable quantities, and mostly owe their colour to copper, though some may contain cobalt. LEPSIUS finally discusses the classical parallels in detail, and concludes that σάπφειρος is lapis, while κύανος is both lapis and its imitations. Egyptian lapis must have come from Bactria, and both *bbr* and *tfrr* are therefore probably the names of trading stations on the route from Badakhshan, the former Babylon, the latter possibly either Tiflis or Tebris to the south of the Caspian sea.

The identification of *ḫsbḏ* as lapis lazuli, so carefully and convincingly argued by LEPSIUS, has gained general acceptance. BRUGSCH, having already proposed lapis in his Wörterbuch[1], follows LEPSIUS closely in the supplement[2] and distinguishes *ḫsbḏ m3ᶜ* 'der Stein σάπφειρος oder κύανος' from *ḫsbḏ iryt* 'das künstliche *ḫsbḏ*, κύανος χυτός oder πεπυρωμένος, κύανος Αἰγύπτιος, ... Kobaltblau, Kupferblau, blaue Kupferlasur, blaue Smalte'. LÜRING[3] also accepts the equation with σάπφειρος, and the translation lapis lazuli, and compares passages in Ebers and DIOSCORIDES for the use of lapis as an eye-salve. WENDEL[4] takes *ḫsbḏ* as a term for the blue stones in general, and lapis in particular, referring also to blue glass and various blue colouring agents. Two kinds, *ḫsbḏ m3ᶜ* and *ḫsbḏ iryt* were recognised, and the genuine material came principally from *tfrr*, which WENDEL thinks may be Tiflis as LEPSIUS suggests. Lapis lazuli was the most important among Egyptian stones, and was used for a variety of objects, and also as a pigment and medicament. The colour of Egyptian blue glass and glaze was generally due to copper, and the pigment was of similar nature, so that unless *ḫsbḏ m3ᶜ* is specifically mentioned, it is better to take references to the latter as indicating artificial lapis. BIRCH[5] also thinks that whenever *m3ᶜ* is not added, *ḫsbḏ* refers to an imitation, and HOFMANN[6] points out that *ḫsbḏ iryt* is unlikely to refer to a cobalt glaze, and must owe its colour to a copper compound.

There has been some discussion as to the nature of *ḫsbḏ nfr n bbr*, which LEPSIUS[7] regards as genuine lapis, perhaps powdered or otherwise treated at Babylon on its way to Egypt. MÜLLER[8] maintains that since there was no lapis in Babylon, the expression must refer to an artificial frit, and this view is repeated by BISSING[9], who also thinks that *ḫsbḏ n tfrr* may be in powder form. Moreover, GARDINER[10] also distinguishes between *ḫsbḏ m3ᶜ*, true lapis lazuli, and *ḫsbḏ nfr n bbr*, which he takes to be blue frit, quoting SMITH[11], and mentioning that lapis is found in the Amarna letters as a present from Assyria.

Of the origin of the name *ḫsbḏ*, little has been written, but MONTET[12] has suggested that it may be a foreign word, derived from the name of the country from which lapis came. He points out that Badakhshan, the chief, if not the only source of lapis, has the same consonantal structure as *ḫsbḏ*, consisting of two groups of two radicals, though in the reverse order, which may be significant. Concerning *tfrrt*, the source most frequently mentioned, he agrees that it must represent a station on the trade route from Bactria to Egypt, perhaps the limit to which the Egyptians themselves had ventured, and suggests identification with Sippar.

The present discussion will be concerned only with *ḫsbḏ* as a precious stone, or the glass and faience imitations substituted for it. The use of *ḫsbḏ* as a pigment is treated below (p. 148–9), where it is shown that as such it must refer either to azurite, or to a blue frit coloured

---

1 Wb. p. 1137.
2 Wb. Erg. p. 973.
3 Med. Kent. p. 95.
4 Bau- und Edelsteine p. 87.
5 ZÄS. 11 p. 153.
6 ZÄS. 23 p. 65.
7 Métaux p. 31; cf. p. 61.
8 Asien und Europa p. 277.
9 Stat. Taf. p. 10; p. 38.
10 Onom. 1 p. 192.
11 Early History of Assyria p. 232.
12 Mon. Piot 41 p. 19.

by a copper compound, since there is no evidence to suggest that powdered lapis was ever used in this connection[1].

ḫsbḏ was undoubtedly the most highly prized of all the semi-precious stones, and regularly follows *nbw* and *ḥḏ* in the enumeration of minerals[2], sometimes even preceding *ḥḏ*[3], though also on occasion giving place to the base metals[4], and at Sinai to *mfkꜣt*[5]. It is depicted as baskets of blue lumps, both large and small[6], and is frequently measured in *dbn*[7], while blocks of *ḫsbḏ* are also mentioned,[8] one weighing 40 *dbn*, ie. about 8 lbs.[9]

ḫsbḏ is brought as tribute from various localities, including Assur[10], Babylon[11], Ḫatti[12], Sengar[13], Syria, including Retenu and Djahi[14], Cyprus[15], Pwenet[16] and Meroe[17], while *ḫsbḏ* is also mentioned among the stones from the region of Elephantine[18], and *ḫsbḏ n sṯt* among the minerals found at Sinai[19]. But more important than any of these are the two primary sources of *ḫsbḏ*, the well known *tfrrt*[20], and *ḫnty-š*[21], which may perhaps be identified with it[22].

Now lapis lazuli was probably obtained from only one locality in the ancient world, namely Badakhshan, to which it is unlikely that the Egyptians ever penetrated[23]. Thus it seems tolerably certain that *tfrrt* represents one of the more important trading stations on the route from Badakhshan, perhaps, as Montet suggests, the furthest point that the Egyptians themselves ever reached. If *ḫnty-š* is in fact to be identified with *tfrrt*, some region near the Caspian sea might be supposed, and Lepsius' Tiflis or Tebris are possible. On the other hand, *tfrrt* may be Sippar as Montet maintains, or possibly even Mt. Bikni, *Dapara*, the Tapures of Ptolemy[24], and Pliny's Tapyres[25] [26], which may have been on the trade route[27], though the *ᵃuḳnu*, normally lapis, which was obtained from there by the Assyrians was probably turquoise[28].

The other regions from which *ḫsbḏ* was obtained either as tribute or otherwise, may be variously explained. Lapis from Badakhshan was common enough in Assyria, whence it was sent to Egypt as is known from the Tell el Amarna letters[29], and from numbers of lapis objects with cuneiform inscriptions found among the Tôd treasure[30] and elsewhere[31]. There is therefore no reason why lapis should not have passed to Egypt as tribute from Assur, Babylon, Ḫatti, Sengar, and even places in Syria, and there is no real ground for supposing that

---

[1] Lucas, Materials p. 394–5.
[2] Eg. Edfou II, 215, 4–5; VI, 165, 15–16; Dend. Mar. I, 22; Urk. IV, 744; Harr. 13a, 1 etc.
[3] Admon. 3, 2; Kamose 13–14; Hier. Ostr. 95, 1 rt. 3.
[4] Karn. I, 27.
[5] Sinai 182; 200; cf. however ib. 275.
[6] Urk. IV, 638; 1098; cf. also ib. 930.
[7] Karn. I, 27; Urk. IV, 701; 708; Harr. 62b, 12; 52b, 3 etc.
[8] Urk. IV, 668; 732; Harr. 14a, 2–3; 33a, 12–13.
[9] Urk. IV, 668.
[10] Urk. IV, 668; 669.
[11] Urk. IV, 668; 701.
[12] Müller Res. II, 91.
[13] Urk. IV, 701.
[14] Urk. IV, 669; 688; 744; 929; 930 etc.
[15] Urk. IV, 708.
[16] Urk. IV, 1098.
[17] Philae ⟨1990⟩ Phot. 241.
[18] Sehel 16.
[19] Sinai 182; 200; 275.
[20] JEA. 4, 36; LD. III, 132c; Med. Hab. ⟨710⟩; Edfou I, 46; Mem. Inst. 24, 108 = Coff. Texts VI, 212–13; Ombos I, 60, 62; Dend. Mar. I, 70; I, 71; II, 12; Düm. G. I. II, 63; II, 69; IV, 102; IV, 170 etc.
[21] Edfou VI, 202, 1; Dend. Chass. IV, 159; Dend. Mar. I, 70; Düm. G. I. II, 77.
[22] Dend. Mar. I, 70.
[23] For other possible sources cf. Hornblower, Man 1949 p. 82.
[24] VI, 2, 6.
[25] HN. 6, 16.
[26] Cf. Campbell Thompson, Dictionary p. 130.
[27] Ball, Roman Book p. 292.
[28] loc. cit. above, n. 26.
[29] Eg. letter 25 passim.
[30] Trésor de Tôd (Doc. de Fouilles IFAO. XI) Ch. 2.
[31] Mon. Piot 41 p. 20.

the much-discussed *ḫsbḏ nfr n bbr*[1] or *ḫsbḏ* from any other of these localities was an artificial product[2]. Similarly, *ḫsbḏ* from Pwenet and Meroe may well refer to lapis brought from Badakhshan by some more devious route, either by sea to the Somali coast, or across southern Arabia. *ḫsbḏ* from Cyprus cannot have been lapis, but may have been azurite from the copper mines, while the occurrence of *ḫsbḏ* in the Sehel inscription is in all probability nothing more than a repetition of a conventional formula. What the significance of *ḫsbḏ n sṯt* from Sinai may be is not altogether clear, but it is just conceivable that it refers to azurite, or even to blue turquoise, which the Assyrians appear to have confused with lapis[3].

ḫsbḏ had a variety of uses, and was employed for making vessels[4], small statues[5], parts of statues[6], scarabs[7], various types of amulets[8], seals[9], collars[10], and other objects[11], including a window frame[12]. Further, parts of the body such as the hair[13], eyebrows[14] etc.[15] are said to be of *ḫsbḏ*, as has been noted by FAULKNER[16], and this may have symbolic meaning as SCHÄFER and others suggest[17], or may be due to an inability to distinguish clearly between blue and black[18]. On the other hand, the hair of statues and also the eyebrows were frequently made of lapis lazuli, and the reference may be to this practice, though the origin of the convention remains to be explained.

Among the many references to objects of *ḫsbḏ* are several which provide conclusive proof that lapis lazuli is the material referred to. These are of particular importance in that the identification has hitherto depended solely on the colour of *ḫsbḏ* and the fact that lapis was the chief, if not the only blue stone known to the Egyptians. Now, however, two pieces of direct evidence may be quoted, one the lapis clasp of a collar on which the description *ḫsbḏ mꜣꜥt* is given[19], the other a lapis figure of Horus inserted in the Berlin amulet board, and accompanied by the label *ḫsbḏ mꜣꜥ*[20]. Moreover, the list of amulets for which *ḫsbḏ mꜣꜥ* is the material in the Dendera amulet table[21] provides striking confirmation of the meaning lapis, when compared with the actual examples cited by PETRIE[22]. Of the scarabs, 13 are lapis and 8 blue 'glaze'; of the figures of Thoth, 7 lapis, 23 blue 'glaze' and 2 blue glass; of the *ḏd* pillars, 37 lapis, 56 blue 'glaze' and 4 blue glass; of the *wꜣḏ* sceptres, 8 lapis, 35 blue 'glaze' and 2 blue glass; and of the *wḏꜣt* eyes, 2 lapis, 46 blue or green 'glaze' and one blue glass[23]. No other material, except the ubiquitous green 'glaze', occurs with such frequency, or is

---

1 Urk. IV, 668; 701.

2 It is, however, clear that at least one form of artificial lapis, *ḫsbḏ iryt*, was obtained from Sengar.

3 Campbell Thompson, Dictionary p. 129f.

4 Medum 13; Urk. IV, 951 = Mem. Miss. V, 356; Cairo 28023.

5 Totb. Nav. 175, 45; Goshen 5, 4; Harr. 40b, 9; ZÄS. 65, 9.

6 P. Tur. PR. 32, 6; Louvre C 15, 8; Urk. IV, 701.

7 Harr. 15b, 4; 33a, 13; 52b, 7.

8 Dend. Mar. IV, 87; Berl. 20600; Harr. 15a, 16; 26, 5; Edfou VI, 299; Dend. Mar. IV, 36, 52–3; Totb. Nav. 26 Ba etc.

9 P. B. M. 10052, 6, 12.

10 Cairo 28027; 28083; Beni Has. II, 7; Mon. Piot 41, 11.

11 Harr. 21a, 8; 52b, 2.

12 Orb. 17, 3.

13 Destr. 2; Hier. Ostr. 95, 1 rt. 9; Düm. G. I. II, 25.

14 Totb. Nav. 172, 13; Sh. S. 65.

15 ZÄS. 11 pl. I, 15; Westc. 10, 11; Düm. G. I. II, 25.

16 JEA. 22 p. 140.

17 Schäfer, V. ä. K. p. 73; Kees, Farbensymbolik p. 466; Borchardt, Porträts p. 6 n. 2.

18 Ransom-Williams, Per-Nēb p. 53f.; Mackay, JEA. 5 p. 113.

19 Mon. Piot 41, 11.

20 Berl. 20600; cf. also Petrie, Amulets p. 48 n. 245 for similar figures.

21 Dend. Mar. IV, 87.

22 Amulets nos. 92; 202; 35; 20; 38.

23 Cf. also several *wḏꜣt* eyes in the Tôd treasure – Trésor de Tôd (IFAO. XI) p. 14; Trésor de Tôd (Cat. Cairo) p. 40 n. 70700.

common to all. Other texts also mention *ḏd* pillars, *wḏ3t* eyes and *w3ḏ* sceptres of *ḫsbḏ*[1], and in addition the figure of a ram[2], for which PETRIE[3] gives 5 of lapis and 26 of blue 'glaze'[4].

There is thus quite definite evidence for the identification of *ḫsbḏ* as lapis lazuli, but it remains to be considered how often the word refers to the genuine stone, and when to the glass or faience imitations used as substitutes for it. It is clear that the Egyptians did not always distinguish between the two, and it would be wrong to assume, as BIRCH[5], that whenever the qualification *m3ʿ* is absent, an imitation is implied. Genuine lapis was used extensively[6], and the Tôd treasure alone included over 200 items of lapis[7], so that there is nothing to suggest a paucity of the real stone.

It seems reasonable to suppose that in all cases where *ḫsbḏ m3ʿ* occurs, except as a pigment, true lapis is indicated, and this may also be the case wherever *ḫsbḏ* is shown as lumps in baskets, though in one instance[8], small irregular lumps of *ḫsbḏ m3ʿ* seem to be contrasted with larger pieces of *ḫsbḏ*, which may therefore be some kind of imitation. Certainly genuine lapis is *ḫsbḏ m ḏw.f*, which is said to be brought from the Coptite nome[9], which, however, can only have been a trade station. Furthermore, it seems likely that the greater part of the *ḫsbḏ* imported from Mesopotamian and other similar sources was the genuine lapis lazuli obtained originally from Badakhshan. Mention has already been made of several amulets said to be of genuine lapis, each of which can be shown to exist in that material, and both *ḫsbḏ m3ʿ* and simple *ḫsbḏ* also occur as the material from which vessels are made[10]. BALCZ[11] was unable to quote an actual vessel in lapis, but LUCAS[12] knows one small one of genuine lapis, and the citation of a block of lapis, *ḫsbḏ m3ʿ*, weighing 40 *dbn*, ie. about 8 lbs.[13], shows that the stone was obtained in sufficiently large pieces for the purpose. Therefore, although chalices and the like of blue faience are frequent, it is not necessary to assume that vessels of *ḫsbḏ* are always imitations.

Specific references to artificial lapis are few, though both *ḫsbḏ iryt* and *ḫsbḏ n wḏḥ* occur once, the latter being discussed below. *ḫsbḏ iryt*, mentioned together with *ḫsbḏ m3ʿ* and *ḫsbḏ(nfr n) bbr*[14], can only be an artificial lapis, though the fact that it was imported may suggest a particular variety, and one distinct from the Egyptian imitations. That *ḫsbḏ nfr n bbr* is also an ersatz lapis, is, as has been pointed out, extremely doubtful. However, the *ḫsbḏ nfr nfr* used as a pigment is almost certainly a frit, and references to *ḫsbḏ nfr nfr*[15] and *ḫsbḏ nfr*[16] elsewhere may also be taken as such. Moreover, artificial *ḫsbḏ* is directly implied in the title *irw ḫsbḏ*, 'maker of (artificial) lapis lazuli'[17], one example being in fact upon a blue glass stela. Other references which can only be to faience or glass include the mention of a window(-frame) of *ḫsbḏ*[18], and the passage noted by NEWBERRY[19], where Pi-Ramesse is said to be 'dazzling with halls of *ḫsbḏ* and *mfk3t*'. Similarly, 'walls filled with *ḫsbḏ*'[20] must indicate

---

[1] Dend. Mar. IV, 36, 52–3; Edfou VI, 299.
[2] Edfou VI, 299. [3] Amulets no. 211.
[4] Cf. also figures of *M3ʿt*, Caminos, Chronicle p. 129–30.
[5] ZÄS. 11 p. 153. [6] Lucas, Materials p. 456.
[7] Cairo 70640–70754; Louvre E. 15214–73 & 15278–318; cf. Trésor de Tôd (Cat. Cairo) p. 111.
[8] Urk. IV, 638. [9] Düm. G. I. I, 46.
[10] Medum 13; Urk. IV, 951 = Mem. Miss. V, 356; Cairo 28023. [11] Mitt. Kairo 3 p. 71.
[12] Materials p. 482. This is the vessel found at Dashur, De Morgan, Dahchour 1894 p. 71; pls. 19; 25. Another, also small, came from Giza, Petrie, Gizeh & Rifeh p. 4; pls. 3; 5.
[13] Urk. IV, 668. [14] Urk. IV, 701.
[15] P. Gurob fr. G vs. 2, 2 ⟨8a⟩ = RAD. 21. [16] Hier. Ostr. 62, 3 vs. 1; 7.
[17] Edin. 449 ⟨15⟩; Cairo Wb. 110 ⟨291⟩. [18] Orb. 17, 3.
[19] JEA. 25 p. 120 – Anast. III, 7, 5; cf. also Anast. IV, 16, 8, where *ḫsbḏ* is said to sparkle.
[20] Lansing 12, 3.

either faience tiles or blue pigment, and a crown of *ḫsbḏ*[1] is also likely to be faience, of which the detachable crowns of statues were often made. But the majority of examples of *ḫsbḏ* are equivocal, and it is impossible to tell whether the real stone or an imitation is meant.

In connection with the imitation of lapis lazuli in glazed faience or glass, it must be noted that in general the colour was due to a copper compound, though LUCAS[2] cites several examples of blue glass coloured by a cobalt compound[3]. In no case does faience or glass owe its colour to powdered lapis, and IVERSEN[4] is wrong in suggesting that this is the case. However, a small inscribed plaque in the Ashmolean Museum, Oxford[5], was found to be composed of clay mixed with powdered lapis, giving it a uniform dull blue colour. Whether this technique was at all frequent can only be established by examination of many more specimens, but it is perhaps unlikely, since the wastage of material must have been considerable, the blue lazurite forming only a small proportion of any piece of lapis. Furthermore, this case cannot be taken to imply the use of powdered lazurite in other instances, since to colour glass or glaze with it is more or less impossible, and to produce pigment from lazurite would be extremely wasteful.

To summarise the evidence for *ḫsbḏ*, it is quite clear that lapis lazuli is meant, though the word also refers to glass and faience imitations, and in certain cases, chiefly as a pigment, to azurite or blue frit, the former being perhaps confused with lapis[6]. That any other blue stone is also embraced by *ḫsbḏ* seems unlikely, though it is just possible that *ḫsbḏ n sṯt* from Sinai may refer to blue turquoise or some other stone, not lapis lazuli[7]. It may, however, simply be azurite, which was certainly found at Sinai. The principal source of lapis in antiquity was Badakhshan, and the frequently mentioned *tfrrt* must indicate a trade station supplying the material to Egypt. That *ḫsbḏ* corresponds to THEOPHRASTUS' σάπφειρος, lapis lazuli, and κύανος, lapis lazuli, azurite and artificial frit (κύανος σκευαστὸς)[8], and to the Latin sapphirus, cyanus and caeruleum, cannot be doubted. Imitations of lapis lazuli were generally coloured by a copper compound, but also occasionally by a cobalt compound; lapis itself was not used to colour glass or glaze, or in the production of blue frit.

ḫsbḏ wḏḥ (Wb. III. 334.11) occurs in only one context[9], where it is accompanied by *mfkȝt wḏḥ*, and the WÖRTERBUCH translation 'blauer Glasfluss' is almost certainly correct. Lapis lazuli was imitated by a dark blue glass[10], coloured by a copper compound, or more rarely by a cobalt compound[11], and it may be to this that *ḫsbḏ wḏḥ* refers. On the other hand, some less common substance may be indicated, possibly a particular species of faience, such as LUCAS' variant D[12]. A similar expression is *ḫsbḏ ỉryt*, discussed above, which was perhaps an imported ersatz not known in Egypt. No precise conclusion is possible, and a translation 'artificial lapis lazuli' is perhaps to be advocated.

---

[1] P. Tur. PR. 32, 6.

[2] Materials p. 217; p. 294.

[3] Cf. Lepsius, Métaux p. 26.

[4] Paints and Pigments p. 18.

[5] No. 1–41 1893; Petrie, Tell el Amarna pl. xiii, 16 (described as 'blue paste').

[6] Cf. Chassinat, P. Méd. Copte p. 226.

[7] Cf. Akkadian ^{a}*uḳnu*, Campbell Thompson, Dictionary p. 129.

[8] Caley & Richards, Theophrastus p. 126, point out that σάπφειρος and κύανος were evidently varieties of the same mineral, the difference apparently being that the former contained specks of pyrites, the latter not.

[9] Harr. 64b, 8.

[10] Lucas, Materials p. 216.

[11] ib. p. 217; p. 294.

[12] Materials p. 187.

shrt (Wb. IV. 208.16–18; 19.16; 207.10; 209.4) is found in a number of different writings, of which *shr(t)* seems to be the most usual[1], though *sꜣhr(t)*[2] and *shy(t)*[3] are also known. Moreover, both *sꜣhwt*[4] and *shwt*[5], accorded separate existence by the Wörterbuch, are to be regarded as writings of *shrt*, with *w* for *r*, which can be proved for the latter by the occurrence of *shrt* in Ebers where Hearst has *shwt*[6]. That *shrr*[7] is also connected with *shrt* seems likely and the possible relationship will be discussed below.

The identification of *shrt* is a considerable problem. Wendel[8] quotes Dümichen for Dakhla (sic. – in fact Baharia) as a source of the stone, but is unable to suggest any meaning on the scanty evidence available, and Gardiner[9] notes that neither the colour nor the true nature of the stone is known. Barguet[10] seems to take *ḥrst* and *shrt* together, and translates 'l'ocre rouge de *shrt*', though with what justification it is difficult to imagine. The colour of the stone is tentatively suggested as green by Caminos[11], who points out that in Anastasi III a baboon is said to be of *shrt*, and that in painting baboons are often shown as being of a green colour, which may be significant. Barns[12] quotes Dawson for the opinion that *shrt* is chalcedony, and notes Gardiner's comparison of *sꜣhwt*, which as a product of the Wadi Natrun might confirm this view. But Gardiner also observes that since heart amulets were made of *shrt*, one might expect it to be pale green, which perhaps does not suit chalcedony, though there is a greenish variety. Thus no very satisfactory suggestion has been made, and the question may be approached afresh.

The textual evidence for *shrt* is not in itself very informative, but reveals a few important facts which, considered in the light of what is known of other stones, may serve to indicate a possible identification. *shrt* occurs among the 24 *ꜥꜣt* at Dendera[13], once as *shr n ꜥr*, and on both occasions followed by *shrr*, and is also mentioned elsewhere in company with other semi-precious stones[14]. It came from Baharia oasis[15], from the Wadi Natrun[16], from 'the mountains'[17] from *hbks*[18], which is not identified, from another locality, now lost[19], and from the region round Elephantine[20]. *shrt* was used medicinally[21], being apparently solid and requiring crushing or powdering[22], and was the material of the heart scarab[23] and other amulets[24], as well as for statuettes of the Nile god[25]. It is also the substance of a decan[26], is used for filling the sound eye[27], and describes a baboon[28], while Hathor is once referred to as *nbt shrt*[29].

---

[1] Eg. Dend. Mar. IV, 36, 49–50; 39, 140–2; Totb. Nav. 29 Bi; Harr. 41a, 9 etc.
[2] JEA. 4 pl. 9 1.11; P. Ch. B. IX vs. B 18, 8; P. Ram. III, A 24.
[3] Mem. Inst. 24, 108 = Coff. Texts VI, 212–3; Sehel 17.
[4] Peas. R 20 (Wb. IV. 19. 16).
[5] Hearst 6, 13; Eb. 31, 18 (Wb. IV. 207.10).
[6] Eb. 44, 16 = Hearst 6, 13. Grapow, Grundriß V p. 256, reads *shrt* in both passages.
[7] Dend. Mar. IV, 36, 49–50; 39, 140–2 (Wb. IV. 209.4).
[8] Bau- und Edelsteine p. 116.
[9] JEA. 4 p. 37.
[10] Stèle de la Famine p. 25.
[11] LEM. p. 89.
[12] Five Ram. Pap. p. 19.
[13] Dend. Mar. IV, 36, 49–50; 39, 140–2.
[14] P. Ch. B. IX vs. B 18, 8; Edfou VI, 202, 10; JEA. 4 pl. 9, 11; Mem. Inst. 24, 108 = Coff. Texts VI, 212–3; Sehel 17; 26; Mammisi 158.
[15] Edfou VI, 202, 10.
[16] Peas. R 20.
[17] JEA. 4. pl. 9, 11.
[18] Mem. Inst. 24, 108 = Coff. Texts VI, 212–3.
[19] Müller Res. II, 92.
[20] Sehel 17; 26.
[21] Eb. 31, 18; 44, 16; 60, 6; 65, 1; 81, 3; Hearst 6, 13; cf. also the occurrence of *shrt* among materials used in mummification, Ann. Serv. 53, 178.
[22] P. Ram. III A 24.
[23] Totb. Nav. 29 Bi.
[24] Dend. Mar. IV, 87.
[25] Harr. 41a, 9.
[26] Thes. 18.
[27] Edfou VIII, 136–7; Philae ⟨1675⟩ Phot. 519.
[28] Anast. III, 5, 1.
[29] Ombos II, 130, 701.

It is therefore clear that *shrt* is a semi-precious stone, obtained from a number of localities and probably of a green colour. This latter fact is suggested not only by the description of the baboon remarked upon by CAMINOS, but also by the name of the stone, *shrt*, which suggests joy and therefore a green colour[1]. Moreover, the heart-scarab of *shrt*, wrongly identified as carnelian by NAVILLE[2] and BUDGE[3], implies a green stone, since the majority are in 'basalt', jasper, felspar, serpentine and other stones of a similar colour[4], and the colour of the various objects of *shrt* in the Dendera amulet table[5] is predominantly green. Of 5 sistra cited by PETRIE, 1 is green felspar and 3 green 'glaze'; of 17 plumes, 3 are obsidian, 3 serpentine, 2 'basalt', and 1 green 'glaze'; of the total *wḏꜣt* eyes, a large proportion are green 'glaze', and some half dozen of different green stones; and of the scarabs well over half are green[6]. Now although none of these indications would in itself be sufficient to warrant the assumption that *shrt* was a green stone, together they strongly suggest that such was the case, and since no alternative evidence of colour is forthcoming, this may provide the basis for an identification.

Of the green stones used by the Egyptians, the principal ones, turquoise, felspar and malachite, are known by name, and of the others, a few, such as chrysoprase, olivine and peridot, are too rare to be considered in connection with *shrt*. Possible alternatives are therefore beryl, jasper, and serpentine, of which the first is unlikely because it was known only at a late period, when it was probably referred to as *wꜣḏ n bꜣẖ*. Green jasper would suit *shrt* quite well, since it is of a bright green colour similar to that which baboons are usually painted, and heart scarabs made of it are fairly common. However, it has been suggested above that this stone may in fact have been known as *nmḥf*, and if this is so, then *shrt* must be some other. Serpentine is less typically green, but heart scarabs were certainly made of it, and it does occur among the materials for one of the amulets in the Dendera table. But no very satisfactory conclusion is possible, and the meaning given to *shrt* must depend to some extent on the identification of *nmḥf*, which may be green jasper, or alternatively basalt or serpentine.

Lastly, something must be said concerning *shrr*, which occurs with *shr(t)* in the lists of the 24 *ꜥꜣt* at Dendera. Had it been found alone in another context, it would undoubtedly have been regarded as a mere variant writing of *shrt*, and this may in fact be the case, since in one of the lists[7] *shrr* follows *shr(t) n ꜥr*, a similar distinction being observed between *mfkꜣt* and *mfkꜣt n ꜥr*, and between *wꜣḏ mḥ*, *wꜣḏ šmꜥ* and *wꜣḏ n ꜥr*. On the other hand, the word is distinctly written *shrr* in both instances, whereas *shr(t)* has only one *r*, and in the other list is not qualified by *n ꜥr*. It seems likely therefore that *shrr* was a variety of *shrt* not recognised before the Graeco-Roman period, but no identification is possible while *shrt* itself is uncertain.

šbn (Wb. IV. 441.12–13) is clearly related to the verb *šbn* (Wb. IV. 440.5), and seems to be a term for 'various materials'. It is found as the designation of the material of collars and the like[8], which, when depicted, are coloured eg. black, white, red, blue and green[9], red, blue and green[10], and red and green[11]. Whenever a precise indication of material such as

[1] Cf. Appendix IV.
[2] Totb. Einl. p. 128.
[3] Book of Dead Vocab. p. 296.
[4] Cf. Petrie, Amulets p. 23–4 nos. 89–90.
[5] Dend. Mar. IV, 87.
[6] Petrie, Amulets nos. 32; 39–40; 138–40; 89–92.
[7] Dend. Mar. IV, 39, 140–2.
[8] Cairo 28024; 28027; 28088; 28118; Ostr. Liv. M. 13, 625 vs. 3; Mem. Miss. 24, 76; 108; 212.
[9] Cairo 28024.
[10] Cairo 28024; 28027.
[11] Cairo 28088.

ḫsbḏ or *mfk3t* is possible, it is given, and it seems that *šbn* describes variegated objects composed of different coloured stones or faience. A somewhat analogous term appears to be *ˁš3*[1], which refers to two collars with differently coloured strings of beads, and *ḫkryt*[2] is once used in a similar way to describe a menat[3].

šsmt (Wb. IV. 539. 1–3) has been variously identified in the past, but is now generally accepted as malachite. Brugsch[4] originally thought that *šsmt* was green natron, but later[5] rejected this in favour of 'mineralische Substanz von grüner Farbe', 'eine Art von grüner Porzellanerde'; and this is accepted by Wendel[6], who rejects Birch's suggestion 'emerald'[7]. Paton[8], too, seems to favour Brugsch's 'Porzellanerde', and Bissing[9] adopts it on the grounds that the representation of *šsmt* as an irregular mass is against its being a stone. Gardiner[10] points out that *šsmt* is connected with *mfk3t*, and was found in Sinai and the Sudan, and Säve-Söderbergh[11], noting the same connection, gives amazon-stone, green jasper, green turquoise, malachite or olivine as possible translations. But Newberry[12] identifies *šsmt* more specifically as the green mineral malachite associated with *mfk3t* at Sinai, the common name for which was *w3ḏ*, and this is accepted by Lucas[13], Sloley[14] and Černý[15].

The textual evidence for *šsmt* is somewhat restricted, but serves to establish that it was a precious stone[16], that it was obtained from Sinai[17], Nubia[18] and Retenu[19], and that it was green in colour. Not only does the label *šsmt* accompany a basket of green stuff[20], and a heap of greenish material[21], but *šsmt* occurs in close parallelism with *w3ḏ* and *mfk3t* in the Pyramid texts[22].

The most significant fact is undoubtedly that *šsmt* was obtained from Sinai, and since, apart from turquoise, the only stone occurring there in any quantity is malachite, there seems no alternative but to accept Newberry's suggestion that it is to this which *šsmt* refers. But such an interpretation raises one serious problem which Newberry dismisses all too easily, namely the relation of *šsmt* and *w3ḏ*. To regard the latter simply as a more common term for *šsmt* is not possible, since the two are found together in the Pyramid texts[23] and elsewhere[24], and *w3ḏ šmˁ* and *šsmt* three times occur in the same list[25]. However, it may be that *šsmt* represents a particular quality of malachite, specifically associated with Sinai, and contrasted with that from Upper Egyptian sources, *w3ḏ šmˁ*, though it must be admitted that *šsmt* also came from Nubian sources. Moreover, *w3ḏ* is not always a precise term, and in the Pyramid texts may well refer to something other than malachite.

On the evidence, therefore, it seems reasonable to suppose that *šsmt* was a green stone, probably a specific kind of malachite, though there are difficulties in such a translation, and the identification cannot be regarded as wholly certain.

[1] Cairo 28034, 44; 28035, 31.
[2] Cairo 28027, 36.
[3] Cf. also *ḫkr* = *ˁš3 iwnw* as the colour of summer, Thes. 528; 634, 21–2.
[4] Dict. géog. p. 1348.
[5] Wb. Erg. p. 1204; ZÄS. 18 p. 13.
[6] Bau- und Edelsteine p. 101.
[7] ZÄS. 15 p. 32.
[8] Records III, 2 p. 58 n. 95.
[9] Stat. Taf. p. 28.
[10] JEA. 5 p. 222.
[11] Äg. und Nub. p. 217.
[12] Griff. Stud. p. 320.
[13] Materials p. 458.
[14] JEA. 34 p. 125.
[15] Sinai 2 p. 125.
[16] Urk. IV, 875 B; P. Ch. B. IV vs. 7, 5; IX vs. B 18, 8; Sinai 200; Harr. 41a, 2.
[17] Sinai 200; 182; cf. 121.
[18] Urk. IV, 1099; Menkhep. pl. 9.
[19] Urk. IV, 706.
[20] Urk. IV, 1099.
[21] Menkhep. pl. 9.
[22] Pyr. 567b.
[23] Pyr. 567b.
[24] Sinai 121.
[25] Urk. IV, 875 B; Dend. Chass. IV, 159; P. Ch. B. IV vs. 7, 5.

ḳꜥ (Wb. V. 18.2) occurs only in texts of the Graeco-Roman period, where it is evidently the name of a valuable mineral, probably a semi-precious stone. In the Dendera tribute tables[1] it is brought from a locality in Nubia, and in the Sehel inscription[2] it is listed among the products of the Elephantine neighbourhood; *ḳꜥ* is also one of the stones used for filling the sound eye[3], and a pair of wings are described as like the colour of gold and *ḳꜥ*[4]. From none of these references can anything be deduced, and were it not for the mention of *ḳꜥ* as the material for a number of amulets in the Dendera table[5], no suggestion as to the meaning of the word would be possible.

Concerning this occurrence of *ḳꜥ*, PETRIE[6] comments '*Qo* or *Qedu* is unknown, another form is *Qy*; being used for the plummet and square which are usually of haematite, this was probably the material', and from a consideration of the amulets said to be of *ḳꜥ*, a translation haematite does indeed seem likely. Of the sixteen represented, the first two are plummets, and of the 21 listed by PETRIE[7], 8 are haematite, while 4 are slate, and 4 more of dark materials. The next two are squares, of which PETRIE[8] lists 60, no fewer than 46 being haematite, and 3 others of dark material. The third pair of amulets, shown as  by MARIETTE, is quite obscure, though they are certainly not staircases as PETRIE assumes[9]. Circumstances have made collation of the text impossible, but it seems not unlikely that  is intended to represent a headrest,  and if so, this is again significant, because of the 86 cited by PETRIE[10], 70 are haematite, and 4 more of dark material. Of none of the other amulets depicted is haematite the most usual material, though scarabs and hearts of haematite are not infrequent, and *wḏꜣt* eyes are also known[11]. For the remainder, no examples of haematite can be adduced, but over the 16 amulets as a whole, the use of haematite exceeds that of any other material except blue and green 'glaze', and an identification of *ḳꜥ* as haematite seems therefore probable.

However, if on this rather slender evidence, a translation haematite is accepted, it can only refer to the black opaque variety with a metallic lustre[12], and not to any of the more amorphous forms. Even so, the question arises as to what was the term applied to haematite before the Graeco-Roman period, since it is hardly possible to suppose that the substance was not recognised earlier. That haematite was confused with any other of the darker semi-precious stones seems unlikely, and the only possible conclusion appears to be that it was one of several analogous substances referred to as *biꜣ*. Its metallic appearance and the fact that it is an iron ore would be sufficient to account for this, and it is certain that the classical authors confused haematite, magnetite, and perhaps also other materials[13]. In Ptolemaic and later texts, *biꜣ* refers exclusively to iron, and the creation of a special term to indicate haematite would therefore be necessary.

kꜣ (Wb. V. 101.13) has been variously read and interpreted in the past, though it is now clear that the substance is *kꜣ* or *kꜣꜣ*, and that it has two varieties, *ḥḏ* and *km*. CHAMPOLLION[14] read *kꜣmt* in one instance, and thought it was sardonyx, while BRUGSCH[15] read *krmt*,

[1] Dend. Mar. I, 70 (*ḳꜥ m irw.f*); I, 71; II, 12 = Dend. Chass. IV, 166, 9 – 167, 1; cf. Düm. G. I. II, 70.
[2] Sehel 16.
[3] Edfou VIII, 136–7; Philae ⟨1673⟩ Phot. 520; Thes. 41–2.
[4] Edfou II, 207, 8.
[5] Dend. Mar. IV, 87.
[6] Amulets p. 6.
[7] ib. p. 16 n. 37.
[8] ib. p. 16 n. 36.
[9] Amulets pl. 48.
[10] ib. p. 15 n. 34.
[11] ib. nos. 89–92; 7; 138–42.
[12] Lucas, Materials p. 452.
[13] Cf. Dioscorides 5, 126 (W); Pliny, HN. 36, 37, 144.
[14] Not. Descr. I p. 506.
[15] Wb. Erg. p. 1235; p. 1285.

comparing Coptic ⲕⲉⲣⲙⲓ, and took another example of the word to be a distinct term *kꜣš*, meaning 'Perlmutter-muschel', Coptic ⲕⲁϣⲁⲃⲏⲗ. Elsewhere[1] he regarded CHAMPOLLION's *kꜣmt* as a variant writing of *ḳmyt*, and this was also the view of VIREY[2]. NAVILLE[3], like BRUGSCH, read *kꜣš* in one instance, and took it to be tortoiseshell, which was accepted by KLEBS[4]. The WÖRTERBUCH correctly identifies *kꜣmt* (sic.) and *kꜣš* (sic.) with a writing *kꜣỉ km*, but knows no examples of *kꜣ ḥḏ* or *kꜣ* alone, concluding therefore that it was 'kostbares Material von brauner Farbe'.

Now in fact the word is known from some half dozen contexts, variously written *kꜣ*, *kꜣꜣ* and *kꜣỉ*, and qualified both as *kꜣ ḥḏ* and *kꜣ km*. It occurs alone in a fragmentary text published by CAMINOS[5], where something is compared to the *kꜣꜣ* stone of stands (*gnw*), from which little can be deduced. But it is also found as *kꜣ ḥḏ* and *kꜣ km* beside two heaps of material[6], as *kꜣ ḥḏ* beside a heap of white material[7], as *kꜣꜣ km* describing brown or red oval pieces in baskets[8], as *kꜣ km* referring to a lump of some material[9], and as *kꜣỉ km* in a fragmentary passage where other semi-precious stones are mentioned[10], three of these references revealing that it was obtained from Pwenet[11]. A similar term *kꜣy*[12] is not in fact this word, but probably a writing of *kꜣw*[13].

To suggest a possible meaning for *kꜣ* is by no means easy, though it was clearly a stone of two varieties, one light, the other dark, either brown or red, and was obtained from Pwenet and Nubia. It might conceivably refer to agate, onyx and sardonyx, which the Egyptians used to a small extent and are likely to have regarded as one stone[14], though if the interpretation of CAMINOS' text as referring to ring-stands of *kꜣꜣ* is correct, this is not altogether suitable. Moreover, if, as seems not improbable, *kꜣ* is to be identified with the *kꜣ* used for vessels in early dynastic times, and discussed above (p. 91), such a translation is impossible. Alternatively, *kꜣ* may refer to some form of quartz, or to a stone not native to Egypt, but in default of further evidence, no very satisfactory conclusion is possible.

knmt (Wb. V. 133.9) seems to occur only once, in Harris[15], and is regarded by both BRUGSCH[16] and the WÖRTERBUCH as a distinct material. That it is a mistake for *ḫnmt* is hardly possible, since the latter is found in the same list of materials for statuettes of the Nile god, nor does it seem likely to be a v o x p e r e g r i n a, for which group-writing might be expected. Beyond the fact, therefore, that *knmt* is a semi-precious stone or similar material, it is impossible to make any suggestion.

tfrr (Wb. V. 300.2–3) seems, as the WÖRTERBUCH assumes, to be an alternative name for lapis-lazuli. MONTET[17], discussing *tfrrt*, the land whence both *ḫsbḏ* and *tfrr* were obtained, notes the writing with *ṯ*, and compares Hebrew *sappīr* and Greek σάπφειρος. He is of the

[1] Thes. 1111.
[2] Mem. Miss. V, i p. 32.
[3] Deir el Bahari III p. 17.
[4] Reliefs NR. p. 126.
[5] Lit. Frag. 12, 1.
[6] Puyem. I, 32.
[7] Menkhep. pl. 9.
[8] Urk. IV, 1096 = Champollion's *kꜣmt*.
[9] Urk. IV, 335 = Brugsch's *kꜣš*.
[10] Kenam. 44.
[11] Puyem. I, 32; Urk. IV, 335; 1096; cf. also Kenam. 44. From Nubia: Menkhep. pl. 9.
[12] Urk. IV, 695.
[13] Cf. Love Tur. 2, 3; Bissing, Stat. Taf. p. 29, identifies with *ḳmyt*.
[14] Cf. classical onychion, which evidently included onyx, sardonyx, and some agate; Caley & Richards, Theophrastus p. 127.
[15] Harr. 41a, 7.
[16] Wb. Erg. p. 1279.
[17] Mon. Piot 41 p. 19.

opinion that as *tfrr* came from *tfrrt*, so *sappīr* came from Sippar, which would suit *tfrrt*, as being a station on the trade-route along which lapis lazuli passed from Bactria to Egypt, perhaps the limit to which the Egyptians themselves had ventured. This view is noted by Černý[1], who, however, concludes that *tfrr* cannot be lapis, since it is mentioned together with *ẖsbḏ* in one of the Sinai inscriptions[2]. Forbes[3] takes *tfrr* to be azurite, though without, it seems, any real grounds for the identification.

tfrr as the name of a material is relatively rare, though the name of the country *tfrrt* is frequent, and *tfrr* as a verbal adjective meaning blue (Wb. V. 300.4) is also known. That *tfrr* was a blue stone, coming from the country *tfrrt* can scarcely be doubted, and there seems little alternative but to identify it with lapis lazuli, usually *ẖsbḏ*. This would be apposite for all the Ptolemaic and later references to the word[4], but it is awkward for the earlier example in the Sinai inscription mentioned by Černý. Here, among *mꜣʿw nb nfr n ẖꜣst n mfkꜣt*, are apparently included both *ẖsbḏ* and *tfrr*[5], though the latter may conceivably be either the name of the country, or the adjective. Satisfactory interpretation of this passage is difficult, but it seems possible that the Egyptians were under the impression that lapis came from more than one source, and may have distinguished *ẖsbḏ* from *tfrr* for this reason. That the word refers to azurite or to an artificial frit it is hard to believe, particularly since it nowhere occurs in connection with other pigments. No clear decision can be reached, but it must be regarded as most probable that *tfrr* is nothing more than a synonym for *ẖsbḏ*, referring specifically to the fact that lapis lazuli came from *tfrrt*.

ṯḥnt (Wb. V. 390.11 f.; 391.16–18) has been the subject of much controversy, and although it is now generally accepted that in the majority of cases *ṯḥnt* refers to faience or glass, the nature of the precious stone *ṯḥnt mꜣʿ* is still in dispute. Brugsch[6] originally thought that *ṯḥnt* was a type of bronze, and so too did de Rougé[7] and Renouf[8], but Brugsch[9] later rejected this in favour of glass. Lepsius[10], on the other hand, takes *ṯḥnt* as a yellow stone, yellow topaz or yellow jasper, and Gensler[11] translates 'crystal', seeing in *ṯḥnt* the origin of the element -ϬⲎⲈⲒⲚ in the Coptic ⲀⲂⲀϬⲎⲈⲒⲚ 'glass'[12]. But Birch[13] thinks it unlikely that the word refers to crystal, and takes the *rnsw* of *ṯḥnt* in Harris as glass beads, glass being also appropriate to vases. Chabas[14], however, points out that *ṯḥnt mꜣʿ* presupposes the existence of imitations, and inclines to the view that *ṯḥnt* was both glass and crystal, though evidently of a colour opposed to red, perhaps yellow, the root meaning of the word being 'sparkling'. Wendel[15] derives *ṯḥnt* from *ṯḥn*, which he takes to mean 'yellow' or 'sulphur-coloured', comparing Coptic ⲐⲎⲚ, and regards it as both a yellow stone and a yellow glass. He accepts Lepsius' yellow jasper as the primary significance, with yellow topaz, yellow glass, and a yellow colour (ochre) as secondary meanings. Paton[16] lists various translations, and favours Erman's[17] faience rather than glass. Newberry[18], however, derives *ṯḥnt* from the

---

1 Sinai 2 p. 208.
2 Sinai 411.
3 Studies III p. 215.
4 Eg. Dend. Mar. I, 69a; Ombos I, 60, 62; Piehl Inscr. II, 47.
5 Sinai 411: 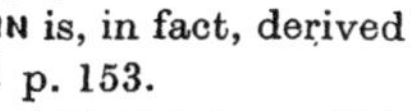.
6 Wb. p. 1589.
7 Mémoire p. 69.
8 ZÄS. 7 p. 66.
9 Wb. Erg. p. 1349.
10 Métaux p. 38.
11 ZÄS. 8 p. 147.
12 ⲀⲂⲀϬⲎⲈⲒⲚ is, in fact, derived from Persian ābgīneh, 'star'.
13 ZÄS 11 p. 153.
14 Études[2] p. 31.
15 Bau- und Edelsteine p. 113.
16 Records III, 2 p. 76 n. 184.
17 Life in Ancient Egypt (1894) p. 458 n. 1.
18 JEA. 6 p. 160; cf. also Hölscher, Libyer und Ägypter p. 17–18.

name of the land *ṯḥnw* whence it was obtained, and sees it as a word for glass, while BREASTED[1] thinks it may be some form of mineral glass, since the word is also applied to glaze or glazed ware, and glass. BARGUET[2] describes *ṯḥnt* as 'pierre de couleur verte', though without giving any reason, and FORBES[3] takes *ṯḥnt* to refer to both faience and coloured glass, and denies any connection with *ṯḥnw*.

In the circumstances, the WÖRTERBUCH is somewhat cautious, and translates 'Fayence und Glas', and 'ein kostbarer Stein aus Nubien', further regarding *ṯḥnt šꜥd* 'Art der Fayence' as a distinct species, and *ṯḥnt* 'als offizinell Verwendetes' as a separate term, though there can be little doubt that both are in fact this word, as will be shown below.

The textual evidence for *ṯḥnt* is extensive, but by no means conclusive, and although there are indications of the nature of the artificial material, there is nothing to suggest what is meant by *ṯḥnt mꜣꜥ*, the significance of which can only be deduced from the former. It therefore seems reasonable to consider first the references to *ṯḥnt* as faience or the like.

Most important of these is a vessel of faience, found at Tell el Yahudiyeh, and bearing an inscription describing it as *ṯḥnt*[4], which can only refer to the actual faience, and not to any blue stone imitated by it, since only lapis would be of the correct colour, and this is clearly inappropriate here. But *ṯḥnt* also refers to glass or glaze, which the Egyptians are unlikely to have distinguished, as is shown by a passage[5] where a statue of *ỉm* is anointed (*mrḥ*) with *ṯḥnt*, which must indicate the glazing of a clay or faience figure. Moreover, *ṯḥnt nbt*[6] may imply the production of either faience or glass, and the *ỉmy-r ṯḥnt*[7] and *sẖm ḥwt ṯḥnt*[8] can plausibly be explained as the controllers of glass or faience 'factories' such as those found at Amarna and elsewhere.

There are also several contexts in which the colour of *ṯḥnt* is implied to be green, and these are most likely to refer to the common green glazed faience, or to green glass. The leaves of a tree[9], a species of wood[10], and a figure of Thoth, probably as a baboon[11] are said to be the colour of *ṯḥnt*, and *ṯḥnt*, like *mfkꜣt*, is used of the pleasing colour of the fields etc.[12], and as a word for joy in opposition to *ḥrst*[13].

The majority of references to *ṯḥnt* in lists of precious stones and minerals, and as the material for amulets or other objects are also perhaps to be taken as faience or glass, except where *ṯḥnt mꜣꜥ* is specifically mentioned. Thus *ṯḥnt* alone is frequently listed among semi-precious stones[14], and is said to be used for small figures[15], amulets of various types, including scarabs[16], *wꜣḏ* sceptres[17], *ḏd* pillars[18], etc.[19], seals[20], sistra[21], beads[22], and, together with gold, for collars and other objects[23].

Moreover, a substance *ṯḥnt* is used in medicine[24], which, despite the doubts of the WÖRTERBUCH, must, as BREASTED[25] and EBBELL[26] tentatively assume, be identified with the word

---

[1] Pap. Smith p. 382. [2] Stèle de la Famine p. 24. [3] Studies V p. 120.
[4] Tell el Yah. pl. 8. [5] P. Salt 825, 18, 3. [6] Totb. Nav. 64, 8.
[7] Tomb. Court. pl. 29. [8] Mar. Mast. C, 1. [9] Love Tur. 2, 4.
[10] Düm. G. I. II, 88, 28. [11] LD. III, 37b; cf. P. Leid. 347, 2, 8; Edfou VI, 299.
[12] Grapow, Bild. Ausdr. p. 54f.; cf. Appendix IV. [13] loc. cit.
[14] P. Ch. B. IV vs. 8, 5; Dend. Mar. I, 22; III, 32; IV, 6–7; Sehel 16 etc.
[15] P. Tur. PR. 124, 14; P. Leid. 347, 2, 8. [16] Edfou VI, 299; Harr. 64b, 3.
[17] Totb. Nav. 125 Schluss. 24. [18] Totb. Nav. 151e; Marseille 42.
[19] Edfou VI, 299; Harr. 64b, 4; Dend. Mar. IV, 83; P. Leid. 347, 12, 10; Ostr. J. Č. 19, 2 [21] etc.
[20] Harr. 15b, 7; 34a, 5; 53b, 3. [21] Dend. Mar. II, 53b; 51a.
[22] P. Bol. 1094, 2, 2; Harr. 15b, 8; 64b, 15.
[23] Harr. 62b, 7; 62b, 8; Philae ⟨1411⟩ Phot. 194; Thes. 21; 28; P. Tur. 2044 rt. col. II, 8 [22]; 9 [23].
[24] Eb. 80, 6; 108, 16; 49, 21; 105, 15; Smith 14, 5.
[25] Pap. Smith p. 382. [26] Pap. Ebers p. 61.

for faience, the compounds *ṯḥnt snwḫt*[1], and *bsn n ṯḥnt*[2] being particularly suggestive, the former perhaps referring to the melting of glaze or glass, the latter to some fundamental constituent of the body material of faience, the nature of which is discussed below (p. 190f.). Only one of the medicinal uses is internal, that of *ṯḥnt snwḫt*, while elsewhere[3] the poisonous effect of glass may be alluded to.

Further evidence of the meaning faience or glass comes from the expression *ṯḥnt šˁd* (sic.), which occurs three times in Harris[4] as a material for *ḥnw* vessels. It may conceivably refer to carved faience, as a reading of  as *šˁd* would seem to imply, but it is also possible that the true reading is *ṯḥnt psḏ(t)*[5], and that the phrase indicates an especially brilliant or glassy faience[6]. In each case, the compound occurs with other items of simple *ṯḥnt*, including beads, seals, and the like, and the genuine stone cannot be meant. The glittering appearance of *ṯḥnt* is also alluded to in other contexts[7], and it seems reasonable to suppose that the comparison is with the brilliance of glass or glazed faience.

In short, there can be little doubt that in very many instances *ṯḥnt* refers to glass, glaze, or faience, and there is no reason to suppose that it means one rather than the other as Erman suggests[8]. It was undoubtedly the glaze of the faience that was uppermost in the Egyptian mind, and it is unlikely that they were careful to distinguish between glaze and glass, the latter being in all probability an accidental by-product of the former[9]. Special words for particular faience or glass imitations of certain semi-precious stones do exist, and the name of almost any stone could equally well refer to its imitation in either of these materials, but *ṯḥnt* stands alone as a general word for faience or glass, referring, perhaps, principally to the green and blue varieties.

The identification of the precious stone *ṯḥnt mꜣˁ* is a more difficult problem, and the few contexts in which it occurs, either among other precious stones[10], or as a material for amulets[11], are of little assistance in establishing the meaning. However, Breasted's[12] observation that since *ṯḥnt* refers to glaze or glazed ware, and glass, it was probably also some form of mineral glass, seems reasonable, and the possibility may be considered. The substances noted by Lucas which might have been regarded as mineral glass include Iceland spar, a transparent variety of calcite, rock crystal, and obsidian[13], to which must be added silica glass and crystalline gypsum[14]. Of these, rock crystal and obsidian may be among the substances referred to as *mnw ḥḏ* and *mnw km*, and Iceland spar is perhaps too rare to merit consideration. Silica glass, on the other hand, is distributed over a large area of some 3000 sq. kms. in the western desert, some lumps being as large as a man's head, and worked flakes being not infrequent[15]. Now *ṯḥnt* came from *ṯḥnw*[16], which lay immediately to the west of the north-

---

[1] Eb. 49, 21.

[2] Eb. 105, 15.

[3] P. Ch. B. III, 10, 2a.

[4] Harr. 15b, 9; 34a, 6; 53b, 2.

[5] *psḏt* is, however, meaningless in the case of *šwb*, Harr. 34b, 1; 71a, 8.

[6] Cf. Lucas, Materials p. 188.

[7] Eg. Totb. Nav. 110 Einl. 17; Mem. Miss. V pl. 2, Harp. 20; P. Leid. 347, 9, 10; Edfou I, 139 etc.

[8] Life in Ancient Egypt (1894) p. 458 n. 1.

[9] Cf. Lucas, Materials p. 207.

[10] Edfou II, 215, 4–5; VI, 165, 15–16; Dend. Mar. IV, 36, 49–50.

[11] Dend. Mar. IV, 87.

[12] Pap. Smith p. 382.

[13] Cf. Pliny, HN. 36, 67, 196; Isidorus 16, 16, 5; 16, 4, 21.

[14] Cf. Theophrastus § 30 for a stone ὑαλοειδής, so called from its similarity to glass.

[15] Bull. Inst. Fouad I du Desert 1 no. 2 (1951) p. 34f.; cf. also Keldani, Bibliography of Geology and Related Sciences nos. 95; 613; 614; 2455; 2456.

[16] Mem. Inst. 24, 108 = Coff. Texts VI, 212–3.

west corner of the Delta according to NEWBERRY[1] and GARDINER[2], and this would perhaps be a possible source of silica glass, so that identification of *ṯḥnt* with this material is feasible[3].

It must here be observed that there is nothing to support the yellow colour suggested by LEPSIUS and WENDEL, and, further, that it is dangerous to assume that because some evidence points to a green colour for what is evidently faience, the same colour indication can be taken to apply to the genuine stone also. In fact, there is no evidence of the colour of real *ṯḥnt*, unless it be in a passage from PLINY[4], quoted long ago by WIEDEMANN[5]. This runs: 'Candidarum dux est paederos . . . haec laudatissima est in Indis, apud quos sangenon vocatur, proxima in Aegypto, ubi syenites/tenites'[6]. Now syenites as the Egyptian name of the stone is quite clearly an intelligible humanistic conjecture in place of the obscure and apparently meaningless tenites, which, however, has better manuscript tradition, and should, on the principle of difficilior lectio potior, be accepted. That tenites is an attempted Latinisation of Egyptian *ṯḥnt*, as WIEDEMANN suggested, seems altogether possible, and, if so, the passage provides evidence for the colour of the latter, although the exact meaning of PLINY's paederos is not quite clear[7].

A white or transparent stone would be consistent with the view of *ṯḥnt* as a natural glass[8], which seems probable for other reasons also, though it would rule out obsidian, which is greenish-black in colour. In default therefore of further evidence, *ṯḥnt* may be regarded as natural glass, principally silica glass, but also perhaps including other transparent stones, as well as being the common term for glazed faience and glass.

ds (Wb. V. 485.16f.) occurs alone, and in three sub-species, *ds ḥḏ*, *ds km* and *ds ṯḥn*, as is recognised by BRUGSCH[9], who translates 'Flintstein', 'Schwerstein', and distinguishes three kinds, 'der dunkle', der 'helle', and 'der leutende', later, however, rejecting this identification in favour of 'weißer Marmor', 'dunkler Marmor', and 'gelber Alabaster'[10]. WENDEL[11] accepts the last translation for *ds ṯḥn*, because he thinks it came from the nome where the alabaster quarries are, but is unable to offer any suggestion as to the meaning of *ds ḥḏ* and *ds km*. The WÖRTERBUCH, however, renders 'Feuerstein', in spite of the fact that the amulet of *ds* in the Berlin board is actually quartz, and this translation is in all probability correct.

The textual evidence for *ds* is not very significant, and there are no definite indications of the meaning of the word. *ds* alone is the label beside an object of quartz on the Berlin amulet board[12], and is stated to be the material for two other amulets in the Dendera table[13]. It is also used for a *wꜣs* sceptre[14], for knives[15], for a box[16], for the limbs of a figure[17], for part of a snake[18], for a statue[19], etc.[20], the majority in magico-religious contexts; it is, furthermore, used medicinally in the form *ꜣst nt ds*, discussed below. *ds ḥḏ* and *ds km* occur together in

[1] JEA. 6 p. 160.
[2] Onom. 1 p. 116f.
[3] Strabo, 16, 2, 25 ch. 758, mentions a γῆ ὑαλῖτις, used in Egypt in the manufacture of glass, and also a ψάμμος ὑαλῖτις.
[4] HN. 37, 46, 129.
[5] Sammlung p. 42.
[6] syenites *D cum U* 856 *ex indice*; senytes L; senites *v*; tenites d, h, *H*; satis F, a; *om.* B.
[7] Cf. HN. 37, 22; 37, 40; 37, 46.
[8] Some silica glass is greenish or greenish-yellow.
[9] Wb. p. 1658.
[10] Wb. Erg. p. 1374.
[11] Bau- und Edelsteine p. 119.
[12] Berl. 20600.
[13] Dend. Mar. IV, 87.
[14] Totb. Nav. 125 Schluß 25.
[15] Dend. Mar. IV, 74; Totb. Nav. 172, 30.
[16] Westc. 9, 5.
[17] P. Tur. PR. 118, 12; cf. Thes. 20.
[18] P. Tur. PR. 119, 6; Totb. Stud. VIII, 16.
[19] Hor. & Seth 9, 10.
[20] Abyd. Mar. I, 52, 17.

the lists of the 24 *ꜥꜣt* at Dendera[1], and the latter also as the material for amulets[2], and medicinally[3]. The third variety, *ds ṯḥn*, is found in the Coffin Texts[4], and in the Edfu mineral list[5], apparently as a product of *spꜣ*.

There is little in these examples to suggest any particular translation, though the use of *ds* in ritual contexts, and especially for knives, is significant. Moreover, it is quite clear that *ds* is closely related to the word *ds* (Wb. V. 486.7), 'knife', which refers particularly to ritual flint implements, and it seems likely that in the early stages of the language 'flint', 'flint-knife' and 'knife' were assimilated, the material perhaps deriving its name from the object most often made from it. Such was probably the origin of *mꜣt*, which SETHE[6] suggests was derived from the name of a characteristic vessel made of granite, and it may also be observed that in Nahuatl *itztli* has the meaning obsidian, obsidian-knife, and knife. Before the introduction of metals, flint was in Egypt, like obsidian in Mexico, the common material for tools and weapons, and its use seems to have survived into historical times, especially in certain ritual and mythological connections.

Flint is normally dark grey or black in colour, which would correspond to *ds km*, the commonest of the three varieties; *ds ḥḏ* might well be chert, an impure species of flint of light grey or light brown colour, and *ds ṯḥn* the pale yellowish flint of the kind from which ripple flaked knives were made in the late predynastic period (Nagada II). That the amulet inserted into the Berlin board is of quartz is unfortunate, since another mistake seems possible, and yet the object is so hard to identify that any check on a more representative selection of examples is impossible.

Concerning *ds ṯḥn*, suggested above to be pale yellowish flint, something more must be said. The identification with alabaster, proposed by BRUGSCH and WENDEL, though ingenious, cannot be maintained, since *spꜣ*, whence *ds ṯḥn* is said to come, cannot now be identified with Alabastronpolis, which may not, in fact, be in the same nome[7]. A possible alternative name for the same material is *ds wbḫ n bšw*[8], which occurs in one variant of a text concerning a snake, the forepart of which is simply of *ds* in eleven of the twelve versions collected by SETHE. The exact significance of *bšw* cannot be determined, but the adjective *wbḫ* suggests that it indicates a specifically bright variety of *ds*, and comparison with *ds ṯḥn* is tempting.

Also of interest is the expression *ꜣst nt ds*, occurring twice in medical contexts[9], once for an eye injury, and once against blood suffusion. GRAPOW's 'Feuersteinsplitter'[10] is perhaps the most apposite translation, though the exact significance of the term is not altogether clear, and the therapeutic virtue of chips of flint is not readily appreciated.

From the evidence as a whole, no very precise indication of the meaning of *ds* is forthcoming, and the translation 'flint' must rest largely upon the connection with *ds*, 'knife', though it is also applicable to all the examples, and satisfactorily explains the three varieties *ds ḥḏ*, *ds km*, and *ds ṯḥn*.

ḏrtt (Wb. V. 598.10) seems to be known from only three examples on coffins[11], where it is parallel to *ḥrst* as the description of a red bead. LACAU[12] renders 'perle', and JÉQUIER[13] 'perle-bracelet', but it is clear from one instance where the inscription runs *swrt nt ḥrst ktt*

[1] Dend. Mar. IV, 36, 49–50; 39, 140–2.
[2] Dend. Mar. IV, 87.
[3] Eb. 59, 20–1; 62, 14; P. Ram. V 69 lower.
[4] Coff. Texts II, 237a.
[5] Edfou VI, 202, 5.
[6] Bau- und Denkmalsteine p. 18; cf. also above p. 72.
[7] Gauthier, Dict. géog. V p. 27–8.
[8] Totb. Stud. VIII, 16.
[9] Eb. 64, 4; 88, 19.
[10] Grundriß III p. 123.
[11] Cairo 28083; 28092; 28123.
[12] Sarcophages 2 p. 170.
[13] Frises p. 50.

n ḏrtt[1] that *ḏrtt*, like *ḥrst*, indicates the material and not the object. Whether it is simply an alternative name for carnelian, or refers to some specific red stone or the like, it is impossible to determine, though the colour is clear. Without further examples, no conclusion can be reached, but that the word is known only from these three similar contexts might tend to suggest that it was merely a variant of *ḥrst*.

Adequately to summarise the results of the foregoing discussions is hardly possible within a small compass, but a few salient facts may be emphasised. Of the six stones which the Egyptians valued most highly, *ḫsbḏ*, *mfkꜣt*, *ḥnmt*, *ḥrst*, *nšmt*, and *ṯḥnt*[2], five can now be identified with certainty as lapis lazuli, turquoise, red jasper, carnelian and green felspar, while the sixth, as well as being the common term for faience and glass, seems to refer also to a natural mineral glass, perhaps silica glass and other apparently similar substances. That these six stones were still esteemed above all others in the Graeco-Roman period would surely suggest that even at that date none of the most noble gems, diamond, ruby, sapphire, and opal, was known, and that emerald, if known, was extremely rare.

Of the secondary stones, *ḥsmn* is undoubtedly amethyst, and several others, such as *wꜣḏ*, *mnw*, *mḫn(m)t*, *nmḥf*, *ḥmꜣgt*, *sḥrt*, *šsmt*, *kꜥ*, *tfrr*, and *ds*, may be identified with more or less accuracy from their colour, provenance, or use. Only *ibhty* and *ꜥꜣt ḥḏt* remain altogether obscure, though it is possible to suggest stones to which they may refer. Among the other terms are many which represent varieties of faience and glass, others which must apply to little known materials, and yet others which are either loan-words or artificial terms used to describe stones better known under other names.

In fact, few semi-precious stones used by the Egyptians cannot be identified lexicographically, and it is always possible that the rarer ones may have been confused with other similar stones, or that green gems such as chrysoprase, fluorspar, jade, olivine and peridot may have been classed generally as *wꜣḏ*. But no name for coral is known, nor for mother-of-pearl, which was used from predynastic times, and the term for the resin employed for scarabs and other objects is also problematic. However, bearing in mind the confusion that exists in the works of Theophrastus and Pliny, and the inconsistencies in jewel terminology even at the present day, it is unreasonable to expect any great accuracy from the Egyptians, who evidently had a quite different system of classification from our own.

Lastly it may be mentioned that there is in Egypt no trace of the lore of precious stones, except in so far as certain amulets were prescribed to be made of certain stones, for reasons which cannot now be appreciated. That turquoise faded was undoubtedly realised, but no superstitions seem to have been connected with this phenomenon, and the common beliefs that stones grow better in hot climates, that poor stones mature into good ones, that stones reproduce, etc., do not appear to have been current.

[1] Cairo 28083.

[2] Edfou II, 215, 4–5; VI, 165, 15–16 etc.

CHAPTER 6

PIGMENTS

The question of Egyptian pigments has already been treated in some detail in a recent essay by IVERSEN[1], and considerable use has been made of this work in writing the present discussion. It is not, however, by any means exhaustive, dealing only with those pigments which are also used in medical prescriptions, and omitting to consider the evidence of the pigment lists, in which other substances occur[2]. There are also other terms referring to painting and writing materials which do not appear in IVERSEN's paper, and are included here for the sake of completeness, among them two which are clearly not minerals[3], but require some mention in an enquiry of this kind.

With the possible exception of *prš*, the pigments used by the Egyptians were either naturally occurring minerals, or artificial frits compounded from mineral substances. These seem to have been divided into two main groups, the hard (*ḏrwy*), consisting of the frits and ground stones, and the soft (*sty*), embracing the ochreous earths and similar materials. The same kind of classification is also adopted by THEOPHRASTUS[4], who distinguishes earths (ochre and minium), sands (chrysocolla and cyanus), and clays (realgar and orpiment), and in Latin limus and harena appear as quasi-technical terms for the soft and hard pigments. PLINY[5] classifies painting materials either by colour, as subdued and vivid, or by composition, as natural and artificial, the latter also being the distinction drawn by ISIDORUS[6].

In Egyptian texts, the hard pigments are normally measured, like metals and precious stones, in *dbn* (ie. according to weight), while the soft, though rarely orpiment and realgar, occur in *hnw*, *ḳbw*, *mnt* or *šnw* vessels, usually the first (ie. according to volume), though the distinction is not rigidly maintained. Another form of measuring pigments is *trỉt*, known from ostraca in the tomb of Senmut[7], which HAYES[8] suggests may mean 'cake', 'lump', 'grain', 'bar' or 'stick', comparing Wb. V. 209.11, and V. 313.9[9]. It is tempting to see in this word the name for the little cakes or discs of pigment found in Egyptian palettes[10], though it may, of course, have a more general significance.

ꜣwt-ỉb (Wb. I. 5.1) was originally known from only a very few examples, which were readily misinterpreted. STERN translates 'nomen granorum quorundam', which is quoted by BRUGSCH[11], and LÜRING[12] thinks *ꜣwt-ỉb* is an aromatic resin. DARESSY[13], noting the relationship to *ḳnỉt*, which he wrongly regards as butter, suggests salt or a special resin, and the

[1] Paints and Pigments.

[2] Iversen quotes only Ostr. Stras. H. 41., though referring also to Ostr. Tor. A. 11, which he was unable to consult. Of Ostr. Cairo 25594 and P. Ch. B. V. rt. 8, 12–14, he makes no mention whatever.

[3] *ḏꜥbt* and *ḳmyt*.

[4] § 40.

[5] HN. 35, 12, 30.

[6] 19, 17, 1f.

[7] Ostr. Senmut 63 rt. 8; 64 rt. 7 (twice); 66, 5.

[8] Ostraca etc. p. 40.

[9] Comparison with Wb. V. 252.6 is less likely.

[10] Lucas, Materials p. 413f.

[11] Wb. Erg. p. 516.

[12] Med. Kent. p. 29.

[13] Ann. Serv. 16 p. 233.

Wörterbuch has 'Art Myrrhe o. ä.'. More recently, however, other references to *ꜣwt-ib* have been noticed on ostraca, from which it is clear that it represents a pigment. Ebbell[1] accordingly translates realgar or orpiment, and Iversen[2] sandarac (ie. realgar). The latter discusses the word at some length, pointing out that as a pigment it is always with, and generally next to, *ḳnit*, which he shows elsewhere to be orpiment. Further, *ꜣwt-ib* is directly explained as a species of *ḳnit*[3], the commoner substance being used to elucidate the lesser known. The medicinal use in Ebers[4] he regards as of particular significance, since the prescription is closely parallel to a passage in Dioscorides[5] describing the use of sandarac in a remedy against coughing. This fact, together with the identification of *ꜣwt-ib* as a form of orpiment, makes interpretation as realgar certain. The two were not properly distinguished in antiquity[6], and since only a colour nuance separates them, some confusion is inevitable. Iversen admits, however, that sandarac has not as yet been found in connection with Egypt.

There can be no doubt that Iversen's reasoning is correct, and that *ꜣwt-ib* is the term for realgar, the red sulphide of arsenic, the sandarac of Pliny and Dioscorides. That it was a mineral should have been recognised earlier, since the word occurs in the Edfu mineral list[7], where its general nature is quite evident. More important, however, are the references to *ꜣwt-ib* in the pigment lists, only one of which[8] is mentioned by Iversen. In two others[9] it is also found next to *ḳnit*, and in the fourth[10] is separated from the latter by two items. From these it is clear that *ꜣwt-ib* was among the regular Egyptian pigments[11].

The relationship between orpiment and realgar is further exemplified in Coptic, where realgar is referred to both as **ⲤⲀⲚⲦⲀⲢⲀⲬⲎⲤ** and as **ⲀⲤⲤⲎⲢⲚⲎϨ ⲈⲦⲦⲞⲢϢ̄** and **ⲀⲤⲤⲈⲢⲚⲎϨ ⲚⲔⲞⲔⲞⲤ**[12], ie. red ἀρσενικόν. The same connection is borne out rather strikingly by some lumps of orpiment from Egypt in the collection of University College London, which are streaked with the characteristic red of realgar[13].

Concerning the use of the word sandarac as a translation of *ꜣwt-ib*, it must be noted that it is not altogether suitable and should be rejected in favour of the more precise realgar. The Greek σανδαράχη may have included several mineral substances of a red colour, such as realgar, cinnabar, vermilion, minium, and perhaps also mineral kermes[14], and was further applied to a gum resin. This dual meaning continued until the Middle Ages, and sandarac is now used principally of the resin of *Tetraclinis articulata*[15]. Realgar, however, is an exact term, referring only to the red sulphide of arsenic, and as such is to be preferred as a translation.

Lastly, it is worth remarking that the identification of *ꜣwt-ib* as realgar provides a useful reminder that an argumentum ex silentio is always to be treated with reserve. Lucas[16] knows no red pigment other than red ochre and, at a late date, red lead, and nowhere mentions realgar. Yet it was evidently in use in the New Kingdom, and was listed among the regular pigments, showing that in this case the archaeological evidence is at fault. This may well be borne in mind when dealing with other lexicographical problems.

[1] Pap. Ebers p. 131.
[2] Paints and Pigments p. 39.
[3] Ann. Serv. 16, 225.
[4] Eb. 54, 18.
[5] 5, 120 = 5, 105 (W).
[6] Cf. Pliny, HN. 34, 56, 1.
[7] Düm. H. I. II, 50b = Edfou VI, 203, 1.
[8] Ostr. Stras. H. 41.
[9] Ostr. Tor. A. 11; Ostr. Cairo 25594.
[10] P. Ch. B. V rt. 8, 12.
[11] For a further example of *ꜣwt-ib* with *ḳnit*, cf. Ostr. Cairo 25819 vs. 2.
[12] Chassinat, P. Méd. Copte p. 82.
[13] Although the lumps of raw material were found in Egypt, there is no evidence that they were native.
[14] Chassinat, P. Méd. Copte p. 82; cf. Pliny, HN. 35, 22, 39.
[15] Lucas, Materials p. 371.
[16] Materials p. 397f.

w3ḏ (Wb. I. 267.3–8 & 9–15) is here considered only as a pigment and in its therapeutic uses, its occurrence as a semi-precious stone having been discussed above (p. 102f.). It must be noted that the WÖRTERBUCH and many previous authorities separate the two, probably because the identity of the stone was misunderstood, but IVERSEN[1] regards them as one, and his view is undoubtedly correct.

Of the early Egyptologists, BRUGSCH[2] and DÜMICHEN[3] both realise that *w3ḏ* is a pigment and eye-paint as well as a valuable stone, though neither identifies it. LÜRING[4], however, also noting that it was the second eye-paint of the Egyptians, points to the apparent connection of *w3ḏ* with copper in Ebers[5], concluding that it must be a copper compound. Of those used in Greek medicine he selects ἰός σκώληξ, artificial verdigris, as the most likely equivalent. FLORENCE and LORET[6] distinguish several species of *w3ḏ*, principally a green powder, which was a copper salt, and the green eye-paint, which was apparently chrysocolla. Similarly, BREASTED[7], discussing *šsyt nt w3ḏw*, notes that *w3ḏ* was both pigment and eye-paint, and takes it to have been a copper salt which the physicians had discovered was astringent. But NEWBERRY[8] regards *w3ḏ* as malachite, and this is also prominent among the translations listed by PATON[9].

More recently, LUCAS' researches into the nature of Egyptian pigments and eye-paints have provided a sounder basis for discussion. He shows that the main natural constituent of both the green eye-paint[10] and green pigment[11] was undoubtedly malachite, which has been found in various forms, though verdigris and chrysocolla were also used. These conclusions are used by IVERSEN[12] to maintain that while *w3ḏ* could perhaps represent practically any green stone or mineral, there was in fact one green substance par excellence to which it primarily referred, and this was malachite. Not only does this apply to the use of *w3ḏ* as an eye-paint and as a pigment, but there are also medicinal uses of *w3ḏ* which correspond to those of malachite and verdigris in DIOSCORIDES and PLINY. He further points out that the *w3ḏ ḥmty* of Ebers is certainly verdigris, and suggests that *ḥsb n w3ḏ* may be ἰός σκώληξ.

The general meaning of *w3ḏ* is thus clearly established, though the nuance may vary somewhat according to the context. As an eye-paint, parallel to *msdmt*[13], it is principally malachite, but also covers chrysocolla which was probably confused with the former; 'green eye-paint' is perhaps the best translation. As a medicament employed for the eyes[14], for boils and the like[15], and for other purposes[16], it may be either malachite or verdigris. In the compound *šsyt nt w3ḏ(w)*[17], which seems to refer to a frit, *w3ḏ* is in all probability malachite, one of the constituents, which gave the frit its colour[18]. In *w3ḏ ḥmty*[19], it can only be verdigris, as IVERSEN points out, and as *w3ḏ n* 𓊛[20] may be either copper rust or verdigris (hydrated carbonate of copper) scraped from boats[21].

---

[1] Paints and Pigments p. 9. [2] Wb. p. 361; Erg. p. 404. [3] ZÄS. 10 p. 100.
[4] Med. Kent. p. 88. [5] Eb. 59, 12; 59, 16; 79, 3.
[6] In De Morgan, Dahchour 1894 p. 160f. [7] Pap. Smith p. 382. [8] Griff. Stud. p. 320.
[9] Records III, 2 p. 54; cf. also Ebers, Die Maasse p. 71, where *w3ḏ* is correctly identified as malachite.
[10] Materials p. 99. [11] ib. p. 396. [12] Paints and Pigments p. 9.
[13] Pyr. 54d; 1681a, b; D. el B. 109; Luxor ⟨561⟩; Goshen 6, 6; Mem. Miss. XV, 60 fig. 156; Mammisi 92; Debod Roeder § 123; Dend. Mar. III, 18; Philae ⟨2731–3⟩ Phot. 995; P. Tur. PR. 118, 4; Dend. Mar. IV, 70 etc.; cf. also Dend. Mar. I, 70; I, 71; II, 12; Medum 13; Urk. IV, 706; Sehel 17; Ann. Serv. 16, 223.
[14] Eb. 55, 12; 56, 2; 56, 9; 57, 14; 57, 16; 60, 4; 60, 9; 61, 8; 62, 2; 62, 20; 63, 4; 63, 7 etc.
[15] Eb. 89, 5; cf. 89, 8.
[16] Eb. 48, 19; 60, 17; 60, 22; 62, 9; Hearst 6, 13; 12, 1; 15, 6; Smith 14, 5; 16, 10 etc.
[17] Smith 14, 5; 16, 10. [18] Lucas, Materials p. 396; cf. p. 392.
[19] Eb. 59, 12; 59, 16; 79, 3. [20] Eb. 43, 6–7.
[21] Cf. Hofmann, Über Mesdem p. 4; Florence and Loret in Dahchour 1894 p. 161.

As a pigment, *wꜣḏ* refers principally to malachite, though such a translation is not applicable in every case. In three of the pigment lists[1], both *wꜣḏ* and *šsyt* occur together, the one almost certainly malachite or chrysocolla, the other green frit. In the fourth[2], however, *šsyt* is omitted, and *wꜣḏ* is qualified as *wꜣḏw šbnw*, 'various green pigments', which must surely include the missing frit. A similar breadth of meaning is also to be assumed in a passage in the 'Book of the Dead'[3], where *dḳw nw wꜣḏ* and *dḳw nw ḥmt wꜣḏt* occur in different versions. In addition, it is clear from some ostraca[4] that the pigment *wꜣḏ* was of different qualities, since *wꜣḏ nfr nfr* is specifically mentioned.

In adducing classical parallels for the use of *wꜣḏ* in medicine, it must be borne in mind that there was confusion between malachite, verdigris, and chrysocolla, which may also have extended to the Egyptian texts. Theophrastus[5] seems to have distinguished smaragdus from chrysocolla, 'which some think is of the same nature, for it is of similar colour', but the references of Pliny[6] and Dioscorides[7] to chrysocolla appear to include both malachite and chrysocolla proper. In Pliny too[8], verdigris is both the basic copper acetate, true verdigris, and the basic carbonate formed when copper is exposed to the air, the same uncertainty being common to other classical authors[9]. It is likely therefore that the Egyptians did not clearly differentiate between the three substances, and may have regarded them as related species of malachite.

Reference has already been made to *ḥsb n wꜣḏ*, identified with ἰὸς σκώληξ by Iversen[10], who cites classical parallels and compares the Egyptian *ḥsbt* (Wb. II. 168.8). The expression is not recognised by the Wörterbuch, where only the Ebers examples are given sv. *ẖpꜣw n wꜣḏ* (Wb. III. 366.4)[11], but Brugsch[12] notes *ḥsbt* as 'Bez. einer Steinart, und möglicherweise Mosaïkstein', and Iversen lists several occurrences from the 'Book of the Dead'.

The reading *ḥsb* for the Ebers examples may be questioned in view of one instance in which *ẖpꜣw* is written in full[13], but is seems possible that the terms are to be assimilated, and since *ḥsb* is well attested elsewhere it may be that *ẖpꜣw* is merely an incorrect expansion of a summary orthography. Identification with ἰὸς σκώληξ, scolecia, is, however, doubtful, since it is clear that this substance, used extensively in classical medicine, was of two kinds, one a natural product, the other a more common artificial preparation[14], the worm-like shape being the result of deliberate rolling[15]. That the Egyptians ever produced either seems improbable, and, moreover, the contexts in which *ḥsb* occurs would rather suggest a species of frit.

Apart from the doubtful Ebers passages[16], *ḥsb n wꜣḏ* is found in the Edfu mineral list[17], as a substance with which a temple is to be sealed[18], and in the 'Book of the Dead'[19] as material

---

[1] Ostr. Stras. H. 41; Ostr. Tor. A. 11; Ostr. Cairo 25594; cf. also Ostr. MMA. 90.6.3. rt. 2 & 4; vs. 1 & 2.

[2] P. Ch. B. V rt. 8, 12.

[3] Totb. Nav. 100, 7–8.

[4] Ostr. Cairo 25247 vs. 8; Hier Ostr. 69, 1 vs. 7.

[5] § 26. The distinction is somewhat confusing since both undoubtedly include malachite. Smaragdus probably embraced the massive and crystalline forms, while the amorphous earthy varieties were covered by chrysocolla.

[6] HN. 33, 26, 86.

[7] 5, 104 = 5, 89 (W).

[8] HN. 34, 26, 110.

[9] Cf. Dioscorides 5, 91 = 5, 79 (W); Theophrastus § 57; Oribasius 13; Vitruvius 7, 12; Isidorus 16, 20, 14.

[10] Paints and Pigments p. 13; p. 17.

[11] Translated 'chip of malachite' by Ebbell, Pap. Ebers.

[12] Wb. p. 995.

[13] Eb. 71, 8.

[14] Cf. Dioscorides 5, 79 (W); Pliny, HN. 34, 26, 110f.

[15] Cf. Bailey, Elder Pliny 2 p. 174.

[16] Eb. 71, 8; 77, 18; 91, 6–7.

[17] Edfou VI, 203, 3.

[18] Fairman, Ann. Serv. 44 p. 274, regards it as something that could be moistened to make a paste.

[19] Totb. Budge 133 (Nu) (p. 291).

for a magical barque, the same to be made simply of *wꜣḏ* in another version[1]. In both these instances a paste or frit seems required, and the latter is specifically suggested by a further rubric from the 'Book of the Dead'[2], where *ḥsb wꜣḏ* is an alternative for *dḳw n ḥmt wꜣ ḏt*. A possible occurrence of *ḥsbt* alone may also be noted[3], though the identification is uncertain.

From these examples it is evident that *ḥsb n wꜣḏ* is something related to frit, possibly the crushed malachite used in making it, and that the translation scolecia cannot in fact be supported.

wꜣ ḏ then is a general term for all green pigments and eye-paints, principally malachite, but also verdigris and chrysocolla. In some instances it appears to have a wider reference, including the green frit *ššyt*, or *ḥmt wꜣ ḏt*, and this applies, too, to the compound *ḥsb n wꜣ ḏ*.

prš (Wb. I. 532.13–15) appears, like *snn* (below p. 181f.) to include two distinct substances, the one mineral, the other vegetable, though the Wörterbuch recognises only one meaning, and translates 'etwas vom Nadelholzbaum'. A somewhat similar view is taken by Brugsch[4], who compares Coptic ⲃⲉⲣϣⲏⲟⲩ, and renders 'coriander', and by Loret, who also favours coriander seed[5], and later juniper berries[6]. But Dawson[7] regards *prš* as a mineral substance, identifiable as minium from the Demotic magical papyrus, where it is used of the red colouring matter of lamps, and surviving in Coptic ⲡⲏⲣϣ. Griffith and Thompson[8] also take *prš* as a mineral, comparing ⲡⲏⲣϣ and Greek μίλτος, which may be either red lead (minium) or red earth, and Iversen[9] thinks there can be no doubt that it means red ochre from the Demotic evidence.

The two views are not, however, irreconcilable, and Chassinat[10] believes that there has been confusion of two distinct terms, similarly written and determined, one a synonym for *prt w῾n*[11], the other a red pigment, Coptic ⲡⲏⲣϣ. Leake[12] points out that a similar confusion exists in the classical 'dragon's blood', which anciently, and in medieval times, covered both cinnabar (red mercuric sulphide) and *Resina draconis*, derived from *Calamus draco*, an East Indian palm tree, both of which are hard, easily powdered, and astringent, and have the same colour.

Whether the references to *prš* conceal two separate words, or a single term the meaning of which was somewhat confused, cannot easily be decided, though the occurrence of both *prš* and *prt w῾n* in a prescription in Ebers[13] might suggest the former. What is clear is that Demotic *prš* and Coptic ⲡⲏⲣϣ refer to a red mineral, and it seems probable that examples of *prš* in Ebers and Hearst[14] are this same word, only the later references[15] being to *prt w῾n*.

The identity of the mineral *prš* rests almost entirely on the passages in the Demotic magical papyrus[16], which concern a lamp used for divination, into which neither *prš* nor gum water must be put. In a similar context in a Greek magical papyrus[17], the lamp is

[1] Totb. Leps. 55, 133, 11.
[2] Totb. Leps. 52, 129, 7; Totb. Nav. 100/129 Te.
[3] Q. Hesy pl. 21, regarded as faience by Quibell; cf. also Medum 15, which may, however, be a place-name, as Edfou VI, 240, 8.
[4] Wb. p. 489.
[5] Rec. Trav. 7 p. 112.
[6] Le Kyphi p. 49f.; Flore² p. 41.
[7] JEA. 20 p. 188; 21 p. 39.
[8] P. Dem. Mag. I p. 44.
[9] Paints and Pigments p. 31.
[10] P. Méd. Copte p. 281.
[11] Düm. G. I. II, 83.
[12] Eg. Medical Papyri p. 45.
[13] Eb. 82, 3.
[14] Eb. 46, 13; 65, 4; 80, 5; 80, 10; 82, 5; Hearst 8, 8; 9, 4; 15, 9.
[15] Düm. G. I. II, 80; II, 83; II, 85B; II, 90; Philae ⟨2883⟩ Phot. 852. (with det.).
[16] P. Dem. Mag. 5, 4; 6, 2; 16, 22; 17, 21; 27, 13.
[17] P. Mag. Gr. Leid. V cols. I, 22; IV, 25; cf. also Peyron, Lexicon, p. 380, sv. ϫⲱⲓ.

required to be ἀμίλτωτος, which shows *prš* to be the equivalent of Greek μίλτος. Now Liddell & Scott translate μίλτος as 1. red earth, red ochre, ruddle, 2. red lead, 3. blight (rust), and 4. blood (in magic), rendering ἀμίλτωτος simply as 'not painted red'[1]. *prš* may therefore be either a red ochre, or red lead (minium), and the former seems the more likely. Minium does not appear to have been known in Egypt before the Roman period[2], and Pliny[3] expressly states that it is unsuitable for use as a medicament except as a styptic[4]; nor does there appear to be any evidence for the use of red lead as the colouring matter of lamps. Red ochre, on the other hand, was regularly used as a slip and as a paint, and the employment of ochres in medicine is well known.

It seems likely therefore that *prš* is a red ochreous earth, one of a variety which occur in Egypt, though the evidence is slight, and it cannot be regarded as one of the regular pigments.

mnšt (Wb. II. 89.12–13) is clearly a mineral substance used as a pigment, though opinions have differed as to its precise identification. Brugsch[5] takes it to be a species of earth, suggesting a connection with Coptic ⲙⲣϣ, ⲙⲉⲣϣ, ⲙⲟⲣϣ, and ⲡⲏⲣϣ, and mistaking *mnšt w3ḏ* as a reference to green colour. Later[6], however, he translates more precisely as 'Mennig' or 'Röthel', and the latter meaning is adopted by Lüring[7], who points out that *mnšt* was a product of the oases, where red ochre is now common, and that the uses of *mnšt* in Ebers are parallel to those of αἱματίτης in Dioscorides. Dawson[8] also regards *mnšt* as an ochre, rejecting the translation minium because the latter is poisonous and therefore could not be taken internally. But he accepts the Wörterbuch suggestion that *mnšt* is yellow in colour, and so prefers 'yellow ochre' to Lüring's 'Hämatit', the other yellow pigment, orpiment, being, like minium, highly poisonous[9].

More recently, the question has been treated in detail by Iversen[10], who concludes that *mnšt* is in fact red ochre, pointing out that there is no evidence for the yellow colour mentioned in the Wörterbuch. He gives instances of the use of *mnšt* as a pigment, and as an astringent and drying agent, though the medical contexts[11] are too general to admit of comparison with classical sources. It is, however, connected with other earthy minerals, particularly *sty*, and cannot be minium, which is not used as a paint until Roman times. Iversen reviews Brugsch's suggested etymological relationship between *mnšt*, ⲙⲉⲣϣ̄ and ⲡⲏⲣϣ̄, and although unable to accept it as proven, thinks that there is circumstantial evidence for such a connection. Red ochre was the common red pigment of Egypt, and occurs both at Aswan and in the oases, the latter corresponding to the source of *mnšt* cited at Dendera[12]. Dioscorides, Pliny and Vitruvius all mention red ochres, and the therapeutic uses of μίλτος coincide very roughly with those of *mnšt*. *mnšt mwyt*[13] would be simply moist ochre, and *mnšt w3ḏt*[14] fresh ochre that has not been burned and washed before use[15]. Lastly, *mnš* is the name of

---

1 μίλτος is clearly an ochre in Theophrastus § 52–3, and Pliny, HN. 33, 36, 111, draws a distinction between red earth (Greek miltos) and minium (Greek cinnabar), both of which were bright red.

2 Lucas, Materials pp. 277; 399; 414.

3 HN. 33, 41, 124.

4 Cf. the use of *prš* as an emollient in Ebers and Hearst.

5 Wb. p. 665.

6 Wb. Erg. p. 610.

7 Med. Kent. p. 94.

8 JEA. 20 p. 188.

9 Artificial orpiment is poisonous, but not the natural mineral; cf. Lucas, Materials p. 400.

10 Paints and Pigments p. 28.

11 Eg. Eb. 62, 13; 62, 15; 69, 8; 69, 13; 87, 16; Hearst 16, 6; P. Ram. V 24 lower; 42 lower; P. Ch. B. VI rt. 5, 10.

12 Dend. Mar. I, 71.

13 P. Med. Lond. 10, 13; Eb. 64, 2.

14 Hearst 16, 6.

15 Cf. Dioscorides 5, 84 = 5, 74 (W).

a kind of boat, which may correspond to the Greek μιλτεῖον, a type used for transporting red ochre.

Iversen's careful discussion can scarcely be improved upon, though one or two points may be noted, and his tentative acceptance of a connection between ⲙⲉⲣϣ and ⲡⲏⲣϣ may be doubted. It is perhaps strange that although *mnšt* is clearly used as a pigment[1], no certain example occurs in any of the pigment lists[2]. The doubtful *pnš* (sic.) of the Strassburg ostracon[3] may be for *mnšt*, and if so would be particularly significant, since it appears to displace *tmḥy* of the other three lists, which seems to be a species of red ochre. *mnšt* is found, together with *tmḥy*, in the Chester Beatty mineral list[4], and as *mnšt wꜣḏ* in the Edfu list[5], where it is used for sealing and must therefore be an earthy material. Like other ochreous pigments, it is measured in *mnt* vessels[6].

mnšt cannot, it seems, be identified with absolute certainly. That it is an ochreous earth used as a pigment is quite evident, but no definite indication of its colour is forthcoming. Since, however, yellow ochre can be shown to be *sty*, and *mnšt* may be etymologically connected with Coptic ⲙⲣⲟϣ, interpretation as red ochre appears likely.

In conclusion it may be mentioned that *mnft* (Wb. II. 80.9), which occurs once in a medical context[7], may be, as Massart[8] suggests, a mistake for *mnšt* rather than a miswriting of *šnft* (Wb. IV. 514.13) or a ἅπαξ λεγόμενον.

ryt (Wb. II. 399.10–12) is undoubtedly the name of the ink used in writing, and of the raw material from which it was made, and the Wörterbuch rendering 'Farbe zum Schreiben' is correct. That it is ink is shown by references to writing or drawing *m ryt*[9], and its occurrence together with papyrus and rush pens[10], and in cakes (*pd*) with other writing materials[11]. When used for writing, often as *ryt wꜣḏ*, it is clearly prepared ink, but the phrase *ryt ḥr mw nw ꜥntyw*[12] and the Demotic *riw ḫl*[13] prove that *ryt* could also refer to the black powder before the gum was added, corresponding to Dioscorides' μέλας ᾧ γράφομεν[14] and Pliny's atramentum[15].

In general, *ryt* must be understood as the common black ink, made largely from soot[16], though *ryt km* is in fact specifically mentioned[17]. A possible reference to the making of ink from charcoal is found in an ostracon from Deir el Medineh[18], where *ryt* occurs with *ḫt ḏḏ(l)*, a particular type of wood from which *ḏꜥbt* seems to have been produced[19], though this is by no means certain. But *ryt* is not only black, but red[20], and in this case must be red ochre[21].

The frequently occurring *ryt wꜣḏ*[22] has also been taken as a colour specification, eg. by Barns[23], who quotes Gardiner in support. Posener[24], however, regards *wꜣḏ* as 'fresh',

---

[1] Abyd. Mar. I, 52, 9; Dend. Mar. IV, 35, 22–3 = Brugsch Wb. Erg. 610; Thes. 1323.
[2] Cf. however Ostr. DM. 609 vs. 2, where *mnšt* occurs together with *wšbt* and *ryt*.
[3] Ostr. Stras. H. 41; cf. also Eb. 77, 12.
[4] P. Ch. B. IV vs. 7, 14.
[5] Edfou VI, 203, 7.
[6] Harr. 65a, 2; 74, 1.
[7] P. Leid. 345, 72, 12.
[8] Leiden Magical Papyrus p. 95 n. 17.
[9] Destr. 78; Emb. Rit. 8, 18; Apoph. 28, 16; Totb. Pl. 164, 16 etc.; cf. Eb. 74, 11 *ryt nt sš*.
[10] P. Ch. B. V rt. 8, 12; cf. Abyd. Mar. I, 31a; Ostr. DM 609 vs. 3.
[11] P. Kah. 20, 26; cf. ib. 19, 23.
[12] Emb. Rit. 9, 16; cf. ib. 8, 18.
[13] P. Dem. Mag. rt. 5, 5; 27, 31; vs. 16, 5.
[14] 5, 162 (W).
[15] HN. 35, 25, 41; cf. Isidorus 19, 17, 1.
[16] Lucas, Materials p. 415.
[17] Emb. Rit. 8, 20; Rit. d'Apis 7, 10.
[18] Ostr. DM 45 vs. 10–11.
[19] Ostr. Len. 2973, 5.
[20] Eb. 10, 11; cf. Emb. Rit. 9, 10, where the red colour is implied.
[21] Lucas, Materials p. 414.
[22] Destr. 78; Emb. Rit. 8, 18; Apoph. 23, 6; 26, 3; 28, 16; 29, 13; 29, 14;Urk. VI, 5, 9–10; Edfou VI, 235, 5; P. Ram. IV C 15 etc.
[23] Five Ram. Pap. p. 26.
[24] JEA. 35 p. 77 n. 8.

rather than 'green', which seems preferable for several reasons. First, green was not a regular colour for writing, and would only have been used in vignettes. Second, in some instances[1] *ryt w3ḏ* is parallel to new papyrus, from which it appears that the newness or freshness of both was magically desirable. Third, 'fresh ink' makes particularly good sense, in that ink which was not fresh would be dried up and require powdering and re-mixing with gum. In fact, in only one instance, in Demotic[2], does *w3ḏ* seem to refer to colour, being there parallel to *km*.

It has been suggested by Breasted[3] that *rwt* (Wb. II. 403.12), found once in Smith[4] may be a variant writing of *ryt*, which is, however, unlikely. The passage is an explanation of the colour nuance *ṯms*, which is described as *dšr irtyw*, ie. reddish purple, like *irtyw rwt* (sic.) *ṯmst*. Taking *rwt* as *ryt*, the latter phrase is interpreted as 'the purplish colour of *ṯmst*', which it cannot mean, since *ryt* has not the meaning 'colour'. 'Ink' is obviously out of place, and it is necessary therefore to regard *rwt* either as a mistake for *prt*, or as an otherwise unknown word.

From the above evidence it is clear that *ryt* represents ink, both as a powder and as a substance prepared for writing. Unless otherwise stated, it is the common black variety made from soot, though it also refers to the red ink for rubrics, which was red ochre.

ḫsbḏ (Wb. III. 334.1–13) has already been examined in detail among the semi-precious stones (above p. 124f.), where it was shown to include a wide range of blue minerals, primarily lapis lazuli, azurite, blue frit, blue glass, and perhaps also cobalt blue. As a pigment, however, its meaning is more restricted, as appears from the evidence collected by Lucas[5]. The principal blue pigment of the Egyptians was an artificial frit, basically a copper calcium silicate, the composition of which has frequently been investigated. The only naturally occurring mineral used was azurite (chessylite), which is known from the Old Kingdom. There is nothing to suggest that powdered lapis lazuli was employed in making blue pigment, and Iversen[6] is wrong in stating that it was. Neither *ḫsbḏ* nor *mfk3t* refers to the actual content of the frit, but to the substance it was intended to imitate[7]. That cobalt was ever used as a pigment is extremely unlikely[8].

Classical sources tend to confirm the archaeological evidence. Theophrastus, who calls true lapis σάπφειρος, refers to a mineral κύανος, used as a pigment, and of different colours[9], one variety being darker and like σάπφειρος[10]. He states[11] that natural κύανος contains χρυσοκόλλα[12], which would be applicable to azurite which often occurs together with malachite or chrysocolla[13]. There were, moreover, several kinds of κύανος, including the Cyprian, probably azurite, the Scythian, possibly powdered lapis[14], and the Egyptian, κύανος σκευαστὸς, which was evidently artificial[15], and therefore presumably the common blue frit[16].

---

1 Apoph. 23, 6; 26, 3; 28, 16; Urk. VI, 5, 9–10.

2 Rit. d'Apis 7, 10. 3 Pap. Smith p. 196. 4 Smith 3, 21.

5 Materials p. 392f.; cf. table on p. 550. 6 Paints and Pigments p. 18.

7 Campbell Thompson, Dictionary p. 131, thinks that Akkadian *ᵃuḳnu*, used as a pigment, probably refers to an artificial frit or to azurite, rather than powdered lapis.

8 Cf. however, Lepsius, Métaux p. 26. 9 § 31.

10 § 37. Probably lapis lazuli without pyrites, ie. lazurite, though it may be azurite. 11 § 39.

12 Lepsius' suggestion, Métaux p. 30, that χρυσοκόλλαν should be read χρυσοκονίαν is ingenious but unnecessary. 13 Cf. Caley & Richards, Theophrastus p. 143. 14 Cf. ib. p. 183.

15 § 55; cf. Dioscorides 5, 91 (W).

16 The grinding of κύανος to produce different colours probably refers to frit — cf. Caley & Richards, Theophrastus p. 186.

PLINY[1] also mentions a blue pigment, caeruleum, probably a frit, though he compares it to chrysocolla (ie. malachite), which might suggest that it was azurite[2]. VITRUVIUS[3], too, knows caeruleum, which he states was an artificial frit invented at Alexandria.

ḫsbḏ is found as the blue pigment in each of the four major lists[4], and also, together with *sty* and *ḳnit*, in certain offering-lists[5], though in others it is absent[6]. It occurs, too, in other ostraca dealing with pigments, notably one published by ČERNÝ[7], where, with *wꜣḏ*, it is clearly a hard pigment, as opposed to the ochres, *sty*, which are friable. In this instance, it is qualified as *ḫsbḏ nfr nfr*, a designation which is known elsewhere[8], and would seem to indicate that different qualities of the blue pigment were recognised. In some of these examples. *ḫsbḏ nfr nfr* is contrasted with simple *ḫsbḏ*, in others with *ḫsbḏ n sš*, a term which occurs frequently[9], and must surely refer to an artificial frit as MACADAM[10] suggests. *ḫsbḏ* is also specifically stated to be the pigment in which things are to be written or drawn[11], in two examples[12] described as *ḫsbḏ mꜣꜥ*, which presupposes a genuine, ie. naturally occurring, substance, and can only refer to azurite, which may perhaps have been confused with lapis lazuli. Azurite may also be indicated by the one early reference to writing in *ḫsbḏ*[13], though this is not necessarily so. A possible example of the use of blue in painting is a passage in Lansing[14], where walls are said to be *mḥw m ḫsbḏ*, but this may equally refer to the use of faience inlay or the like.

It may be concluded that as a pigment *ḫsbḏ* refers to either one of two substances, azurite or blue frit, more generally the latter. *ḫsbḏ mꜣꜥ* is to be regarded as azurite, and *ḫsbḏ n sš* as frit, while *ḫsbḏ nfr nfr* may represent either azurite or a superior quality of frit. Apart from *ḫsbḏ*, no other name for blue pigment is known.

sšw (Wb. III. 481.10–13) constitutes something of a problem, and cannot be identified with any degree of certainty. BRUGSCH[15] transliterates the word as *nꜥꜥ*, and renders 'bunte Farben' or 'Emaillen', evidently on the basis of a passage in Auswahl[16]. STERN, on the other hand, translates 'atramentum', and this is accepted by LÜRING[17], because of the palette ideogram, and the comparison between the medicinal uses of *sšw* and those of Greek μέλαν. He can quote no chemical analysis of Egyptian ink, but cites various classical recipes from which he concludes the basic constituent to have been soot or charcoal mixed with gum or water. EBBELL translates 'ink-powder'.

Now apart from the use of the palette hieroglyph in writing the word, there is nothing to suggest that *sšw* is a term for ink, and the WÖRTERBUCH's 'Tusche' cannot be justified. *sšw* occurs in several medical contexts[18], none of which is very suggestive, and in sundry

---

1 HN. 33, 56, 158.

2 Azurite is 2 $CuCO_3 \cdot Cu(OH)_2$, chrysocolla $CuCO_3 \cdot Cu(OH)_2$.

3 7, 11, 1.

4 Ostr. Stras. H. 41; Ostr. Tor. A. 11; Ostr. Cairo 25594; P. Ch. B. V rt. 8, 13; cf. also Ostr. MMA. 90.6.3. rt. 1; vs. 1.

5 Wolf Opet p. 44; Mem. Miss. XV pl. 70; Luxor ⟨453⟩; ⟨472⟩.

6 Mem. Miss. XV pl. 26; pl. 28–9.

7 Ostr. Len. 2973, Arch. Or. 3 p. 395.

8 Ostr. Cairo 25649; 25247 rt. 6; vs. 8; Ostr. DM 280 rt. & vs.; P. Gurob fr. G vs. 2, 2 = RAD. 21; Hier. Ostr. 69, 1 vs. 7; cf. also ib. 62, 3 vs. 1 & 7.

9 Ostr. Cairo 25247 rt. 6; edge; Hier. Ostr. 69, 1 rt. 7; vs. 7; Kawa VI, 12 etc.

10 Kawa I p. 39.

11 Urk. I, 39; Totb. Nav. 148, 16; 172, 13; Totb. Leps. 165, 12 etc.

12 Totb. Nav. 64, 51–2; Yuya 30B (pl. 16).

13 Urk. I, 39.

14 Lansing 12, 3.

15 Wb. Erg. p. 229.

16 Auswahl 12, 56.

17 Med. Kent. p. 102.

18 Eb. 52, 9; 57, 7; 61, 11; 66, 13; 68, 13; 92, 10; 105, 14.

lists of minerals[1], where it cannot be identified. In two instances[2] it is imported from Retenu, and, with *ỉsmr*, follows the base metals in the enumeration. In a third[3], it is included among tribute from Syria, in the form of oblong bars, which are apparently white in colour.

sšw is thus clearly a mineral substance, probably of a valuable nature, and perhaps not native to Egypt. These considerations, and the fact that it was imported in bars, seem to rule out the translation 'ink'. As pointed out above, Egyptian ink was largely soot mixed with gum[4], and this was known as *ryt*; iron ink seems to have been used only at a comparatively late date[5]. Nor can *sšw* be a generic term for pigment, since the majority of Egyptian pigments were native, and are in any case covered by the term *ḏrwy*. Further, the apparently white colour of *sšw* is not suited to either ink or coloured pigments.

However, if *sšw* is any species of pigment, the colour may be significant. The common white paint was either chalk or gypsum, and no other is known archaeologically[6]. But it is suggested below (p. 178) that ceruse, white lead, was being sent to Egypt in the New Kingdom, and it is just possible that this is the mineral to which *sšw* refers. It is true that both *sšw* and *ḫrr*, evidently an attempt to transcribe Akkadian *ᵃḫulālu*, which may be ceruse, occur in the Chester Beatty mineral list, and that *ḏḥty ḥḏ*, which may be white lead, is found in a medical context[7], which raises some difficulty. On the other hand, the Chester Beatty list is not scientifically compiled, containing several repetitions, and a large number of outlandish words, and *ḏḥty ḥḏ* may well mean tin, so that the objections are not insurmountable.

In fact, no conclusion seems possible, and *sšw* cannot certainly be regarded as a pigment. It may equally well be a metal or other valuable mineral, and, from its limited occurrence, one which is not as yet known from archaeological sources.

sty (Wb. III. 488.3–6) was for a long time misread as *ḫnty*, the recognition of the true reading, and the explanation of the mistake being due to Dévaud[8]. That *sty* was a species of earthy mineral used as a pigment has never been doubted, though different identifications have been advanced. Brugsch[9] regards *sty* as a coloured earth used in painting, probably green, as opposed to *mnšt*, red, and *šsyt*, white (sic.), and Lüring[10] takes it to be the same colour, identifying *sty* with Greek χαλκῖτις, which he thinks is malachite. But Gardiner[11] sees *sty* as the raw material from which the ordinary red paint was made, ie. red ochre, and Caminos[12] follows this interpretation, while Roeder[13] and Daressy[14] favour yellow ochre. Dévaud[15], on the other hand, translates etymologically 'terre de Nubie', and thinks the colour may have been white, the same view also being taken by Černý[16], who regards *sty* as a nisbe, and points to the opposition between *sty* and *ḏrwy*, 'colours', and by Barguet[17].

---

1 P. Ch. B. IV vs. 7, 6; Urk. IV, 706; 731; 744; 1729.
2 Urk. IV, 731,; 744.
3 Urk. IV, 1101.
4 Lucas. Materials p. 415.
5 ib. p. 413f.; cf. however, p. 419 *re* marking ink. Forbes, Studies III p. 220, appears to regard *sšw* as iron oxide ink, following the Wörterbuch.
6 Lucas, Materials p. 399.
7 P. Med. Lond. 6, 7.
8 Rec. Trav. 39 p. 22.
9 Wb. p. 1113; Erg. p. 948; ZÄS. 18 p. 6.
10 Med. Kent. p. 93.
11 Hier. Texts I p. 41.
12 LEM. p. 442.
13 Urk. z. Relig. pp. 282; 283; 286.
14 Ostraca p. 64.
15 Rec. Trav. 39 p. 22.
16 Arch. Or. 3 p. 396.
17 Stèle de la Famine p. 25.

The question has been carefully reconsidered by IVERSEN[1], who accepts the fact that *sty* is connected with *tꜣ-sty*, and points out that Demotic proves the reading to be *sty*[2]. *sty* is found among the pigments, and in close connection with *mnšt*. In the Leningrad ostracon it is opposed to *ḏrwy*, not from its colour, but apparently as the term for soft ochreous clays as distinct from the hard pigments (frits). In medical contexts it is generally used to cool and reduce swellings[3], as are the other pigments, and it seems that all were thought of as related. But *sty* is specifically used for the liver[4], and so is DIOSCORIDES' sinopic earth, a species of ochre[5]. Of the colour of *sty* there is no direct evidence, but vignettes of the 'Book of the Dead' show the boat of Rēᶜ, supposed to be painted with *sty*[6], as yellow. If this is significant, then *sty*, being an earth, must be yellow ochre as suggested by DARESSY, since only two yellow pigments were known in Egypt, the other being orpiment, which is evidently *ḳnỉt*. PLINY's account[7] of the therapeutic uses of ochres corresponds generally to the employment of *sty* in Ebers, and the provenances of ochre given by LUCAS[8], ie. Aswan and the oases, accord with those of *sty*. It must be remembered that ὤχρα covers a variety of colour nuances, from yellow and brown towards red, and *sty* likewise seems primarily to denote a consistency rather than a colour.

This is an undoubtedly satisfactory, and almost exhaustive summary of what is known concerning *sty*, though there are one or two points which may be added. *sty* is found in three out of the four pigment lists[9], while in the remaining one[10] it is apparently replaced by *ṯr*, which would seem to show that *sty* included red ochres as well as yellow. The same impression is given by certain offering lists[11], where *sty* occurs with other pigments, and no term for the red ochres or any other red pigment appears. In the Leningrad ostracon[12] the same is true, and *sty* is undoubtedly the ochreous pigments as opposed to the more solid frits, *ḫsbḏ* and *wꜣḏ*. In the Strassburg list, it is the only item measured in *hnw* rather than *dbn*, and the same distinction is observed elsewhere[13]. *sty nfr nfr* is once mentioned[14], which seems to imply that different qualities were recognised.

Various things are drawn or painted with *sty*, including the boat of Rēᶜ[15], figures of Horus, Atum, Shu and Tefnut[16], a figure of Wepwawet[17], figures of *Ḥḥ*[18] etc.[19]. Of these, the boat of Rēᶜ appears to be yellow, and so also the figures of *Ḥḥ*, which are to be drawn in *ḳnỉt* and *sty*, where parallelism rather than contrast seems likely[20].

The expression *sty bꜣk*[21] is of particular interest, since it evidently refers to ochre prepared in some way for use. It may of course mean simply ochre that has been burnt to make the

---

[1] Paints and Pigments p. 19.

[2] P. Dem. Mag. 23, 3; 29, 28; vs. 22, 7.

[3] Refs. Paints and Pigments p. 22–4.

[4] Eb. 67, 10.

[5] 5, 111 = 5, 96 (W).

[6] Totb. Nav. 130, 44.

[7] HN. 35, 14, 34.

[8] Materials p. 269.

[9] Ostr. Stras. H. 41; Ostr. Tor. A. 11; Ostr. Cairo 25594; cf. also Ostr. MMA. 90.6.3. vs. 4; rt. 5, the latter doubtful.

[10] P. Ch. B. V rt. 8, 13.

[11] Wolf Opet p. 44; Mem. Miss. XV pl. 26; pl. 28–9; Luxor ⟨453⟩; ⟨472⟩; cf. also Ostr. Cairo 25247; Ann. Serv. 16, 225.

[12] Ostr. Len. 2973.

[13] Ostr. Cairo 25247 vs. 17; Ostr. Gard. 135, 3 [24]; Hier. Ostr. 69, 1 vs. 8; Ostr. DM 280; 183, 6; cf. also measurement in *šnw*, Harr. 65a, 3; 74, 2; Koller 4, 3; Ostr. Cairo 25618.

[14] Ostr. Cairo 25247 vs. 17; cf. Emb. Rit. 9, 9.

[15] Totb. Nav. 130, 44.

[16] Totb. Leps. 134, 16.

[17] P. Leid. 347, 12, 9.

[18] Coff. Texts II, 44h; Lacau Sarc. I p. 81.

[19] Totb. Nav. 125; Totb. Leps. 125, 67; Edfou VI, 300, 4; P. Ch. B. III rt. 9, 13 (a house plastered with *sty*).

[20] Cf. Emb. Rit. 8, 19; 9, 9, where *sty* and *ḳnỉt* are parallel, and Theophrastus § 51.

[21] Ostr. DM 183, 6.

powder finer, and then washed, though it could conceivably indicate roasting of yellow ochre to produce red. This latter process was known to the Assyrians[1], and THEOPHRASTUS[2] mentions the production of μίλτος from burning ὤχρα until it changed. PLINY[3], too, refers to the practice, observing that the longer it has been roasted, the better is the colour. Whether the Egyptians ever heated yellow ochre until it changed colour seems, however, doubtful, and the red ochre used by them was probably all the naturally occurring substance[4].

From the evidence, it appears that *sty* is primarily a general word for ochre, and as such is opposed to *ḏrwy*. Like Greek ὤχρα, it includes both yellow and red ochres, but seems sometimes to refer to the former in particular, while several words for species of red ochre are known.

šsyt (Wb. IV. 545.7–8) and *šsꜣyt* (Wb. IV. 545.5–6), though separated by the WÖRTERBUCH, are without doubt the same word. This is proved by the pigment lists, two of which have *šsy(t)*[5], while a third has *šsꜣy(t)*[6] referring to the same pigment. The identification is apparently assumed by BRUGSCH[7], who takes *šsyt* as a coloured earth, perhaps the origin of Coptic ϣⲓϣ ϣⲟⲉⲓϣ ϣⲁⲉⲓϣ[8]. He tends to favour yellow ochre[9], but elsewhere thinks the colour is white[10]. MÜLLER[11], however, regards *šsyt* as a variant of *ḫsꜣyt*, and this seems also to be the view of SÄVE-SÖDERBERGH[12]. BREASTED[13], on the other hand, sees in *šsyt* the form in which a mineral is used, ie. powder, crystals, or the like, but points out that the occurrence of the word alone indicates that it was used exclusively in connection with one mineral, the name of which could be omitted without sacrificing the meaning. But against such a translation as 'powder of green pigment' IVERSEN[14] objects that *šsyt* must be solid, since it has to be ground in Ebers[15].

The identification of *šsyt* is not, however, so very difficult, due largely to its appearance in three of the four pigment lists[16], where it is clearly a mineral substance used as a pigment. Significantly, in the one list in which *šsyt* is not included[17], the common item *wꜣḏ* is expanded into *wꜣḏ šbnw*, 'various greens', implying that *šsyt* was a species of green pigment, and this conclusion is further supported by the absence of *wꜣḏ* in another list[18], where there is no mention of a green pigment unless *šsyt* be taken as such. Again, this agrees with the occurrence of *šsyt* in the phrase *šsyt nt wꜣḏw*[19], which relates the two, and with the doubtful description of *šsyt* as green in a Cairo ostracon[20]. Now if *šsyt* is a green pigment distinct from *wꜣḏ*, malachite, it can only be the name of the common green frit, and *šsyt nt wꜣḏw* then means 'green frit of malachite', the latter being the constituent which gave the frit its colour[21].

---

1 Campbell Thompson, Dictionary p. 31.
2 § 53.
3 HN. 35, 13, 31; cf. Isidorus 19, 17, 1.
4 Lucas, Materials p. 398; Forbes, Studies III p. 207.
5 Ostr. Stras. H. 41; Ostr. Tor. A. 11.
6 Ostr. Cairo 25594.
7 Wb. Erg. p. 1204.
8 ϣⲟⲉⲓϣ in fact derives from *ḫꜣḫꜣ* 'worfeln' (Wb. III. 233. 17).
9 ZÄS. 18 p. 6.
10 Wb. Erg. p. 948.
11 Rec. Trav. 22 p. 104.
12 Äg. und Nub. p. 218.
13 Pap. Smith p. 381.
14 Paints and Pigments p. 13.
15 Eb. 78, 5.
16 Cf. also Ostr. MMA. 90.6.3. rt. 4 (*šsyt* of Kush); vs. 1.
17 P. Ch. B. V rt. 8, 12f.
18 Ostr. Tor. A. 11.
19 Smith 14, 5; 16, 10; Eb. 78, 5.
20 Ostr. Cairo 25677 rt. 21–3.
21 Lucas, Materials p. 396; cf. p. 392.

Whether all the apparent examples of *šs(ꜣ)yt* refer to this frit seems doubtful, and some at least may be writings of *ḫsꜣyt* as has been suggested. It is probable, however, that the *šsꜣ* measured in *dbn* in Harris[1] is in fact this word, and so too the *šsyt* of Ebers[2] and the *šsꜣy* of a Cairo ostracon[3]. Less certain are the *šsꜣy* of Koller[4], and the baskets of *šsꜣ* mentioned in Harris[5], while the *ssꜣyt* (sic.) of Ebers[6] seems altogether dubious.

At all events, it is clear that a distinct word *šs(ꜣ)y(t)* does exist, and that it can only refer to the green frit used as a pigment.

ḳnit (Wb. V. 52.10–16) is recognised by Brugsch[7] as a yellow pigment, though without more precise identification, and the Wörterbuch, while translating 'ein mineralischer Stoff von gelber Farbe' also offers no suggestion. The true nature of *ḳnit* is explained by Iversen[8], who points out that the word occurs in offering lists together with other pigments, and is closely related to *ꜣwt-ib*; realgar, the latter being explained as *ḳnit* in one instance[9]. Its yellow colour is established beyond doubt by a passage in the 'Book of the Dead'[10] where 'dwellings' of the colour of *ḳnit* are bright yellow in the vignettes. But a more precise translation than simply 'yellow pigment' is possible. The Egyptians used only two yellow pigments, yellow ochre and orpiment[11], and of these the former was known as *sty*, so that *ḳnit* is practically certain to be orpiment. This is supported by the fact that neither orpiment nor *ḳnit* appears before the New Kingdom, and by the connection with *ꜣwt-ib*, both being sulphides of arsenic. *ḳnit* is apparently connected with *ḳn*, 'to be fat', and in some instances seems to have a somewhat vague meaning, as 'yellow colour', or 'yellow stuff'.

Iversen's argument, though based almost entirely on the negative proof that no other substance is possible, is none the less convincing, and there seems no reason to doubt that *ḳnit* is orpiment. It occurs in each of the four pigment lists[12], accompanied by *ꜣwt-ib*, and with other pigments in certain offering lists[13]. On several occasions it is parallel to *sty*[14], where parallelism of colour seems to be implied, and it is the colour of 'dwellings'[15], a figure of Shu[16], figures of *Ḥḥ*[17], and of other gods[18] etc. Strangely, *ḳnit* occurs only once, or possibly twice, in medical contexts[19], although arsenicum was a well known medicament in classical times. Lastly, the connection of orpiment with gold and silver ores[20] may be reflected in the expression *nbw n ḳnit*[21], though *ḳnit* may refer simply to yellow colour, as it apparently does in one or two other instances.

In conclusion, an interesting circumstance already alluded to by Iversen[22] must be noted. It is that although orpiment was frequently used, and *ḳnit* occurs in a large number of texts, Lucas knows of no source of the material in Egypt. This would suggest that it was imported, yet no reference to *ḳnit* as tribute can be quoted. But the Periplus implies that orpiment

---

[1] Harr. 65a, 5. [2] Eb. 56, 12. [3] Ostr. Cairo 25677 vs. 11.
[4] Koller 4, 2. [5] Harr. 65a 4; 74, 4. [6] Eb. 84, 6.
[7] Wb. Erg. p. 1254. [8] Paints and Pigments p. 34.
[9] Ann. Serv. 16, 221. [10] Ch. 149.
[11] Lucas, Materials p. 399.
[12] Ostr. Stras. H. 41; Ostr. Tor. A. 11; Ostr. Cairo 25594; P. Ch. B. V rt. 8, 13.
[13] Wolf Opet p. 44; Mem. Miss. XV pl. 26; pl. 28–9; pl. 70; Luxor ⟨453⟩; ⟨472⟩; cf. Kawa VI, 12.
[14] Emb. Rit. 8, 19; 9, 9; Coff. Texts II, 44h; Lacau Sarc. I p. 81.
[15] Totb. Nav. Einl. p. 57. [16] Destr. 46. [17] Coff. Texts II, 44h.
[18] P. Leid. 346, 3, 3; Emb. Rit. 7, 13.
[19] P. Med. Berl. 7, 5; cf. P. Tur. unpublished [25].
[20] Theophrastus § 51. [21] P. Tur. PR. 32, 5. [22] Paints and Pigments p. 37.

was an Egyptian export article[1], and Pliny[2] states that realgar was found on Topazos, St. John's Island, so that it may after all have been among the natural resources of Egypt[3].

The close connection of orpiment and realgar has already been referred to above, but it may here be added that orpiment, like yellow ochre, can be made to change colour if heated twice and cooled. That the Egyptians ever discovered this is unlikely, but if they did, it would serve to explain why *ꜣwt-ib* was regarded as a species of *ḳnỉt*, and not the reverse.

ṯmḥy (Wb. V. 369.4), regarded by the Wörterbuch as 'Art kostbarer Stein aus Nubien', can now be identified with greater precision, owing to its undoubted occurrence as a pigment in three of the four lists[4]. In one of these[5] it is measured in *mnt* vessels, while the other items, except *sty*, are in *dbn*, and this would seem to suggest an ochreous earth[6]. Now it is significant that in none of the lists is there a specific word for red ochre, which one would expect, and that in the one from which *ṯmḥy* is absent[7], its place is taken by a word *pnš* (sic.), probably a miswriting of *mnšt*. *ṯmḥy* and *mnšt* also occur together in the Chester Beatty mineral list[8], and it seems likely that the two are related, and that *ṯmḥy* is a species of red ochre.

This is to some extent confirmed by the other examples of the word. *ṯmḥy* is found among the 24 precious minerals at Dendera[9], in association with other earthy substances such as *wꜣḏ*, *msdmt* and *bḳs*, and it occurs again in a broken context in the Chester Beatty list[10]. It is three times referred to as *ṯmḥy (n) wꜣwꜣt*[11], and, as *tꜣ-mḥy* (Wb. II. 127.1), is listed among the minerals found near Elephantine[12], both of which correspond to known sources of ochre. It also occurs with other minerals in the geographical papyrus[13], but the context is too fragmentary to admit any of conclusion. That *ṯmḥw* (Wb. V. 369.5), listed among garden produce[14], has any connection with *ṯmḥy* seems highly unlikely, since the former is evidently some kind of plant.

It may be concluded, therefore, that *ṯmḥy* is a species of red ochre, corresponding to one of the varieties mentioned by Pliny[15], and that it came from Nubia and Elephantine. It may perhaps refer to the common reddish-brown pigment used for the flesh colour of male figures, rather than to one of the more vivid reds, but this is merely a suggestion, and the position in which it occurs in Harris[16] might point rather to its being of some value.

ṯr (Wb. V. 386.11–12) and the later *wtry* (Wb. I. 381.5) are certainly to be identified, as is shown by the existence of similar terms *ṯr* (Wb. V. 386.13) and *wtry* (Wb. I. 381.3) both with the meaning 'blood'. Indeed, Blackman and Fairman[17] suggest that the reading of the later word is not in fact *wtr(y)*, but *t(w)r*, and if this is so, identification is inevitable.

---

[1] § 56, though it is not stated directly.

[2] HN. 35, 22, 39.

[3] Cf. also orpiment from Tanis, Petrie, Tanis I p. 39, though this may have been imported.

[4] Ostr. Tor. A. 11; Ostr. Cairo 25594; P. Ch. B. V rt. 8, 13 (damaged); cf. also Ostr. MMA. 90.6.3. vs. 3.

[5] Ostr. Cairo 25594.

[6] Cf. Harr. 65a, 2; 74, 1, where *mnšt* is in *mnt*.

[7] Ostr. Stras. H. 41.

[8] P. Ch. B. IV vs. 7, 14.

[9] Dend. Mar. IV, 36, 50; 39, 141–2.

[10] P. Ch. B. IV vs. 8, 1.

[11] Harr. 62b, 14; 70a, 11; Dend. Mar. IV, 39, 141–2.

[12] Sehel 16. – Brugsch, Hungersnoth p. 129, read *mḥy*, regarding it as a species of serpentine, but *tꜣ-mḥy* is certain.

[13] Geog. Pap. pl. 10, fr. 15.

[14] ZÄS. 38, 37.

[15] HN. 35, 13, 31.

[16] Harr. 70a, 11 (following *ḫsbḏ*); 62b, 14 (following *mfkꜣt*).

[17] JEA. 29 p. 7 n. h.

BRUGSCH[1] assumes that the two words are the same, translating 'die dunkel- oder blutrote Farbe', and suggesting a close connection with *ḏrwy*. Whether this latter can be substantiated seems doubtful, though there is certainly some parallelism of meaning in a few instances[2].

The references to *ṯr* are diverse, and precise identification of the substance is somewhat difficult. Of particular interest is its occurrence in the Chester Beatty list of pigments[3], where GARDINER[4] translates 'red', which would seem to be justified by the obvious connection with the similar word for blood mentioned above. On the other hand, it is significant that in this list there is no reference to *sty*, which appears in each of the others, notably at the head of the Toronto list[5], apparently as a general designation of ochres, and it may be that *ṯr* has the same meaning. This would accord well with later examples, where it appears to signify little more than 'colour'[6].

A translation ochre would also be applicable in two medical contexts[7], in one of which *ṯr* is linked with *di͗di͗*, and would be suited to the Nubian provenance given in the Edfu mineral list[8], where *ṯr* is described as *ṯr m ḏt.f*, and *ṯr šmᶜ m i͗rw.f*, which implies that different species were recognised.

But other references to *ṯr* seem to point to something rather more solid, in particular its occurrence as the material for a headrest and other objects[9], and for a statuette of the Nile god[10]. In the former case it is parallel to *ti͗i͗ꜣꜣt* and *kꜣ*, both names of stones, and is presumably of similar nature itself. Specimens of ochreous haematite vary considerably in texture, some being quite hard, and both sandstone haematite, a form of red ochre, and the oolitic variety might be carved into small statues, but none seems suitable for a headrest. *ṯr* also follows *ḏḥty* in a partially effaced list of minerals[11], and occurs both in the geographical papyrus[12], and in an Old Kingdom offering list[13], in none of which is the meaning clear. Lastly, it is found three times in a somewhat obscure passage[14] where one official sends his servant to another to fetch *ṯr n š n mi͗ ḳd.s* and a hide (*mskꜣ*).

From the meagre evidence, no very satisfactory translation emerges, though it is at least clear that *ṯr* is a mineral substance, some form of which was used as a pigment. It may be a general term for ochres, and sometimes has the wider meaning 'colour', though the existence of a similar word, used of blood, would suggest that it was primarily red. Some species of *ṯr* was, however, a solid substance, perhaps a stone. Certain types of ochreous haematite might conceivably be applicable to all examples of *ṯr*, though the identification is only tentative.

di͗di͗ (Wb. V. 421.9–11) was originally thought by BRUGSCH[15] to mean mandrake, by comparison with the Hebrew *doudaim*, but this is specifically rejected by KEIMER[16] who nevertheless regards it as a Semitic loan-word, and suggests no alternative translation. That *di͗di͗* is a mineral substance is, however, abundantly clear, and EBBELL's[17] 'dragon's blood' (from *Dracaena Ombet Kotschy*) may therefore be rejected also.

---

[1] Wb. Erg. pp. 364; 1315; 1334.

[2] Compare, for example, Düm. T. I. II, viii, 3 and xii, 15, and other similar examples of *ṯr* and *ḏrwy* at Dendera. [3] P. Ch. B. V rt. 8, 13. [4] Hier. Pap. B. M. p. 49. [5] Ostr. Tor. A. 11.

[6] Dend. Mar. I, 39a; 39b; III, 34d etc.; cf. Kuban 3.

[7] P. Ram. V 21 lower; 72 upper. [8] Edfou VI, 204, 1.

[9] Cairo 1385 = Saqq. Mast. I pl. 1; cf. Selim Hassan, Giza VI(2) p. 462.

[10] Harr. 41a, 10. [11] LD. III, 200d. [12] Geog. Pap. pl. 10 fr. 16.

[13] Giza J. IV pl. 9; cf. also Macramallah, Un Cimetière Archaique p. 17 for another obscure example.

[14] Hier. Pap. Berl. III pl. 7, l. 3; 4; 6.

[15] ZÄS. 29 p. 31. [16] Gartenpflanzen p. 137. [17] Pap. Ebers p. 133.

Gauthier[1] devoted a long article to *didi*, 'l'argile rouge d'Éléphantine', collecting together all the then known examples of the word. In Ebers, he points out, it is employed as an astringent or emollient[2], and as the former also in the Berlin medical papyrus[3], while in the myth of the destruction of mankind[4] it is brought from Elephantine, ground, and added to beer, to which it imparts the colour of blood. Gauthier maintains that Brugsch is wrong in identifying *didi* as mandrake because 1. the Coptic rendering of μανδραγόρας, ie. ⲚⲞⲨⲦⲈⲘ, is not connected with *didi*, 2. neither Loret nor Schweinfurth has found evidence of mandrake in Egypt, 3. the Egyptians were so unfamiliar with mandrake that in the Song of Songs μανδραγόρας was not understood, and was simply transcribed as ⲘⲀⲚ̄ⲀⲢⲀⲔⲰⲢⲞⲤ in the Coptic version, and 4. Dioscorides says that the Egyptian word for mandrake was ἀπεμουμ. He adds, further, that mandrake would not colour beer red, and that the determinative of *didi* suggests a mineral.

Considering what *didi* may then mean, Gauthier argues that *di* is Coptic ϯ, and that *didi* would therefore give Coptic ϯϯ, a word which in fact exists. Peyron[5] translates it 'lapis super quem fullones vestes excutiunt', and quotes the compound ⲢⲈϤϯϯ as 'fullo'. *didi* is therefore fuller's earth, a reddish variety of which is mentioned by Dioscorides[6] and others. It could be used as a salve – *didi* is never used internally (sic.) – and Pliny[7] specifically mentions its calming properties. It would easily powder and dissolve in water. But although primarily fuller's earth, *didi* may have acquired an extended meaning, and 'Aswan clay' is perhaps a better translation.

Dawson[8] also rejects the identification of *didi* with mandrake, and gives Gauthier credit for proving that it was a mineral. He re-examines the passages in which the word occurs, including some not known to Gauthier, and deduces that *didi* was obtained from Elephantine[9] and Nubia[10], that it was a bright red colour[11], that it was hard and had to be pounded for use[12], that it was valuable[13], that it was medicinal, generally used externally[14], though it was not poisonous and could be swallowed[15], and that since it was a red pigment, its medicinal uses may have been merely colourative. Dawson admits that Gauthier's translation is possible, but thinks that *didi* was something more distinctive than fuller's earth, being rare and bright red. He proposes red ochre, the soft form of haematite, which is known to occur in Nubia, is relatively rare, and was used as red ink and red paint. He draws a distinction between the common haematite, used for eg. the flesh tints of men, and the rarer, brighter, *didi*, for which 'Nubian haematite' is a convenient translation.

Since Dawson's article, few examples of *didi* have come to light, and they add but little to our knowledge. *didi* is finally established as a mineral by its occurrence in the Chester Beatty list[16], where it is called *didi n tꜣ sty*, confirming the Nubian provenance in Koller, and new medical references[17] where it occurs with *ṯr* and *mnšt*, suggest that it was an analogous material.

Despite the ingenuity of Gauthier's argument, and the existence of Demotic *tt*[18] as a link between *didi* and Coptic ϯϯ, Dawson's translation seems the more probable. Fuller's earth

---

[1] Rev. Ég. 11 p. 1f.
[2] Eb. 74, 14; 80, 20; 82, 19; 83, 20. Twice more internally – Eb. 39, 10; 86, 2.
[3] P. Med. Berl. 7, 5.
[4] Destr. 17; 18 etc.
[5] Lexicon Supp. p. 14.
[6] 5, 176 = 5, 156 (W).
[7] HN. 28, 66.
[8] JRAS. 1927 p. 497.
[9] Eb. 39, 10; 74, 15; Destr. 17–18; Urk. IV, 55.
[10] Koller 4, 2.
[11] Destr. 17f.; Rec. Trav. 13, 166.
[12] Destr. 18; Eb. 39, 10; cf. *dḳw n didi*, Eb. 74, 15.
[13] Urk. IV, 55; Koller 4, 2.
[14] Eb. 74, 14; 80, 20; 82, 19; 83, 20; Hearst 8, 11; 8, 14; 8, 17; 11, 10; 12, 14; 15, 11; 15, 12.
[15] Eb. 39, 10; 86, 2.
[16] P. Ch. B. IV vs. 10, 15.
[17] P. Ram. V 44 lower; 71 upper; 72 upper.
[18] P. Dem. Mag. 19, 40.

is not well known in connection with Egypt, and though some species may be reddish, the majority are olive green. The chief characteristic of *didi* is its blood red colour, and it is clearly something more valuable than an ordinary clay[1]. Bearing in mind PLINY's citation[2] of various different kinds of ochre, distinguished by their colour, it is perhaps possible to see in *didi* a particularly vivid, somewhat rare, variety, found at Aswan and in Nubia, and different from those normally used as pigments. Whether the word is originally of Semitic origin it is impossible to say, though Akkadian *ṭiṭṭu* refers to clay and similar earths[3], and may have been adopted by the Egyptians in reference to a particular kind.

ḏrwy (Wb. V. 601.5–10) has long been identified as the most general word for pigment, though it is clear that it also has a secondary and more restricted meaning. BRUGSCH[4] tentatively compares *ḏrwy* with *ṯr*, and there are passages in late texts which seem to suggest a similar meaning for both. More important, he also relates *ḏrwy* and *tìt* (Wb. V. 238.9)[5], and Prof. FAIRMAN confirms, in a letter, that the two are in fact the same word, and that wherever *tìt* (sic.) refers to colour, it is to be read *ḏrwy*[6]. This is proved by exactly parallel passages in Ptolemaic texts[7], and there can be no doubt that the entry in the WÖRTERBUCH is to be eliminated. The majority of the examples are in fact *ḏrwy*[8], all dating from the Greek period, while the only earlier reference to *tìt*[9], which is fully written and cannot be *ḏrwy*, is capable of alternative explanation as the well known word for image (Wb. V. 239.1f.).

Despite its form, *ḏrwy* is a masculine singular, as has been pointed out by HAYES[10], who notes the variant writings. It is a word of fairly frequent occurrence, particularly in Ptolemaic texts[11], where it evidently means 'colour', without any more precise connotation. This also seems to be the meaning in the phrases *mḥ m ḏrwy*[12], which in all probability refers to the filling in of outlines with colour, and *wꜣḥ ḏrwy*[13], the application of colour. The former may, however, have a more literal significance, indicating the filling of sunken hieroglyphs etc. with a coloured paste, the WÖRTERBUCH's 'Farbpaste', though this seems somewhat unlikely since the technique is not common. Similarly, the *ḏrwy* which occurs as the last item in one version of the list of 24 precious minerals at Dendera[14] is probably a general term including all pigments.

It is, however, clear that in addition to the general meaning of colour or pigment, *ḏrwy* has also a more restricted significance, referring specifically to the hard frit pigments as opposed to the more friable ochres. This is evident from the Leningrad ostracon[15], where both *sty* and *ḏrwy* are mentioned, the latter being the blue and green pigments, *ḫsbḏ* and *wꜣḏ*, probably in this instance frits, while the former seems to include all ochreous pigments[16].

---

[1] Cf. Urk. IV, 712.
[2] HN. 35, 13, 31.
[3] Campbell Thompson, Dictionary p. 18.
[4] Wb. p. 1561; Erg. p. 1334.
[5] Wb. Erg. p. 154; p. 1310. Originally, Wb. p. 1528, he thought *tìt* was a metal.
[6] Cf. the interchange of signs in other similar words, eg. *ḏryt*, 'shrine'.
[7] Compare eg. Dend. Chass. III, 62, 4 (*tìt*) with Dend. Chass. III, 46, 10; 140, 11 (*ḏrwy*).
[8] Dend. Mar. II, 61b; IV, 36; Düm. H. I. II, 56; T. I. II, xii, 10; Ombos I, 60, 62 etc.
[9] Louvre C 12 (Lesest. 76, 10).
[10] Ostraca etc. p. 41.
[11] Edfou I, 23; 553; Dend. Mar. II, 8; II, 10; II, 57d; II, 73a; II, 77b; Düm. Baug. 33; 40; Mammisi 6; Thes. 266, 14 etc. Cf. above n. 7 & 8.
[12] Louvre C 12; Thes. 1442; P. Tur. Leps. Grundplan Xb2 (plan H.); Philae ⟨311⟩ etc.
[13] Thes. 266, 14 cf. also *m sš m ḏrwy*, Düm. G. I. II, 18.
[14] Dend. Mar. IV, 36, 50; cf. also the *ḏrwy nbt* of P. Tur. PR. 32, 6.
[15] Ostr. Len. 2973.
[16] Cf. also Ostr. MMA. 90.6.3. vs. 4, where *ḏrwy* and *sty* occur together, though the significance is not clear, since other pigments are listed.

Again, in another ostracon[1], *ḏrwy* embraces *ḫsbḏ nfr nfr* and *wꜣḏ*, and reference is made to their being crushed[2], which implies lumps of naturally occurring mineral, ie. azurite and malachite, or frit. The same distinction is also observed in listing *ḫsbḏ* and *wꜣḏ* in *dbn*, while *sty* is always in *hnw* vessels or the like, and it may be that *ḏrwy* measured in *dbn*[3] refers only to the hard pigments.

Other references to *ḏrwy* are ambiguous, and may bear either meaning. *ḏrwy* is apparently to be used for sealing[4], and this might indicate either a frit or any other pigment, Similarly, the *ḏrwy ʿšꜣ* brought from Retenu[5] may refer to a miscellaneous consignment of pigments accepted as tribute, or to some specific kind produced outside Egypt. The translation of *ḏrwy* in the Senmut ostraca[6] depends rather on the meaning of *trỉt*. If the latter is simply the name of the small cakes into which pigment was made, ready for use, *ḏrwy* will have a general meaning. If, on the other hand, it refers to lumps or sticks of pigment, *ḏrwy* is more likely to specify the hard pigments[7].

Included in the Wörterbuch examples of *ḏrwy* are several which may not, in fact, be the word for pigment, and should perhaps be eliminated. Among these are references to *ḏmʿ ḏrww*[8] and *ḫt ḏrww*[9] in Harris, where Piehl, Brugsch, Breasted and others regard *ḏrww* as a distinct word, which is altogether probable. If *ḏrww* is pigment, then either two commodities are lumped together, which is unlikely, or it is used adjectivally, as 'coloured papyrus', 'coloured wood', neither of which makes particularly good sense. More probable is a translation of *ḏrww* as bark or rind, as Breasted proposes[10]. Equally doubtful is a supposed occurrence of *ḏrwy* in connection with a statue[11], which Weill regards as a mistake[12]. Similarly, the Wörterbuch example of *ḏrwy* 'als Bez. für bemalte Stelle'[13] is translated 'wall, side wall' (Coptic ϪⲞ) by Caminos[14], following Gardiner, which is undoubtedly correct, and the same translation seems also to apply in another instance[15], so that both may be disregarded.

To sum up, the evidence for *ḏrwy* points to either of two closely related meanings. In the first instance it is the most general term for pigment, sometimes to be translated colour, and secondly it refers more specifically to the hard pigments as distinct from the ochreous clays. As a general term in late texts, *ḏrwy* seems to resemble *ṯr*, but identification of the two words is unlikely.

In concluding this chapter, some mention must be made of two substances which are not mineral, and cannot have been regarded as such by the Egyptians. Since, however, they occur in close connection with the various pigments, they are relevant to the present discussion and require brief notice.

ḳmyt (Wb. V. 39.3–15), Coptic ⲔⲞⲘⲘⲈ, Greek κόμμι, gum is applied to both true gums and gum-resins[16], and several different varieties are specifically mentioned, including *ḳmyt nt ʿntyw*[17], *nt snṯr*[18], *nt šndt*[19], *nt hbny*[20], *ḳmyt ḥḏt*[21], *ḳmyt n pwnt*[22], and also *mw nw ḳmyt*[23]. A

[1] Hier. Ostr. 69, 1 rt. 6. [2] ib. rt. 6; 9; vs. 2. [3] P. Tur. 1881 rt. 5, 10 [26].
[4] Ostr. DM 198 rt. col. 2. [5] Urk. IV, 670. [6] Ostr. Senmut 63, 8; 64, 6; 66, 5.
[7] Cf. Hayes, Ostraca etc. p. 41, where *ḏrwy* is taken to refer to blue-grey background colour.
[8] Harr. 17a, 12; 20a, 1; 72, 13. [9] Harr. 34b, 3.
[10] AR. IV p. 134. [11] Urk. I, 294. [12] Dec. Roy. p. 88.
[13] Lansing 12, 3. [14] LEM. p. 415. [15] P. Tur. PR. 42, 6.
[16] Lüring, Med. Kent. p. 160; Lieblein, Rec. Trav. 16 p. 147; Loret, Flore² p. 84 etc.
[17] Urk. IV, 329, 3; 346, 13. [18] Eb. 87, 6–7. [19] Eb. 54, 16; 77. 19.
[20] Emb. Rit. 9, 17. [21] Eb. 10, 10; 43, 11; P. Med. Berl. 3, 12.
[22] Düm. H. I. I, 32; Edfou I, 435. Cf. also *ḳmyt n tꜣ-nṯr*, Düm. H. I. I, 32.
[23] Eb. 68, 4; 75, 1; 107, 20; Hearst 15, 15 etc.

detailed consideration of the evidence would here be out of place, and discussion will be confined to those examples where *ḳmyt* is closely connected with pigments, and has a particular meaning.

ḳmyt occurs in the Toronto pigment list[1], following *sty*, on a Cairo ostracon[2], with *ḫsbḏ n sš*, on an ostracon from Deir el Medineh[3], with *ḫsbḏ* and *sty*, and on an ostracon now in New York[4]. What is also, it seems, a writing of *ḳmyt* is found with *ḫsbḏ* and *wꜣḏ* on another Cairo ostracon[5], and what may also be *ḳmyt* heads the Strassburg pigment list[6]. The latter is in fact written *bꜣy* (sic.), which makes no sense, and it is not impossible that  is a mistake for , though it must be admitted that the item is measured in *dbn*, while one would expect *ḳmyt* to be in *hnw* or the like.

Now the function of *ḳmyt* in connection with Egyptian painting is quite clear, since Lucas[7] points out that in executing tempera painting some adhesive comparable to modern size or gum is necessary, and that without it azurite, malachite and the frits at least could not be made to adhere to any surface. Such an adhesive can only have been gum, glue, or, less probably, albumin, and it is clearly the first to which *ḳmyt* refers.

But there is direct textual evidence of the use of gum as a vehicle, as was recognised by Hofmann[8]. In a passage in one version of the 'Book of the Dead'[9], a scarab is to be drawn *m ḫsbḏ ḥr mw nw ḳmyt*, which is quite explicit, and there are also references to *ryt ḥr mw nw ʿntyw*[10], and *ri͗w ḫl*[11], which probably refer to the practice of making up cakes of ink with gum and water[12].

There can be no doubt therefore that the mention of *ḳmyt* in conjunction with mineral pigments, particularly *ḫsbḏ* and *wꜣḏ*, refers to gum used as a paint vehicle in the Egyptian tempera technique.

ḏʿbt (Wb. V. 536.8–17), Coptic ϫⲃⲃ(ⲉ)ⲥ is the regular word for charcoal as a fuel, the Wörterbuch's 'Holzkohlen', and is only considered here because it seems possible that in some instances it refers to the black pigment used by the Egyptians, for which no other term is known. Concerning this black pigment, Lucas[13] says that it was almost always carbon in some form, generally soot, though occasionally fairly coarse, and quotes one example of the use of powdered charcoal. Discussing the analogous black ink[14], he quotes comparative cases of the production of carbon by burning incense and date stones, and elsewhere[15] quotes Lane for the burning of frankincense and almond shells to make black eye-paint[16].

Now it is true that *ḏʿbt* does not occur in any of the pigment lists, where neither black nor white pigment appears to be mentioned. It is, however, found in certain offering lists[17] in the form *ḏʿbt ʿn(t)* (Wb. V. 537.3), and accompanying various pigments, and in one

---

[1] Ostr. Tor. A. 11. [2] Ostr. Cairo 25246 [27].
[3] Ostr. DM. 280. [4] Ostr. MMA. 90.6.3. rt. 5; vs. 2.
[5] Ostr. Cairo 25247 vs. 8 is in all probability a mistake for .
[6] Ostr. Stras. H. 41. [7] Materials p. 401. [8] ZÄS. 23 p. 67.
[9] Totb. Leps. 165, 12.
[10] Emb. Rit. 9, 16; cf. ib. 8, 18.
[11] P. Dem. Mag. rt. 5, 5; 27, 31; vs. 16, 5. [12] Lucas, Materials p. 414.
[13] Materials p. 391. [14] ib. p. 415. [15] ib. p. 101.
[16] Cf. also Dioscorides 5, 162 (W); Pliny, HN. 35, 25, 41; Vitruvius 7, 10; Isidorus 19, 17, 1.
[17] Wolf Opet p. 44; Mem. Miss. XV pl. 26; pl. 28–9; Luxor ⟨453⟩; ⟨472⟩ [28]; Ann. Serv. 16, 223.

instance[1] *ḏꜥbt* is described as *ꜥnt pw*. The interpretation of this expression is doubtful, but it seems at least possible that *ꜥnt* is *ꜥntyw*[2], and that it refers to carbon black made by burning frankincense. *ḏꜥbt* is also mentioned in the Leningrad ostracon[3], where, in addition to the pigments, the scribes receive *ḏḏ*–wood and *ḏꜥbt*, the implication of which is not clear, though *ḏꜥbt* may well be the black pigment, and *ḏḏ*–wood a species particularly suitable for burning to produce a fine soot[4]. In other instances[5], *ḏꜥbt* is referred to in connection with work on a tomb, where Černý[6] suggests that fuel for a fire is meant, though again it is possible that the raw material of black pigment is indicated.

Lucas[7] explains the coarseness of some specimens of carbon black by suggesting that it may have been soot collected from masonry or plaster surfaces, and this is perhaps confirmed by the phrase *ḏꜥbt nt inb* (Wb. V. 537.1–2)[8], which Stern, Joachim, Wreszinski and Ebbell all regard as soot, though the Wörterbuch is less certain. An analogous expression *ḏꜥbt nt ḥmty* also occurs[9], where, however, *ḏꜥbt* may have a somewhat different meaning, perhaps verdigris[10].

ḏꜥbt is not the only known word for soot, to which *wbd* (Wb. I. 297.9) also seems to refer, though there is nothing to suggest that the latter was ever used as a pigment. Further, the word *ryt*, discussed above, refers not only to the prepared ink, ie. carbon black mixed with gum and water, but also to the sooty powder itself before it has been made into cakes. Thus, if *ḏꜥbt* does indeed represent the raw material for black pigment, as appears possible, it cannot refer to ink, but merely to the black used for outline drawing and painting. if indeed the Egyptians distinguished the two.

The lexicographical evidence for Egyptian pigments may now be summarised, and compared, as far as possible, with what is known from purely archaeological sources. On the whole, it seems to present a fairly coherent picture, though there are some points where difficulty exists, and which require further elucidation, and others where the textual and archaeological data do not appear to correspond. In such cases, comparative information from other ancient and classical sources is of considerable importance, and cannot lightly be disregarded.

The most general word for pigment is undoubtedly *ḏrwy*, which is applied to all kinds, regardless of their nature and origin, particularly in texts of the Graeco-Roman period. In some earlier examples, however, it has also a more specialised meaning, and refers specifically to the hard pigments, azurite, malachite, and the blue and green frits, as distinct from the softer and more friable ochreous clays. The latter are known collectively as *sty*, which may also have included the arsenic compounds, though there is no definite evidence of this, and their precise standing is uncertain. A third word, *ṯr*, seems likewise to have a somewhat general significance, in some cases parallel to *ḏrwy* as a broad term for pigment or colour, and in one instance apparently replacing *sty* as a designation of the ochres. But it also refers to some particular mineral, evidently of a red colour.

---

[1] Ann. Serv. 16, 225. [2] Cf. writings of *ꜥntyw* (Wb. I. 206.7f.). [3] Ostr. Len. 2973, 5.

[4] For this interpretation cf. Ostr. DM. 45 vs. 10–11, where *ryt* and *ḫt ḏḏl* are found together, Hier. Ostr. 26, 2, 3–4, where *ḫt ḏnḏr(t)* follows *ḫt n šmw*, and also the term *ḫt n ḏꜥbt* of P. Mayer 12, 18.

[5] Ostr. Cairo 25266, 1; 25269, 1; Giornale 50, 2.

[6] Arch. Or. III p. 398–9. [7] Materials p. 391.

[8] Eb. 46, 17; Hearst 16, 1; P. Ram. V 30 upper. [9] Eb. 106, 2.

[10] Cf. Coptic [S]ϣⲏⲃⲉ (Crum 551a), parallel to [B]ϫⲉⲃⲥ, used to render ἰός in Ezek. 24, 6, and apparently rust or verdigris. Akkadian 'dust of the wall' was evidently a form of sal murale – cf. Campbell Thompson, Dictionary p. 10.

A similarly basic term is *ryt*, ink, which covers both the raw material, soot or ochre, and the resulting substance compounded with gum and water. In most cases it refers to the common black ink used in writing, though occasionally to the red employed for rubrication. Of iron oxide ink, used inter alia as a marking ink on linen, there is no trace in the texts, since it was certainly not *sšw* as has been suggested. It seems likely, therefore, that it was also included under *ryt*.

The hard pigments were the blues and greens, the naturally occurring copper carbonates azurite and malachite, and the analogous frits, copper calcium silicates, made in imitation of them. Only one name for the blue pigments is known, *ḫsbḏ*, which must therefore refer to both azurite and blue frit, though the latter may be more specifically indicated by the expression *ḫsbḏ n sš*, the former by *ḫsbḏ mꜣꜥ*. For the greens, however, there are two quite distinct terms, *wꜣḏ* and *šsyt*, the one representing malachite, chrysocolla or verdigris, and also in some cases the frit, the other applied exclusively to the frit. A third term, *ḥmt*[1], evidently the name of a type of (green) frit, is also upon occasion used with reference to a green pigment.

The yellow pigments were two in number, namely yellow ochre and orpiment, the yellow sulphide of arsenic. Yellow ochre is represented by the same word *sty* that is used of ochreous earths in general, while orpiment is *ḳnἰt*, the two words not infrequently occurring together in parallel phrases.

The red pigments present a somewhat greater problem, since the lexicographical evidence does not altogether coincide with the archaeological information collected by Lucas. The one most widely used was undoubtedly red ochre, for which a number of names appear to exist, while red lead seems to have been employed to some extent at a late date. Of the latter, however, there is no apparent trace in the texts, while another red pigment, realgar, the red sulphide of arsenic, which is not known archaeologically, is clearly referred to in *ꜣwt-ἰb*.

mnšt, *ṯmḥy*, *dἰdἰ*, and probably also *prš* and *ṯr*, are all names for species of red ochre, while *mny*[2] is perhaps recognisable as ruddle, a preparation of the same substance. This multiplication of superficially similar terms is at first sight rather disturbing, but can be parallelled in Assyrian terminology[3], and in the accounts of red earths given by classical authors.

A detailed examination of references to varieties of red ochre by Theophrastus, Dioscorides and Pliny is not desirable at this juncture, but a few salient points may be mentioned[4]. In Theophrastus, red ochre is clearly μίλτος, which he states could be made by burning ὤχρα, and of which he mentions three varieties, one bright red, one pale, and one midway between the two[5]. He also refers to a red pigment cinnabar, which is, however, in all probability minium[6]. Dioscorides likewise mentions Sinopic μίλτος, and other species of the same substance which are of inferior quality[7], and cites an analogous form of haematite[8], while carefully distinguishing both cinnabar and minium, which are undoubtedly not ochres[9].

The treatment of red ochres by Pliny is somewhat more explicit, though he mentions several kinds which cannot now be identified. He draws a distinction between the red earth miltos, and minium, called cinnabar by the Greeks, both of which, however, are of a vivid

[1] Discussed elsewhere, cf. above p. 117–8.
[2] Discussed elsewhere, cf. below p. 172.
[3] Cf. Campbell Thompson, Dictionary p. 19.
[4] Cf. Bailey, Elder Pliny 1 p. 217; 2 p. 208f.
[5] § 52–3. μίλτος was probably a generic term for all pigments owing their colour to ferric oxide; cf. Caley & Richards, Theophrastus p. 172.
[6] § 58; cf. Caley & Richards, Theophrastus p. 193.
[7] 5, 96 (W).
[8] 5, 126 (W).
[9] 5, 94 (W).

red colour, and differ in that respect from the more subdued rubrica and sinopis[1]. Each of the latter has three varieties, rubrica being either Lemnian, Egyptian or African, while sinopis, which some regard as an inferior rubrica, is either bright red, dull red, or an intermediate shade, the African species being brighter than the rest[2].

It is quite evident from these passages, particularly the account of PLINY, that in classical times several varieties of red ochre, distinguished either by colour or provenance, were recognised, and it seems not unlikely that similar distinctions should have been observed by the Egyptians. In paintings, numerous shades of red and reddish brown can be noticed, and there is an obvious difference between the dull colour of male bodies and the brighter reds used elsewhere. It is therefore quite possible that there was more than one term for red ochre, and each of the words mentioned above may well refer to a particular species recognised by its colour or some other characteristic. That any one of them should represent minium seems most improbable, since all occur in texts long before that substance was apparently first used, though the archaeological evidence may be misleading on this point, as it is with regard to realgar. However, *prš* and *ṯr* present certain problems of interpretation which are not readily resolved.

Neither of the two remaining 'primary' colours of the Egyptian palette, white and black, is represented in any of the known pigment lists, and the evidence for each is largely inferential. White was normally either whiting or gypsum[3], but no word for these occurs in any context which might suggest its employment as a pigment. On the other hand, there is the possibility that *ḫrr*[4] refers to white lead, the use of which has not, however, been supported from examination of surviving pigments. Similarly, it seems just conceivable that *sšw* too conceals a white pigment, perhaps also ceruse, though this identification cannot be pressed[5].

Of black likewise there is no certain indication in the texts, but *ḏ'bt* does appear in a number of contexts where a translation of soot or black pigment seems possible, though in no case is interpretation as charcoal fuel ruled out. From the absence of both black and white from the pigment lists and ostraca, it seems likely that the Egyptians, though using them among their six 'primaries', did not regard them as pigments or colours in the same sense as the other minerals employed.

Lastly, the presence of *ḳmyt* in connection with the pigments would seem to suggest that a supply of gum, the vehicle with which the colours were applied, and without which they would not adhere, was regarded as an indispensable part of the Egyptian painter's equipment.

---

[1] HN. 33, 38, 115f.
[2] HN. 35, 13, 31f.; cf. also Isidorus 19, 17, 3 and Vitruvius 7, 7, 2.
[3] Cf. Theophrastus § 65; Pliny, HN. 35, 18, 36; Vitruvius 7, 7, 3.
[4] Discussed elsewhere, cf. below p. 178.
[5] Ebers' suggestion, Die Maasse p. 70, that *sty* is white lead carbonate may be disregarded.

CHAPTER 7

MISCELLANEOUS MINERALS

The foregoing chapters have been concerned almost exclusively with mineral substances which fall naturally into certain more or less clearly defined groups. Beside these, however, there are others which are not so readily classified, friable minerals and miscellaneous chemicals in no way related to the items treated above. For convenience these are collected together here, and with them are also discussed a number of other terms which clearly refer to minerals, but which cannot be identified more precisely. Among the latter are several words occurring in the mineral lists with the ▭ determinative, suggesting that they are perhaps building or monumental stones. Since, however, the same determinative is found with eg. *ỉbnw*, it is unwise to assume this, and such words are therefore included here rather than in chapter 4.

ỉꜣꜣ (Wb. I. 27.8) occurs only three times, and in the same text[1], once simply as *ỉꜣꜣ*, following *ꜣwt-ỉb*, and once as *ỉꜣꜣ šmꜥ* and *ỉꜣꜣ mḥ*, *ḥr sḥb mnḏty.k*. By its appearance in the Edfu list, *ỉꜣꜣ* must be a mineral, evidently of two varieties (cf. *wꜣḏ*, *bỉꜣ*, *sỉꜣ* etc.), and the contexts may perhaps suggest a pigment. By analogy with other words for green colour[2], the use of the verb *sḥb* might possibly point to green as the colour of *ỉꜣꜣ*, though this is the merest hypothesis. In fact, the nature of the mineral *ỉꜣꜣ* cannot satisfactorily be determined.

inḫꜥbt (no Wb. ref.) is apparently ἅπαξ in the Chester Beatty mineral list[3], where it occurs in a somewhat broken context in the phrase *inḫꜥbt n tꜣ sty*. The determinative and the Nubian provenance might suggest an ochreous earth, but no identification is possible.

ỉḫr (no Wb. ref.) is found only once[4], in the phrase *ỉḫr šmꜥ r ḫtm inb.k*, which would perhaps suggest a clay or a frit used for sealing, though the evidence is insufficient to admit of any conclusion.

ỉsmr (Wb. I. 132.18–19), Ptolemaic *smr* (Wb. IV. 139.11) is clearly a loan-word, and the fanciful etymology proposed by Gensler[5] may be disregarded. The Semitic origin of *ỉsmr* was first pointed out by Müller[6], who favours enamel as a translation, and so rejects any connection with Hebrew *ḥšml* or Assyrian *ešmaru* (sic.). Brugsch[7], however, compares Hebrew *šāmîr* and Greek σμυρίθης λίθος, regarding *ỉsmr* as emery, and this is the view taken by Wendel[8], Maspero[9], Bissing[10], Paton[11] and others, though alternative translations have also been proposed[12].

[1] Edfou VI, 202, 9; 203, 1.
[2] Cf. Appendix IV.
[3] P. Ch. B. IV vs. 7, 8.
[4] Edfou VI, 203, 4.
[5] ZÄS. 8 p. 148.
[6] Asien und Europa p. 306 n. 11.
[7] Wb. Erg. p. 146.
[8] Bau- und Edelsteine p. 120.
[9] Struggle p. 203 n. 6.
[10] Stat. Taf. p. 12.
[11] Records III, 2 p. 53 n. 73.
[12] Cf. Paton's list.

The translation of *ỉsmr*, like that of *ỉbnw* discussed below, depends not upon Egyptian evidence, but upon that from other languages, and it is the meaning of the cognate words that has first to be established. There can be no doubt that *ỉsmr*, *smr* is a loan-word derived from Sumerian ^{a}AŠ.MUR and Hebrew *šāmîr*, and connected with Greek σμίρις, σμυρίθης λίθος, or that the latter, through Latin smiris, is the origin of emery, émeril, Smirgel, smeriglio and other similar words. The obvious inference is that *ỉsmr* means emery, but, as will be shown below, archaeological evidence is against such a translation, and the range of meaning of the Sumerian, Hebrew and Greek words must be reconsidered.

The Sumerian ^{a}AŠ.MUR is probably the origin of the Hebrew and hence the Greek words, as well as of the Egyptian, and so must be investigated first. The meaning is not necessarily emery, and CAMPBELL THOMPSON[1] gives pumice-stone, powder of clay, powder of lead, corundum and emery as possible translations, pointing out that ^{a}AŠ.MUR is prescribed for eye diseases, where anything gritty is clearly inappropriate. Elsewhere[2] he translates ^{a}*algamišu* and ^{a}AŠ.MUR as corundum, emery or amethyst, so that even as an abrasive, ^{a}AŠ.MUR cannot be specified as emery alone. Further, there was confusion between ^{a}AŠ.MUR and ^{a}AS.ḪAR, an arsenic compound the precise meaning of which is not known[3], which also suggests that ^{a}AŠ.MUR was susceptible of various renderings.

Turning to *šāmîr*, it appears that the Hebrew word is equally capable of varied interpretation. It occurs in only a few instances[4], and the traditional translation is diamond. KÖHLER's Lexicon, however, renders emery, referring to HARRIS[5], and suggests a loan-word, comparing ^{a}AŠ.MUR. DRIVER[6] quotes the Jeremiah reference, and says, of a pen or stylus, 'with a point of diamond – or rather emery'. But if *šāmîr* is derived from ^{a}AŠ.MUR, corundum or amethyst would be equally applicable, and emery is not the only possible translation.

The same also applies to Greek σμύρις, σμιρίτης (σμυρίθης) λίθος, which translates *šāmîr* in the Septuagint. LIDDELL & SCOTT render 'emery-powder used by lapidaries', quoting DIOSCORIDES and HESYCHIUS as their authorities. But DIOSCORIDES merely says that σμύρις is a stone with which jewellers polish gems[7], and refers to σμίρις used by the lapidaries[8], while HESYCHIUS explains σμύρις as a form of sand by which the hard parts of stone are polished. To these descriptions may be added that of ISIDORUS[9] who comments 'smyris lapis asper et indomitus et omnia adterens, ex quo lapide gemmae teruntur'. While these passages make it clear that σμύρις was a form of corundum or a ferruginous sand, they do not necessarily suggest emery.

Thus it is evident that although emery may be a convenient translation of the words derived from ^{a}AŠ.MUR, it is not the only possible rendering, and certainly cannot be pressed for *ỉsmr* in the face of conflicting archaeological evidence.

ỉsmr is known to have been obtained from a number of sources. It is listed among tribute from Retenu[10], as part of a ship's cargo captured on the return from Djahi[11], and as a product of *tꜣ sty*[12], *bhn(t)*[13], and *kns(t)*[14], all regions of Nubia. Now whereas emery might have been imported from Syro-Palestine, there is no certain evidence of its having been found in Egypt or Nubia, or of its use by the Egyptians[15]. Sand at Aswan is stated by WAINWRIGHT[16] to

---

1 Dictionary p. 52. 2 ib. p. 167. 3 ib. p. 52.
4 Job 41, 15; Jer. 17, 1; Ezek. 3, 9; Zech. 7, 12.
5 Development of the Canaanite Dialects p. 33f.
6 Semitic Writing p. 84. 7 5, 147 (W). 8 ib. 9 16, 4, 27.
10 Urk. IV, 731; 744; cf. 1729. 11 Urk. IV, 686. 12 Philae ⟨1980⟩ Phot. 265.
13 Brugsch, Dict. géog. 1034; Wb. Erg. 146; Gauthier, Dict. géog. II p. 26.
14 Brugsch, Wb. Erg. 146; Gauthier, Dict. géog. V p. 205.
15 Lucas, Materials p. 294. 16 Balabish p. 38.

contain 15 per cent. of emery, and BRUGSCH[1] quotes QUATREMÈRE (Mém. géog. II) for the observation that the Arab geographers give Lower Nubia as the source of 'l'émeril qui sert à polir les pierres précieuses', but neither of these statements is confirmed. PLINY[2] calls the sand-covered region of Ethiopia Zmiris, which seems to connect smiris with Nubia, but since smiris may have a wider meaning than emery, the reference is of no great importance. In fact, there is no certain evidence for emery in Egypt or its immediate environs, and LUCAS may be right in doubting its occurrence.

Nor is the evidence for its use by the ancient Egyptians at all convincing. Various small objects stated to be of emery have proved on more careful examination to be ferruginous sandstone[3], and LUCAS thinks it highly unlikely that emery was used as an abrasive in cutting hard stones[4], for which quartz sand would have been adequate. Indeed, emery has not been found at all in connection with Egypt, and it seems most improbable that it was ever used.

The meaning of *ỉsmr* is therefore placed in doubt, and the contexts are of no assistance in this respect. The derivation from Sumerian ^{a}AŠ.MUR is confirmed by a somewhat tentative phonetic writing[5], and from the Assyrian evidence some abrasive seems likely. Emery is a possible translation of *ỉsmr* brought as tribute from Retenu, but some other rendering must be adopted for the Ptolemaic references to *smr* as a product of Nubia. In these later examples, *ỉsmr* must represent some form of corundum, or a ferruginous or quartz sand, perhaps a native substitute for the rarer product brought from abroad. No single translation is entirely satisfactory, though corundum or ferruginous sand are to be preferred to the more specific emery. Even if, however, the latter is accepted in the Urkunden examples, it cannot be taken as evidence that emery was used as an abrasive, since *ỉsmr* was evidently valued sufficiently to be taken as tribute, and would scarcely have been put to such a use.

ỉkbw (no Wb. ref.) is known from only two occurrences on the same ostracon[6], where it is included in what are evidently lists of pigments. In one case the quantity is given as 'one large *kp* vessel', which would seem to indicate a soft pigment of some description, and the writing of the word itself might suggest a vox peregrina. Since the ostracon can be dated to the latter half of Dynasty 18, before the sulphides of arsenic were in common use, and since neither *ḳnỉt* nor *ꜣwt-ỉb* occurs in the lists, it is just possible that *ỉkbw* is a loan word representing either or both of these pigments. But no Semitic original suggests itself, and in the absence of more definite evidence no certain conclusion can be drawn, though it is extremely probable that *ỉkbw* is a mineral pigment.

ỉdbw (Wb. I. 153.11) is evidently a mineral, probably a semi-precious stone, the nature of which cannot be determined. It is found parallel to *mꜣṯ*[7], and as the material for the two eyes of a figure[8], in a context very similar to that in which the ἅπαξ *wꜣḏwꜣḏt* (sic.) occurs. The latter example might perhaps suggest rock crystal or the like, though this cannot be insisted upon in view of the probable meaning of the parallel *wꜣḏwꜣḏt* (sic.), discussed above (p. 104–5). The word may be connected with the common *ỉdb*, 'bank', (Wb. I. 153.2–10), perhaps by some Ptolemaic word-play, but the meaning is not apparent, and not even the colour can be deduced.

[1] Wb. Erg. p. 146.
[2] HN. 36, 25, 129; cf. ib. 36, 9.
[3] Lucas, Materials p. 294–5.
[4] ib. p. 91–2; p. 295.
[5] P. Tur. PR. 66–7, 11.
[6] Ostr. MMA. 90.6.3. rt. 3; vs. 3.
[7] Dend. Mar. IV, 38, 127.
[8] Dend. Mar. IV, 39, 146.

wꜣy (no Wb. ref.) appears to be confined to one example in the Edfu mineral list[1], in the phrase *wꜣy nfr m snm*, the significance of which is obscure since *snm* too is otherwise unknown. *wꜣy* may possibly be a mistake for *ḫt-ꜥwꜣ(y)*, with one group, 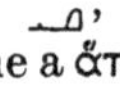, omitted, but emendation is not altogether satisfactory, and it is perhaps better to assume a ἅπαξ λεγόμενον.

wꜣḫ (no Wb. ref.) occurs only once, in the Chester Beatty mineral list[2], in a broken context from which nothing can be gleaned. The determinative suggests a stone, but no conclusion is possible[3].

wbḫt (no Wb. ref.) is found only in the Chester Beatty list[4], as *wbḫt n ḫtꜣ*, following *ꜥꜣt ḥḏt*. The determinative suggests a stone, and connection with *wbḫ*, 'hell sein' (Wb. I. 295.12–20) and other similar words might point to a stone of white colour, or a natural glass. The juxtaposition with *ꜥꜣt ḥḏt* may be significant, and *wbḫt* is perhaps a mineral of similar nature obtained exclusively from Asia Minor (Ḫatti).

wšk (no Wb. ref.) is also known from a single reference in the Chester Beatty list[5], the context being, unfortunately, broken. *inr nmḥ* precedes a lacuna, which appears to be followed by *m-ꜥ wšk m [sm]dt*, and this would suggest that *wšk* is an otherwise unknown precious stone. Alternatively, *mꜥ* may be part of the name, the first group being lost in the lacuna.

biꜣ (Wb. I. 438.8–11) is undoubtedly the same word as that used to denote iron. It may, however, conveniently be separated here, since it is clear that while in some cases *biꜣ* refers to iron, in others it cannot, and a distinct but related meaning must be assumed. But in dividing up the references to *biꜣ* in this way, there arises the difficulty that some are quite equivocal, and may be regarded as examples of either use, since they spring from the primary connection of *biꜣ* with heaven.

Nevertheless, it is quite evident that whereas *biꜣ* represents iron as far as the *nwꜣ*, *msḫtyw* and *mḏtft* of *biꜣ* are concerned, and probably also in the case of certain weapons of *biꜣ* occurring in late religious texts, the *biꜣ* which is found with some frequency in offering lists as the material of the two blocks used in 'opening the mouth'[6], and of the *ḥnt* vessels[7], must bear a somewhat different meaning. Nor does iron seem applicable to the *ḏbt* of *biꜣ* which are occasionally mentioned[8], or to the *biꜣ ḳsy* of Ebers and elsewhere[9], while references to fingers of *biꜣ*[10], *biꜣ* which opens the mouth[11], ropes of *biꜣ* by which the ascent to heaven was made[12], and *biꜣ* as the material of walls[13], of the heavenly throne etc.[14], are of doubtful interpretation.

[1] Edfou VI, 203, 2.
[2] P. Ch. B. IV vs. 8, 2.
[3] Perhaps compare *wꜣḫ nḫbt* (Wb. I. 257.12; II. 293.13).
[4] P. Ch. B. IV vs. 7, 7.
[5] P. Ch. B. IV vs. 7, 13.
[6] Amenemhet pl. 21; D. el B. 110; Beni Has. I, 17; I, 35; Mem. Miss. V pl. 8; Pyr. 30b etc.
[7] Pyr. 40a; Dendereh 4; Beni Has. I, 17; I, 35 etc.; cf. Selim Hassan, Giza VI, 2 p. 187.
[8] Totb. Nav. 65, 52; 183, 14; 148, 16 etc.
[9] Eb. 48, 19; 59, 18; Totb. Leps. 25, 31; cf. Kenam. 44; P. Ch. B. IV vs. 7, 10.
[10] Pyr. 1983e.
[11] Pyr. 14; Petos. 82, 71; cf. also Petos. 82, 34; Lib. Fun. I p. 103.
[12] Cairo 28092; Pyr. 138c; Coff. Texts I, 271c; Rec. Trav. 29, 149; Text. Relig. 22.
[13] Totb. Nav. 109, 7; 149,7.
[14] Pyr. 1992c; Pyr. 1562b–c; Totb. 153A Nu 20, 13.

The undoubted connection of *bỉ3* with the heavenly vault, and with the meteorites that came from there, has been examined in great detail by WAINWRIGHT[1], who concludes that it was principally iron, but could refer to meteoric material in general. Now this latter meaning seems quite applicable to some of the examples of *bỉ3* to be discussed here, and it may be suggested that the Egyptian conception of meteoric material was not very exact, and that *bỉ3* therefore included other analogous materials, particularly iron ores, the nature of which was not properly understood.

That such confusion might have existed is not unlikely, and may perhaps be inferred from the absence, until the New Kingdom at least, of any precise terms for haematite and magnetite, and from the substitution of *bỉ3 ḳsy* for *bỉ3 šmʿ* in a passage in the 'Book of the Dead'[2]. Moreover, the mention of *bỉ3* as the material for a figure of Tewosret[3] is unlikely to refer to iron, though among those mentioned by PETRIE[4] are three made of haematite, which may be significant. PETRIE himself thinks that *bỉ3* included haematite[5] and notes that it was almost two-thirds the weight of iron. In its natural state, haematite, with its metallic lustre, is particularly like iron, and specular haematite even more so.

Haematite, magnetite, and other iron ores were universally confused in antiquity, and DIOSCORIDES[6] lumps together earthy haematite, the hard black variety, and magnetite. PLINY[7] cites SOTACUS' classification of magnets into five varieties, some with magnetic power, and some without, and while the majority are undoubtedly magnetite, one at least, the haematites magnet, is not. Whether this uncertainty extended to meteoric materials is doubtful, though there was certainly considerable ignorance concerning the true nature of the latter.

Speaking of ceraunia, PLINY[8] mentions varieties resembling axes, a black kind called baetuli, and another species which is found only when a thunderbolt has struck; he also refers to ombria or notia, which like ceraunia and brontea, fall out of the heavens[9]. Commenting on this, BALL[10] points out that the baetuli were evidently meteorites, but that some cerauniae at least were in all probability nothing more than prehistoric arrows and axes, which many primitive peoples consider to be of supernatural origin, and connected with lightning, a belief which was current until the beginning of the 17th century. The same point is made by COOK[11], who says that other things besides meteorites might be reckoned as thunderbolts, eg. flint implements[12].

That such a belief was ever held in Egypt, cannot, of course, be proved, but it may be significant that the blocks of *bỉ3* superseded the *psš-k3f* as the means of 'opening the mouth', and it could be that the Egyptians regarded certain prehistoric artifacts as of meteoric origin, and used them for a purpose to which their shape and supposed nature was suited[13].

In short, it seems that the meanings of *bỉ3* cannot satisfactorily be established, but there is some evidence to suggest that it was originally a word of fairly general meaning, referring to substances which the Egyptians thought were of meteoric origin. It thus included iron, and meteoric agglomerate, which was, however, rare, and perhaps also other minerals which

---

[1] JEA. 18 p. 3f.
[2] Totb. Nav. 64 Nach. 4 Aa.
[3] Edfou VI, 299.
[4] Amulets p. 47.
[5] Anc. Eg. 1917 p. 47.
[6] 5, 126 (W).
[7] HN. 36, 25, 126; cf. Isidorus 16, 4, 1.
[8] HN. 37, 51, 134.
[9] HN. 37, 65, 176.
[10] Roman Book p. 297f.
[11] Zeus III p. 885; cf. ib. II p. 505; p. 643.
[12] For an interesting example of this mistake cf. Ann. Serv. 22 p. 157f.
[13] It is not suggested that all prehistoric implements were minsunderstood in this way, since cer tain of them, eg. ripple-flaked knives, would easily be recognised as being of human manufacture.

were falsely identified with them. Among these would be magnetite and the haematite of metallic appearance, and perhaps also other iron ores, which were later recognised as distinct materials and re-named accordingly. It seems likely also that the hippopotamus bones from Qau, 'mineralised, heavy and black, and presenting a metallic lustre and appearance', were thought to be of meteoric origin or were in some way confused with *bi͗ꜣ*[1]. Whether certain worked flints were also thought of as *bi͗ꜣ* one cannot tell, but the suggestion may be offered for what it is worth.

Lastly, a word must be said concerning the use of the epithets *mḥ* and *šmꜥ* in connection with *bi͗ꜣ* in those cases where it refers to the two blocks, or to the *ḫnt* vessels. In other instances, eg. with *nṯry* and *wꜣḏ*, these seem to have geographical significance, but if *bi͗ꜣ* were simply meteoric material, such an interpretation would have no validity. Nor is it very satisfactory to dismiss them as ritual repetitions, one for Lower Egypt, the other for Upper Egypt, though this is, of course, possible. It is interesting that the one example of *bi͗ꜣ šmꜥ* outside these contexts occurs in the 'Book of the Dead'[2] in connection with a brick, which is elsewhere described as of *bi͗ꜣ ḳsy*[3]. It is a possible inference that the two are identical, and this would support the interpretation of *bi͗ꜣ* here proposed. Iron ores occur in Sinai and in the eastern desert, and others, including haematite, at Aswan, and it may be to these deposits that *bi͗ꜣ mḥ* and *bi͗ꜣ šmꜥ* refer, though the point cannot be pressed.

br (no Wb. ref.) is known from only two examples, one in the Chester Beatty mineral list[4], which indicates its general nature, the other on an ostracon from Deir el Medineh[5], where it occurs with *sty bꜣk*, which might conceivably suggest a pigment. The writing in the latter instance seems to point to a vox peregrina, but no Semitic original suggests itself. One may perhaps compare *br* (Wb. I. 465.7), used for filling the sound eye, though this appears doubtful, and no worthwhile conclusion is possible.

brt (no Wb. ref.) is apparently confined to the Chester Beatty list[6], where it is described as *brt n i͗sr*. Both the writing and the provenance, evidently Phoenicia[7], point to a foreign word, and it is tempting to compare Akkadian ^{a}*birûti*, which may be a name for turquoise[8], though it only occurs once and cannot certainly be identified.

bḳs-ꜥnḫ (Wb. I. 480.6–7; V. 18.1) constitutes one of the most involved problems of Egyptian mineralogical lexicography, since the word is known under at least five different forms all of which, however, seem to be related. The earliest, and the most explicit, is *bi͗ꜣ n ḳst*[9] or *bi͗ꜣ ḳsy*[10], literally *bi͗ꜣ* of Cusae, which directly connects the word with *bi͗ꜣ* discussed above, and is apparently parallel to *bi͗ꜣ šmꜥ*[11]. From this appear to develop the later *bḳs*[12] and *bḳs-ꜥnḫ*[13],

[1] Cf. JEA. 18 p. 3f.
[2] Totb. Nav. 64, 52; cf. 148, 16.
[3] Totb. Leps. 64, 31 (pl. 25).
[4] P. Ch. B. IV vs. 7, 13.
[5] Ostr. DM. 183, 6.
[6] P. Ch. B. IV vs. 7, 8.
[7] Gauthier, Dict. géog. I p. 105.
[8] Campbell Thompson, Dictionary p. 134.
[9] P. Ch. B. IV vs. 7, 10.
[10] Eb. 48, 19; 59, 18; Totb. Leps. 64, 31; perhaps also Kenam. 44 (damaged).
[11] Totb. Leps. 64, 31 = Totb. Nav. 64, 52 Aa.
[12] Philae ⟨1991⟩ Phot. 241; ⟨1986⟩ Phot. 245 (twice); Brugsch, Dict. géog. p. 1033.
[13] Dend. Chass. IV, 177, 2; Edfou VIII, 136; Philae ⟨1664⟩ Phot. 523; Düm. Baug. 29.

the latter once written *bỉ3 ḳsy-ʿnḫ*[1], perhaps archaistically. But the *b* is also omitted, and both *ḳs*[2] and *ḳs-ʿnḫ*[3] are known.

At first sight this multiplicity of forms might suggest that more than one word is involved, and that the WÖRTERBUCH's separation of *bḳs* and *ḳs-ʿnḫ* is justified, but closer examination points to their being only one name, variously written. The most important fact is that at Philae, *bḳs*[4] and *ḳs*[5] are both brought by the same region in similar lists of tribute, and may therefore be assumed to be the same. Further, two of the variant spellings, *bḳs* and *ḳs-ʿnḫ*, are confined to particular temples, Philae and Dendera respectively, while only *bḳs-ʿnḫ* occurs at Dendera, Edfu and Philae, and that, apart from one reference, in parallel texts of the filling of the sound eye. There is thus a certain system in the Ptolemaic variants, and there can be little doubt that they are all to be identified. Nor does it seem probable that the earlier examples are to be separated, particularly in view of the form *bỉ3 ḳs(y)-ʿnḫ* in the Sehel inscription, which has the appearance of an attempt to write the current *bḳs-ʿnḫ* in a way which would indicate its origin in *bỉ3 ḳsy*. *bḳs-ʿnḫ* is therefore a substance connected with *bỉ3*, first distinguished in the New Kingdom, and fairly common in the Graeco-Roman period.

Previous discussion of the question has followed two main lines of thought, either that *ḳs* is a writing of *šs*, and that *ḳs-ʿnḫ* is therefore alabaster, or that *bḳs* is connected with *bỉ3*, and that something akin to iron, eg. magnetite, is meant. No treatment has endeavoured to reconcile all the references, and in these circumstances, the WÖRTERBUCH decision to recognise two distinct words is perhaps understandable.

The identification of *ḳs* with *šs* seems to have originated with PLEYTE[6], who regarded *ḏbt nt bỉ3 ḳsy* as 'une pierre taillée des rochers de Ksu', and took *bḳs* to be *bỉ3 ḳs*, *bỉ3* meaning 'rocher', and *ḳs* being Coptic, ϭⲱⲥ, ⲕⲱⲥ alabaster, also known as *šs*[7]. The compound *ḳs-ʿnḫ*, however, he was unable to explain. PLEYTE's view is accepted by BRUGSCH[8], who reads *ḳs-ʿnḫ* as *šs-ʿnḫ*, and translates 'der lebendige (sic.) Alabaster', and by LÜRING[9], who rejects STERN's 'iron', and regards the writing of *ḳsy* as the name of Cusae as a secondary refinement. *ḏbt nt bỉ3 ḳsy* is simply a block of alabaster, and LÜRING refers to a stele of alabaster in Hamburg which is inscribed in blue.

On the other hand, STERN took *bỉ3 ḳsy* to be iron, and this lead is followed by DÉVÉRIA[10], who denies that *ḳs* and *šs* are necessarily to be identified, and points out the impossibility of a translation 'fer d'albâtre', 'fer alabastrite'. He suggests that *bỉ3 ḳs* is the same as DIODORUS' 'Ethiopian stone', i.e. 'quelque matière ferrugineuse, comme la sidéritine ou fer arseniaté dont l'éclat est résineux; la sidérose ou fer spathique; l'hématite ou sanguine, qui pourrait très-bien porter un texte tracé en bleu'. Similarly, EBBELL[11] and FORBES[12] are in favour of some connection with iron, and both translate magnetite.

Considering first the proposed translation alabaster, it is to be noted that the reading *ḳs* for ȣ has now been destroyed by GARDINER[13], so that any direct equation must be abandoned. Moreover, none of the known examples of *bḳs-ʿnḫ* is particularly suggestive of alabaster, since the majority group it with earthy minerals[14] or semi-precious stones[15], and only Nubian

[1] Sehel 17. [2] Düm. G. I. IV, 180; Thes. 19 A'12 (Dend.).
[3] Dend. Mar. IV, 87; IV, 36, 49–50; IV, 39, 140–2.
[4] Düm. G. I. IV, 188 = Philae ⟨1986⟩ Phot. 245. [5] Düm. G. I. IV, 180.
[6] ZÄS. IV p. 17. [7] Brugsch, Recueil I p. 23. [8] Hungersnoth p. 131; Wb. p. 1474.
[9] Med. Kent. p. 96. [10] Mél. Arch. 1 p. 10. [11] Pap. Ebers p. 132.
[12] Metallurgy p. 430. [13] BIFAO. 30 p. 161.
[14] Dend. Mar. IV, 36, 49–50; 39, 140–2; Sehel 17; Eb. 48, 19; 59, 18.
[15] Philae ⟨1991⟩ Phot. 241; ⟨1664⟩ Phot. 523; Edfou VIII, 136; Dend. Chass. IV, 177, 2; Dend. Mar. IV, 87; Thes. 19 A'12.

sources are given[1]. In the one instance where objects of *bḳs-ʿnḫ* are mentioned[2], no correspondence with the use of alabaster can be observed. Indeed, of all the specimens of each of the sixteen types of amulet collected by PETRIE, only three are of alabaster, ie. 2 of 134 *wḏȝt* eyes, and 1 of 193 hearts, which is not altogether convincing. Lastly, it is difficult to see what connection there could be between alabaster and *biȝ šmʿ*[3] or *biȝ n pt*[4], to both of which *biȝ ḳsy* is evidently related.

The second suggestion, connecting *bḳs-ʿnḫ* with iron is far more plausible, though whether the precise translation magnetite is correct may perhaps be doubted. Magnetite is the naturally magnetic oxide of iron, ferroso-ferric oxide (Fe_3O_4), usually iron black in colour with a metallic lustre, though the appearance varies somewhat. In classical times it was to some extent confused with haematite[5], though μάγνης ζῶν and ferrum vivum[6] seem to refer specifically to magnetite. LUCAS makes no mention of magnetite as a natural product of Egypt, and only once refers to its use[7], but PLINY[8] cites one variety of magnes, the best, which comes from Ethiopia in the region called Zmiris, and adds that the haematites magnes was found in the same area.

Now various Nubian sources are given for *bḳs-ʿnḫ*[9], which is evidently either a mineral ore[10] or a semi-precious stone[11], and is clearly connected in some way with both *biȝ šmʿ*[12] and *biȝ n pt*[13]. This would suit magnetite well, and the epithet *ʿnḫ* would be particularly apposite as referring to the apparently live nature of the mineral in attracting iron[14].

On the other hand, magnetite does not occur at all among the materials cited by PETRIE for the various amulets of *ḳs-ʿnḫ* listed at Dendera[15], though haematite is known for three, the scarab, the *wḏȝt*, and the hawk. Further, *ḳs-ʿnḫ* occurs in the Demotic magical papyrus[16], where, however, magnetite is represented by the Greek μάγνης, written *mʿknis*, *mʿnis* and *mʿknisyʿ*, and evidently to some extent confused with haematite[17].

It is thus difficult to accept the translation magnetite in all cases, though the undoubted connection with *biȝ* would seem to suggest that *bḳs-ʿnḫ* was in fact a species of iron ore[18]. It is perhaps possible that the word has a somewhat wider meaning, and, like the classical magnes, includes certain varieties of haematite with a metallic appearance, though this would not altogether explain the inconsistency in the Demotic papyrus, and would conflict with the apparent existence of a Ptolemaic term for haematite, *ḳʿ*, discussed above (p. 133). No very satisfactory conclusion can be reached, but it seems reasonable to assume that *bḳs-ʿnḫ* was an iron ore of attractive appearance, regarded as a semi-precious stone in the Greek period.

---

1 Philae ⟨1986⟩ Phot. 245; Brugsch, Dict. géog. p. 1033.
2 Dend. Mar. IV, 87.
3 Totb. Nav. 64, 52 Aa = Totb. Leps. 64, 31.
4 P. Ch. B. IV vs. 7, 10.
5 Cf. Dioscorides 5, 126 (W); Pliny, HN. 36, 25, 126; Isidorus 16, 4, 1; P. Dem. Mag. vs. 2, 7f.
6 Cf. Pliny, HN. 34, 42, 147.
7 Materials p. 100 — as an eye paint.
8 HN. 36, 25, 128.
9 Philae ⟨1986⟩ Phot. 245; Brugsch, Dict. géog. p. 1033.
10 Dend. Mar. IV, 36, 49–50; 39, 140–2; Sehel 17 etc.
11 Edfou VIII, 136; Dend. Chass. IV, 177, 2; Philae ⟨1664⟩ Phot. 523; ⟨1991⟩ Phot. 241; Dend. Mar. IV, 87.
12 Cf. note 3 above.
13 P. Ch. B. IV vs. 7, 10.
14 Cf. Demotic *mʿkns nt ʿnḫ* (= μάγνης ζῶν) P. Dem. Mag. vs. 2, 11.
15 Dend Mar. IV, 87.
16 P. Dem. Mag. rt. 4, 6; 10, 12 (*wʿ ȝny n ḳs-ʿnḫ*); 10, 32; 27, 10; vs. 22, 4.
17 ib. vs. 2, 7f.
18 Cf. also the expression *bḳs ḥr mn.f*, Junker, Große Pylon p. 269, 5.

bk (no Wb. ref.) occurs only in the Chester Beatty mineral list[1], described as *bk n imntt*, from which nothing can be deduced. It is possible that the Toronto variant[2] may have had an alternative writing, but even this is doubtful.

mꜣd (Wb. II. 35.1–2) is apparently known only from the Ebers papyrus, where both Stern and Joachim identify it with *mꜣṯ*, granite[3]. This seems unlikely, particularly in the compound *mꜣd n ꜥnḫ*, and Ebbell[4] regards it as a distinct word, translating 'pumice'. The latter has been found in Egypt, where it occurs native in small quantities[5], and Pliny[6] mentions therapeutic uses of pumice for eye ailments, ulcers and emollient plasters. The contexts in which *mꜣd* occurs are of little assistance in establishing the meaning. In one prescription[7] it is used with two other mineral substances and a liquid to dispel the marks of a beating, and in another[8] is finely ground and mixed with *ꜥntyw* to counter an obscure ailment. In a third[9], *mꜣd n ꜥnḫ* is pounded together with a large number of other ingredients, and employed to prevent suffusion of blood. From the orthography of *ꜥnḫ* it seems that the word for mirror is meant, and 'pumice of the mirror' might therefore refer to the use of pumice as an abrasive to polish the face of a mirror and remove verdigris. Of this particular use there is no confirmation, but the abrasive properties of pumice were well known in the ancient world, and it is probable that the lumps of pumice found in Egypt were for smoothing or abrading. However, the evidence for *mꜣd* is so slight that a translation 'pumice' must be purely conjectural.

mn (Wb. II. 68.1) is known to the Wörterbuch from only two examples, to which others may now be added, without, however, providing any very clear indication of the nature of the substance. Of the two original references, one[10] suggests that *mn* was of a dark colour, though not necessarily black as the Wörterbuch has it, while the other[11] reveals that it was brought as tribute from Syria. In the tomb of Puyemrēꜥ, what is evidently the same material, though written *mni*[12] and *imn*[13], is imported from Pwenet and a region in Nubia, no colour indication being given in the publication, though the contents of the baskets appear still to be visible. Elsewhere[14], *mn* seems to follow *ḫsbḏ* and *mfkꜣt* in a list of temple donations, though the reference is included under *mnw* in the Wörterbuch, and may even be a mistake for *mḫn(m)t*. A further apparent example of *mn*[15] may be disregarded, as it is clearly a mistake for *šꜥy*[16].

Discussion of *mn* is somewhat complicated by the existence of a number of similarly written words, *mnit*, *mny*, and *mnw*, all apparently minerals, and only one identified with any degree of certainty[17]. Equation of *mn* with *mnw* seems unlikely, but it is quite possible that either *mnit* or *mny* is to be assimilated with it, particularly in view of the writing *mni* of Puyemrēꜥ. In the one case *mn* would then be a semi-precious stone or even a resin, and in the other perhaps a form of red ochre.

---

[1] P. Ch. B. IV vs. 7, 10. [2] Ostr. Tor. A. 4.
[3] Ebers, Die Maasse p. 39, suggests haematite, which is hardly possible.
[4] Pap. Ebers p. 132. [5] Lucas, Materials p. 91. [6] HN. 36, 42, 155.
[7] Eb. 69, 20. [8] Eb. 72, 3. [9] Eb. 88, 17.
[10] LD. III, 163. [11] Urk. IV, 1101. [12] Puyem. 32. [13] Puyem. 43.
[14] Karn. R. III. pl. 90. [15] Düm. H. I. I, 8, 16.
[16] Cf. Möller, Paläographie nos. 540 and 335, and 331 and 99.
[17] Chassinat, Man. Mag. Copte p. 66, regards *mn* as a writing of *mnnn*; cf. below p. 173.

From the few certain examples of *mn*, however, all that can be gathered is that it was a substance of a dark colour, obtained both from Syria and regions to the south of Egypt. A mineral seems likely from the depiction of *mn* as a mass of lumps in a basket, and if the Karnak reference is correct, it may further be identified as a semi-precious stone. The writing is, however, doubtful, and although it is probably not *mnw*, it may be an obscure corruption of *mḫn(m)t*, which occurs in two similar lists, also at Karnak[1]. Altogether, no very clear appreciation of *mn* can be made, and although it would appear to be a mineral substance, it might equally well be a resin or the like[2].

mnỉt (Wb. II. 76.9–10), if it is not a mere variant of *mn* discussed immediately above, occurs only three times. In the Harris papyrus[3] it is found, with *wꜣḏ*, as the material for a statuette of the Nile god, and may perhaps be a semi-precious stone, and this could also apply to *mnỉt* as the label beside a necklet[4], though the latter might be the name of the object rather than its material. A third example of the writing *mnỉt* appears in a context[5] which would suggest that it was a species of resin, and therefore possibly a different word. But Lucas[6] cites various instances of the use of resin for beads and other small objects, and in view of this, it may be possible to relate the references, and regard *mnỉt* as a species of resinous material occasionally employed in the same manner as the precious stones.

mny (Wb. II. 76.13) was originally known from only one example[7] which was thought by Wreszinski and others to be a mistake. Lüring[8] took it to be a writing of *mnnn*, which he regarded as 'eines der aromatischen Harze', and Ebbell[9] evidently favoured a similar identification, translating as bitumen. Iversen[10] remarks that such a translation seems very doubtful, but proposes no alternative, and in fact, the Ebers reference alone is of no significance.

Two more recently published examples of the same word are, however, of greater interest. In one, a love-song[11], the lover compares his position to that of a (domesticated) ox who is easily caught and branded by his mistress. Gardiner[12] translates as follows: 'She casteth a noose (?) upon me with her hair; she will catch me with her eye; she will subjugate (me) with her ruddle (?); that she may brand (me) with her seal'. In the other[13], an obscure context, reference is made to 'stone cartouches, like *mny* upon the flesh of the gods', and Caminos[14] accepts Gardiner's 'ruddle' as a probable translation. Now both examples are apparently concerned with marking the flesh, and ruddle, a species of red ochre used for marking sheep and cattle is quite appropriate for *mny*. Klebs[15] discussing the marking of cattle, suggests that royalty were perhaps marked in some way to show their divinity[16],

---

[1] Karn. I, 27 = Urk. IV, 1668; LD. III, 127b.

[2] This might be confirmed if *mny* of Eb. 54, 19 were to be taken as an example of this word; cf. Grapow, Grundriß IV (1) p. 166.

[3] Harr. 40b, 15. [4] Berl. 45. [5] Amonhymn. Leid. 1, 8. [6] Materials p. 444.

[7] Eb. 54, 19, regarded as a species of resin by Grapow, Grundriß IV (1) p. 166. Chassinat, Rec. Champ. p. 463 n. 8; Man. Mag. Copte p. 66, takes this example and Eb. 68, 10 as *mnnn*, though the latter is *mnḥ*

[8] Med. Kent. p. 29. [9] Pap. Ebers p. 132. [10] Paints and Pigments p. 40.

[11] P. Ch. B. I rt. 17, 13. [12] P. Ch. B. I p. 37.

[13] Lit. Frag. 12, 2. [14] Lit. Frag. p. 34. [15] Reliefs NR. p. 69.

[16] The question of such marks is discussed at length by Keimer, Remarques sur le Tatouage, Mem. Inst. Ég. 53 pp. 47–53; 64–72, where the views of many other authors are cited.

and this might find support in the reference to 'cartouches like ruddle upon the flesh of the gods', though the point cannot be pressed.

In default, therefore, of further evidence, *mny* may be regarded as ruddle, a preparation of red ochre, used not as a pigment, but in marking the hides of animals.

mnnn (Wb. II. 82.9–14) occurs principally in the embalming ritual, where *ḏ3hy*[1], *dpwr*[2], Coptos[3] and Pwenet[4] are mentioned as sources, and in the texts relating to the making of the nine oils at Edfu[5]. LÜRING[6] takes it to be 'eines der aromatischen Harze', and JÉQUIER[7], referring to the Edfu examples, regards it as 'probablement ... une huile ou une autre substance grasse'. LORET[8], on the other hand, identifies *mnnn* as bitumen of Judaea, the Coptic ⲁⲙⲣⲏϩⲉ, comparing the sources of bitumen given by DIODORUS, STRABO, DIOSCORIDES and HERODOTUS with those of *mnnn* in the embalming ritual. He points out that it was solid and could be crushed, and that it became liquid or nearly so on heating, which would suit bitumen, and further identifies *mnnn* with Coptic ⲙⲓⲟⲗⲱⲛ, translated by the Arabic for 'mummy' in the Scalae[9]. A similar view is also taken by CHASSINAT[10], who discusses *mnnn* in detail, reviewing LORET's arguments, and casting doubt on the identification with ⲙⲓⲟⲗⲱⲛ[11]. He cites passages to show that it was black[12], that it was solid but could be powdered, or liquefied by heat, and that it was used in mummification, concluding that there were two species of *mnnn*, one the natural product pissasphalt, the other an artificial preparation of pissasphalt and other substances[13].

Now the treatment of the evidence by LORET and CHASSINAT is exhaustive, and there can be little doubt that *mnnn* was a material used in embalming and thought to be bitumen by the Copts and Arabs. That it was in fact bitumen is, however, unlikely, as has been shown by LUCAS[14], who discusses the question in some detail. Classical and modern authors alike have been of the opinion that bitumen was among the common embalming materials, but recent work by RUFFER, DAWSON, and LUCAS himself has shown that this was not so before the Graeco-Roman period, and that the black resinous substances from various earlier mummies are not bitumen. LUCAS specifically mentions the passage in the Rhind papyri[15], and remarks that it is more likely to refer to resin.

There is thus good reason to suppose that *mnnn* is not bitumen but some other material of similar appearance, probably wood tar or wood pitch, though a coniferous oil or resin is also possible[16]. Such an identification would accord with all that is known from the texts, and would be particularly apposite to the Edfu references, where *mnnn* is a constituent of various oils.

mrḥ (Wb. II. 111.13) is separated from *mrḥt* (Wb. II. 111.1–10) in the WÖRTERBUCH, though the two are treated together by CHASSINAT[17] and may well be the same word. The

---

[1] Emb. Rit. 3, 1; cf. also Q. Saqq. III, 90, 7.
[2] Emb. Rit. 3, 7.
[3] ib. 3, 6; 7, 7.
[4] ib. 6, 16; cf. also Harr. 33b, 12; 70b, 14.
[5] Edfou II, 210, 3–6; cf. also II, 226, 7–8 etc.
[6] Med. Kent. p. 29.
[7] Frises p. 147; cf. also Selim Hassan, Giza VI, 2 p. 250.
[8] Rec. Trav. 16 p. 157.
[9] The change of *n* to ⲗ can be parallelled in the case of *mnḥ* and ⲙⲟⲩⲗⲁϩ.
[10] Man. Mag. Copte p. 65; cf. also Rec. Champ. p. 463.
[11] But cf. Daumas' note, Man. Mag. Copte p. 111.
[12] Assuming that *mn*, discussed above p. 171, is a writing of *mnnn*, which seems improbable.
[13] Cf. also Forbes, Mnemosyne 3e Ser. 4 (1936) p. 67; Ambix 2 p. 69.
[14] Materials p. 384f.
[15] P. Rhind I, 3, 9.
[16] Cf. Lucas, Materials p. 376; p. 355.
[17] Rec. Champ. p. 463.

distinction is, however, convenient, since it is evident that in some few instances *mrḥ(t)* has a more specific meaning, akin to that of *mnnn*, and is the origin of the Coptic ⲁⲙⲣⲏϩⲉ as pointed out by Loret[1]. This latter connection might suggest that it refers to a species of bitumen or asphalt, and this is to some extent confirmed by the term *mrḥ(t) ḫ3st*[2], which Chassinat[3] takes as an indication that *mrḥ(t)* was a mineral substance. On the other hand, the parallelism between *mrḥ* and *mnnn* in the Rhind papyri[4] and the occurrence of the expression *mrḥ(t) n ḫt*[5] would rather indicate wood tar or resin, which is also more appropriate in other instances. Thus *mrḥ* is used for covering or coating coffins[6], which points to either a resinous varnish or the black anointing material discussed by Lucas[7], and is shown being daubed on the (?) bandages of a mummy[8], where either would again be applicable.

Without more definite evidence, no consistent translation is possible, though it seems likely that *mrḥ* is to be identified as wood tar or the like and that the single reference to *mrḥ(t) ḫ3st* indicates a mineral of similar appearance, possibly petroleum or liquid asphalt, Chassinat's 'pissasphalte'.

msdmt (Wb. II. 153.8–15), Coptic ⲥⲧⲏⲙ, Greek στίμμι, Latin stibium, is undoubtedly derived from the verb *sdm* (Wb. IV. 370.3–8), as is shown by the earlier form *sdm* (Wb. IV. 370.9–10) which survives collaterally with it[9], and the existence of an analogous term *sdm* (Wb. IV. 370.11), 'Schminkmittel', which occurs in a number of headings in Ebers. The original meaning of *msdmt* was probably eye-paint, and as such it may once have included the older green paint[10] as well as the black, but already in the Pyramid texts[11] and in other early examples[12] it refers specifically to the black eye-paint as distinct from the green, *w3ḏ*.

The nature of this black eye-paint has been widely misunderstood, owing to the fact that στίμμι and stibium refer principally to antimony compounds, stibnite etc., from which it was assumed that Egyptian *msdmt* represented the same substances. Thus Brugsch[13] consistently renders *msdmt* as 'Antimon', Lüring[14] regards it as antimony sulphide, and Wiedemann, in an early work[15], also refers to it as the latter. A glance at the translations collected by Paton[16] will show how this misconception has persisted.

But as early as 1889, Wiedemann[17], in possession of analyses by Fischer, realised that *msdmt* was not primarily, if at all, antimony sulphide, and points out that it represents 'eine ganze Reihe ähnlicher Substanzen, die schwarze Farbe besaßen und als Schminke dienen konnten', the chief among them being undoubtedly lead sulphide (galena). A similar conclusion is also reached by Hofmann[18], and by Florence and Loret[19], the latter specifically condemning the uncritical acceptance of antimony as a translation of *msdmt*, which is, on the contrary, chiefly galena, other substances used being 'des sophistications'. They also note that although Coptic has the word ⲥⲧⲏⲙ, the term for antimony is not that, but ⲃⲁⲥⲟⲩⲣ.

[1] Rec. Trav. 16 p. 157. [2] Eb. 22, 8. [3] Rec. Champ. p. 463.
[4] P. Rhind I, 3, 9; cf. also B. v. Durchw. 28–9. [5] Sin. B. 295.
[6] Hier. Ostr. 21, 1 vs. 2–3. [7] Materials p. 352; p. 359. [8] JEA. 13 pl. 17; p. 47.
[9] Cf. Ann. Serv. 16, 225, where *sdm* is described as *msdmt*.
[10] Cf. Lucas, Materials p. 99; Wiedemann, Aeg. Stud. p. 43. [11] Pyr. 54d.
[12] Eg. Cairo 1330; 1392; Urk. I, 21; Medum 13.
[13] ZÄS 29 p. 30; Hungersnoth p. 131; Wb. p. 713. [14] Med. Kent. p. 86.
[15] Sammlung p. 41. [16] Records III, 2 p. 70 n. 153.
[17] Aeg. Stud. p. 25f.; cf. also Medum p. 41–4. Ebers, Die Maasse p. 74f., had already suggested that *msdmt* included both antimony sulphide and galena, and was inclined, ib. p. 200f., to take galena as the primary meaning.
[18] Über Mesdem — I have been unable to consult this at first hand.
[19] In De Morgan, Dahchour 1894 p. 153f.

More recently, LUCAS[1] has re-examined the question, showing that the idea that antimony has frequently been found, and was commonly employed, in Egypt is erroneous. Only one case of metallic antimony and a very few instances of the employment of antimony compounds can be quoted, while only one, or at most two, examples of its use as an eye-paint are known. Discussing eye-paints in particular[2], he points out that the common black variety is galena, which, together with malachite, has been found in graves in several conditions. Of 61 specimens of ancient kohl that have been analysed (including those known to WIEDEMANN and FLORENCE and LORET), no fewer than 40 have proved to be galena, only one being an antimony compound. He concludes, therefore, that 'the general idea ... that ancient Egyptian kohl, except when it was the green malachite or chrysocolla, always either consisted of or contained antimony or an antimony compound is wrong, and hence it is most misleading to term it stibium (an early Latin name for sulphide of antimony transferred later to the metal) as is sometimes done'.

A substantially similar argument is also put forward by JONCKHEERE[3], who maintains that *msdmt* was galena, not antimony, though it did not always have a precise chemical meaning, and could be applied generally to all black coloured substances used as eye-paint.

References to *msdmt* are very numerous, the majority exemplifying its use as an eye-paint, parallel to *wꜣḏ*[4], often as labels to small bundles, the colour of which is sometimes shown[5]. In some instances, *wꜣḏ* is specified for one eye, *msdmt* for the other[6], though *msdmt* is also for both[7], and in one example *msdmt* is for under the eye, *ḥsbḏ* for over it[8]. In addition to these there are a few references to *msdmt* simply as eye-paint, where no particular distinction is implied[9], and the meaning is akin to that of *sdm*, 'Schminkmittel'.

msdmt is also widely used as a medicament[10], sometimes in conjunction with *wꜣḏ*[11], sometimes with *ḥtm*[12], and sometimes with *gsfn*[13], the two latter being varieties of *msdmt*, as is clear from the Dendera tribute tables[14]. In lists of minerals *msdmt* is normally found with the other ores and earthy substances[15], but at least once occupies a peculiarly prominent position[16].

Several different sources of *msdmt* are mentioned, including Syria[17], Pwenet[18], Coptos[19], Elephantine[20], and perhaps Sinai[21], concerning which LUCAS[22] remarks that there was no need to import *msdmt*, since all the materials used as black eye-paint, except the very rare antimony compounds, occur in the country. *msdmt* from Coptos is probably galena from the Red Sea coast, and *msdmt* from Pwenet may also be galena, which is found in Arabia, and may have been imported via that region; most of the substances used as *msdmt* are also native to Asia.

---

[1] Materials p. 222. [2] ib. p. 99f.

[3] La Mesdemet, in Histoire de la Médecine 1952 No. 7 p. 2f.

[4] Eg. Berl. 9; 1107; 1108; 13772; Cairo 1330; 1392; 28024; Beni Has. I, 35; II, 7; Mem. Miss. V, 8, 49; 457; 462; XV, 26; 60, 156; 61, 167; Huy 35; Deshasheh 29; D. el B. 109 etc.

[5] Eg. Beni Has. II, 7.

[6] Dend. Mar. III, 181; Philae ⟨2732⟩ Phot. 995; Edfou III, 143, 17, 18; V, 191, 12–14 etc.

[7] Urk. VIII, 85. [8] ZÄS. 11 pl. I, 15f.

[9] Love Harris 7, 4; Love Beatty vs. C 3, 1; P. Tur. PR. 118.

[10] Eg. Eb. 33, 3; 61, 7; 96, 19; Hearst 10, 1 etc.

[11] Eb. 48, 19; 63, 4. [12] Eb. 59, 4; 61, 5; 61, 11.

[13] Eb. 56, 4; 57, 16; 58, 4. [14] Dend. Mar. I, 70; I, 71.

[15] Dend. Mar. IV, 36, 49–50; 39, 140–2; Sehel 17; Ostr. Tor. A. 4 etc.

[16] Lesest. 96, 10. [17] LD. II, 133; Urk. IV, 715.

[18] Urk. IV, 335; cf. also Dend. Mar. I, 71. [19] Rec. Trav. 16, 51.

[20] Sehel 17. [21] JEA. 4 pl. 9. [22] Materials p. 103.

That there were different species and qualities of *msdmt* is evident from such expressions as *msdmt mꜣꜥt*[1], which probably refers specifically to galena, *msdmt nfrt*[2], and *msdmt mꜣꜥt nfr nfr nfr*[3], and from the compounds *ṯꜣy n msdmt*[4], and *ḫpr msdmt*[5]. Of these the former is probably the male stibium of Pliny[6], which is coarser, rougher and less dense, and was possibly sulphide of antimony, though the Egyptians may have made the distinction according to colour or some other criterion[7], and the assumption of antimony sulphide is not inevitable. The nature of *ḫpr msdmt* is, however, obscure, though it presumably indicates some form of the substance, perhaps a solid lump[8] such as must have been used to fashion the statuettes of the Nile god mentioned in Harris[9]. The other varieties of *msdmt* known as *ḥtm* and *gsfn* will be discussed in detail below.

Finally, a word must be said concerning the Wörterbuch entry *dsds* (sic.) (Wb. V. 487.7), which is clearly to be deleted. Fairman[10] has pointed out that the apparent examples of this word are nothing more than graphic variants of *msdmt*, as is shown by certain instances where the two directly alternate[11]. In almost every case the writing occurs parallel to *wꜣḏ*[12], though it is also found in two versions of the text of filling the sound eye[13]. No examples are known before the Ptolemaic period, and the majority seem to occur at Edfu and Philae.

Summarising the evidence for *msdmt*, it is reasonable to conclude that it was the most general word for eye-paint, and, like Sumerian ŠIM.BI.ZI.DA[14], originally applied to all varieties of whatever colour. In the earliest known examples, however, it is already a more specific term for black eye-paint as opposed to green (*wꜣḏ*), though the more general significance has not entirely been lost. Since the Egyptian black eye-paint was made of a number of somewhat similar substances, *msdmt* can refer to any one of these, but it must normally mean galena, which was by far the most common. Any connection with antimony compounds is purely incidental, though *msdmt* is undoubtedly the origin of στίμμι and stibium, which do refer principally to these[15].

rnṯt (Wb. II. 399.5) is apparently known from only one passage[16], where the accompanying materials, including *ḥsbḏ*, *mfkꜣt*, *shrt*, *ḥt-ꜥwꜣ* and *msdmt*, might suggest that it was of mineral nature. The location from which it is said to come, *bꜣwḳ dšrt*, is unknown, but may perhaps lie in the eastern desert. No suggestion can be made as to the meaning of the word.

ḥtm (Wb. III. 199.1–2) and the similar word (Wb. III. 199.4) representing the material of the *ḥnt* vessel, are hardly to be distinguished as they are in the Wörterbuch, or by Balcz[17], who prefers to read the latter *ḥsꜣ*, comparing *ḥsꜣ* (Wb. III. 160.6f.), 'dough or paste', and regarding it as a form of clay. That the reading is *ḥtm* is shown by one example where

[1] Eb. 59, 6; 59, 18; P. Berl. 11247; Louvre Wb. 22 ⟨65⟩; Cairo 44548 = Bénédite, Objets de Toilette p. 42; Schiaparelli, Relazione II p. 90 fig. 60. [2] Louvre N. 1363 Wb. ⟨Var. 5a⟩.

[3] BIFAO. 1, 231; Ann. Serv. 2, 143; also on two receptacles seen with a dealer in 1903, cf. Grapow, Grundriß III p. 54. Note in particular the writing of *mꜣꜥt*.

[4] Eb. 58, 5; 61, 6. [5] Eb. 56, 21. [6] HN. 36, 33, 101.

[7] Cf. Campbell Thompson, Dictionary p. xxxvi.

[8] Cf. Lüring, Med. Kent. p. 86; Wiedemann, Aeg. Stud. p. 35. [9] Harr. 41a, 8.

[10] BIFAO. 43 p. 120 n. 1. [11] Edfou III, 143, 17–18; V, 191, 12–14.

[12] Edfou III, 144, 5; VII, 277, 10–12; Mammisi 92; Philae ⟨2731⟩; ⟨2732⟩; ⟨2733⟩; (Phot. 995); Dend. Mar. III, 181; Debod (Roeder) § 153.

[13] Edfou VIII, 136; Philae ⟨1663⟩ Phot. 523. [14] Campbell Thompson, Dictionary p. 49.

[15] Cf. Wiedemann, Aeg. Stud. p. 25f. [16] JEA. 4 pl. 9, 11.

[17] Mitt. Kairo 3 p. 75.

the word is fully written[1], and by comparison of the writings of the verb *ḥtm* in the Coffin Texts[2], and there is, moreover, no clear differentiation between the forms of the two words[3].

It has long been realised that *ḥtm* is closely related to *msdmt*[4], and indeed this fact provides the most important clue to the meaning of the word. BRUGSCH[5] describes *ḥtm* as 'Bez. eines Minerales ... durchaus identisch mit dem sonst *msdmt*, seltener *ksfn*, genannten Minerale', and a similar observation is made by both WIEDEMANN[6] and LÜRING[7], who point out that the three substances are not absolutely identical, since in Ebers they must have distinct meanings.[8]

For the meaning of *ḥtm*, LÜRING compares Greek καδμεία, Latin cadmia, calamine or cadmia, claiming that the Greek word is of foreign origin[9], and suggesting that it may have come from Egyptian. He supports the identification by comparison of passages in Ebers and DIOSCORIDES illustrating the uses of *ḥtm* and καδμία. JOACHIM also takes *ḥtm* as cadmia, and the question is discussed at some length by CHASSINAT[10] in dealing with Coptic ΚⲀⲖⲘⲒⲀⲤ. The latter falsely separates the *ḥtm* of the medical texts from the Ptolemaic *ḥtm*, which he takes to be antimony, and notes that it is used for the eyes and other ailments, and that the Greeks, Romans and Arabs alike use cadmia for the eyes and for ulcers. But cadmia was not used raw[11], and *ḥtm* is therefore not calamine (Galmei), but cadmia (Ofenbruch).

Before examining the textual evidence for *ḥtm*, it will be as well to distinguish clearly between calamine and cadmia, both of which have been proposed as translations, and are included in the classical cadmia. Calamine is zinc carbonate, the natural product from which brass was extracted, and which is found associated with carbonate and sulphide of lead at Gebel Rosas[12]. Cadmia, on the other hand, is zinc oxide, a kind of soot produced by the smelting of copper ores with zinc impurities, or by roasting pyrites.[13] DIOSCORIDES[14] mentions both natural and artificial cadmia, and PLINY[15] distinguishes several different species. That the Egyptians should have had any more precise system of classification seems unlikely, though CHASSINAT may be right in suggesting that the cadmia used medicinally – if cadmia it is – was the artificial product.

The earliest references to *ḥtm* are as the material of the *ḥnt* vessels[16], for which *biꜣ* is also used, though how far these are to be taken literally may perhaps be questioned. It is later used medicinally, for the eyes[17], and for other ailments[18], and occurs as a variant for *msdmt*[19], with which it is also found in another context[20]. The connection with *msdmt* is particularly significant, since it shows that *ḥtm* was a substance used as a black eye-paint and probably associated with galena, though not identical with it as is proved by passages in Ebers[21] and elsewhere where the two occur together. That *ḥtm* was of a black or dark blue-purple colour is confirmed by a context in which it is evidently described as *irtyw*[22].

---

[1] P. Abusir rt. 19 unpublished [29].
[2] Coff. Texts I, 118b–c; 138a; 259a; II, 273c; IV, 169h etc.
[3] Compare Eb. 59, 4 with P. Abusir rt. 19; Eb. 59, 19 with Meir II, 6; Q. Saqq. III, 91, 2 with Pyr. 40.
[4] Dend. Mar. I, 70.
[5] Wb. Erg. p. 869.
[6] Aeg. Stud. p. 27.
[7] Med. Kent. p. 90.
[8] Sauneron, BIFAO. 57 p. 159 stresses the connection with *msdmt* and is inclined to regard *ḥtm* as a simple synonym.
[9] Liddell & Scott sv. καδμία say nothing of this.
[10] P. Méd. Copte p. 132.
[11] Pliny, HN. 34, 22, 100.
[12] Lucas, Materials p. 276.
[13] Cf. Campbell Thompson, Dictionary p. 72.
[14] 5, 74 (W).
[15] HN. 34, 22, 100; cf. Isidorus 16, 20, 12.
[16] Pyr. 40; Beni Has. I, 17; I, 35; Meir II, 6; Düm. H. I. II 5e etc.
[17] Eb. 59, 4; 59, 19; 61, 5.
[18] Eb. 61, 11; 79, 21; 95, 6; Hearst 8,8; 15, 9.
[19] Dend. Mar. I, 70.
[20] Q. Saqq. III, 91, 2; cf. also P. Ch. B. IV vs. 7, 6 and 7, 11 (supplied from Ostr. Tor. A. 4).
[21] Eb. 59, 4; 61, 11; 61, 5; cf. also refs. in note 20 above.
[22] Sinai 200, 8 – noted by Černý.

Now calamine might conceivably be used for making small vessels, though none are known of such material, and cadmia would undoubtedly be suitable for the eyes, while either would serve as an eye-paint, and might be associated with *msdmt*. It is interesting that zinc carbonate and lead sulphide are found together in nature, though zinc ores as a whole are rare in Egypt, and no zinc compounds occur in the specimens of black eye-paint listed by LUCAS[1]. Thus, although *ḥtm* could be calamine or cadmia, there is no direct evidence to suggest that it is, and the identification is therefore suspect.

Alternatively, *ḥtm* may represent another substance, known to have been used as an eye-paint, namely oxide of manganese[2], which occurs widely in Egypt and Sinai[3]. Manganese compounds were used to colour glass and glazes purple[4], and one, pyrolusite, has been reported as a pigment[5], while 6 of the 61 specimens of eye-paint cited by LUCAS[6] were manganese oxide. The evidence for the use of manganese oxide is therefore more satisfactory than that for calamine or cadmia, and its dark colour and occurrence in Sinai are especially suited to *ḥtm*. But again, no vessels of the material are known, and this constitutes a difficulty.

In short, no translation can be proposed which satisfactorily covers all the examples of *ḥtm*, though for the medical and later references alone either manganese oxide or calamine and cadmia would be suitable. The instances of *ḥtm* as the material for *ḥnt* vessels are, however, against either, and it may be that, despite the strikingly similar orthography, they represent a separate word or another material later confused with the eye-paint.

ḫrr (no Wb. ref.) is known from the Chester Beatty mineral list[7], where its occurrence in *hnw* vessels perhaps suggests an earthy mineral. KUENTZ[8] conjectures that it may mean ceruse, white lead, and compares Akkadian *ᵃḫulalu*[9], and this seems not unlikely, though LUCAS[10] does not mention its use in Egypt. However, *ᵃḫulalu* was among the substances recorded as being sent to Egypt in the Tell el Amarna letters[11], and so must have been known, and *ḫrr* may well be an attempt to render the word in Egyptian. It may be noted that a similarly written word occurs three times in another text[12], once alone, once as *ꜣḫt ḫrr*, and once as *grg* (?) *ḫrr*. In each case it is apparently the name of a field in the desert, and it seems very doubtful whether there is any connection with the term here discussed. On the whole, no very satisfactory conclusion is possible, though identification of *ḫrr* with *ᵃḫulalu* seems very probable, and translation as ceruse is feasible.

ḫt—ʿwꜣ (Wb. III. 340.8) is regarded as a species of wood by the WÖRTERBUCH, which compares a reference to 'wood which the *kk*-worm has eaten'[13], and translates 'faulendes Holz'. This is substantially the view taken by early commentators on the medical texts, where the word occurs with some frequency. STERN has 'lignum quoddam fortasse madidum seu putridum', and JOACHIM 'fauliges Holzpulver', while LÜRING[14] compares DIOSCORIDES'

[1] Materials p. 100. [2] ib. p. 100. [3] ib. p. 296.

[4] It is perhaps just possible that *ỉrtyw* refers to this purple, ie. the colour produced by *ḥtm*.

[5] Materials p. 392. [6] ib. p. 100. [7] P. Ch. B. IV vs. 7, 11. [8] BIFAO. 36 p. 182.

[9] Campbell Thompson, Dictionary p. 135f. But *ᵃḫulālu* cannot certainly be identified as white lead. Ass. Dict. VI p. 226 takes it to be a precious stone, moth-coloured and of the colour of a feather, and cites two varieties, genuine *ᵃḫulālu* and mountain *ᵃḫulālu*.

[10] Materials p. 399. [11] Eg. letter 25; cf. Campbell Thompson, Dictionary p. 139.

[12] LD. III, 229c, 16; 17; 18. [13] Anast. III, 5, 12. [14] Med. Kent. p. 109.

description of styrax[1]: δολίζουσι δὲ αὐτὸ τοῖς ἐκ τοῦ δένδρου πρίσμασιν ἅπερ ὑπὸ τῶν σκωλήκων ἀνατίτραται. Latterly EBBELL[2] identifies *ḫt-ʿwꜣ* as aloë socotrina, though without explanation.

But the medical references[3] are of no assistance in establishing the meaning, and there is in fact nothing, apart from the presence of *ḫt* in the name, to suggest that *ḫt-ʿwꜣ* is a species of wood, or even a vegetable substance. On the contrary, in at least two instances it appears to be otherwise, since it occurs both in the Edfu mineral list[4], and in another short list of products, the majority of which are mineral[5]. In the former it is described as *ḫt-ʿwꜣ* of Pwenet, and in the latter the source is given as the mountains of *ḥstyw*, both of which, incidentally, tend to rule out aloë socotrina, which comes from only one region, namely the island of Socotra.

If then *ḫt-ʿwꜣ* is in fact a mineral, the name must surely be figurative, indicating that it has the appearance of rotten or worm-riddled wood, and there is in fact a substance which answers this description well, namely silicified wood. This is, to quote LUCAS[6], 'petrified or fossil wood the substance of which has been removed by natural agencies and replaced by silica in such a manner that the structure of the wood has been preserved'. It occurs plentifully in Egypt[7], and examples of its use are known, including a statuette of New Kingdom date, which looks exactly as though it were worm-eaten[8].

What the medicinal application of silicified wood can have been it is not easy to see, but he fact that *ḫt-ʿwꜣ* is evidently a mineral looking like rotten wood seems to make the identification with silicified wood almost inevitable[9].

ẖʿwꜣyt (no Wb. ref.), if it is indeed so to be read, occurs only in the Chester Beatty mineral list[10], following a lacuna, in which part of the word may have been lost. It may perhaps be described as *ḥmꜣgt pw n* . . ., though the *pwn* . . . may equally well be the name of yet another mineral, the last signs of which are missing. Alternatively, *ẖʿwꜣyt* (sic.) could be a writing of *ḫt-ʿwꜣ* discussed above, (*ẖt*) being for *ḫt* by mistake, though this is not altogether likely. In the circumstances, no conclusion seems possible.

sꜣ-wr (Wb. III. 415.21) is apparently confined to medical contexts[11], with one notable exception[12]. STERN has 'nomen speciei granorum', JOACHIM 'Bleivitriol', WRESZINSKI σῶρυ, and LÜRING[13] the same, pointing out that *sꜣ-wr* is used against eye ailments in Ebers, while PLINY says that sori is 'oculorum curationi utile'. EBBELL[14] also renders sory, but BARNS[15] quotes DAWSON for the opinion that *sꜣ-wr* is not in fact the mineral σῶρυ, but rather a resin. BRUGSCH[16] compares *sꜣ-wr* with Coptic ⲂⲀⲤⲞⲨⲢ, antimony, though the latter is clearly related to other words of the same root ⲂⲀ-.

---

[1] 1, 79 = 1, 66 (W). [2] Pap. Ebers p. 133.
[3] Eb. 57, 7; 57, 16 etc., the majority in prescriptions for the eyes.
[4] Edfou VI, 202, 9. [5] JEA. 4 pl. 9. [6] Materials p. 515.
[7] Cf. Bull. Inst. Eg. 25 p. 161. [8] Aldred, New Kingdom Art no. 174.
[9] Theophrastus § 17 refers to a stone, like rotten wood in appearance, which may be brown fibrous lignite, though *ḫt-ʿwꜣ* can hardly be equated with this.
[10] P. Ch. B. IV vs. 7, 6.
[11] Eg. Eb. 9, 13; 25, 18; 105, 7; Hearst 8, 8; P. Med. Lond. 3, 5; P. Med. Berl. 6, 3.
[12] P. Dem. Mag. 3, 25. [13] Med. Kent. p. 92.
[14] Pap. Ebers p. 133. [15] Five Ram. Pap. p. 16. [16] Wb. Erg. p. 990.

The medical references are not in themselves conclusive as to the nature of *sꜣ-wr*, but the occurrence of the word in Demotic[1] suggests that it is a stone, and the identification with σῶρυ (sori) of Dioscorides, Pliny and Galen may perhaps therefore stand. But it is by no means certain to what the latter itself refers. Liddell & Scott give 'a kind of ore, perhaps ferrous sulphate, melanterite', while Bailey[2] thinks it was probably disintegrating pyrites. In view of this, it is hardly possible to propose an exact translation, and the classical sori is the most apposite rendering.

Whether sori was borrowed from Egyptian, or whether both came from a common Semitic source, it is difficult to determine, However, it is probably significant that μίσυ, misy, closely connected with sori in classical texts, was derived from Akkadian ^{a}*muṣu*[3], and a similar origin for sori may therefore seem likely. This would no doubt account for the writing *sꜣw-wr* in Egyptian, a possible attempt at a phonetic equivalent.

sꜣ-šmꜥ (Wb. III. 416.3) is known from only two late texts[4], where, however, it may be a variant writing of *siꜣ šmꜥ* as Lüring[5] suggests, though whether he is correct in regarding it as a form of natron merely because it comes from *Nḫb* and is parallel to *bd*[6], seems doubtful. No more satisfactory interpretation can, however, be offered.

siꜣ (Wb. IV. 38.8–10) occurs with some frequency in medical contexts[7], though not, apparently, elsewhere, unless Lüring is right in regarding *sꜣ-šmꜥ* as a variant writing. Stern has 'semina plantae cuiusdam', Joachim 'sa-korn', quoting also Ebers' 'Natron oder Salpeter', while Ebbell does not translate. Lüring[8] notes the two varieties *siꜣ mḥ* and *siꜣ šmꜥ*, and identifies the latter with *sꜣ-šmꜥ* which comes from the nome of which Nekheb is the capital and is therefore in all probability a kind of natron. Dawson[9], on the other hand, concentrating on the medicinal uses of *siꜣ*, notes that it is never used internally and was therefore 'impalatable or poisonous', and thinks it may have been one of the arsenic salts orpiment or realgar, which were known to the Egyptians and used by the Assyrians in medicine[10].

The true nature of *siꜣ* cannot, perhaps, be determined, but Dawson's interpretation may be rejected, because orpiment and realgar can be shown to be *ḳnit* and *ꜣwt-ib*, both of which occur, though rarely, in the medical papyri. Further, no obvious sources of arsenic salts are known to correspond to the northern and southern *siꜣ*. That *siꜣ* is a term for natron or the like is not impossible, though, if so, it might be expected to occur in other texts beside, and to have the more explicit determinative [hieroglyph]. Lüring's hypothesis is in fact too tenuous to form the basis of any conclusion.

The word occurs alone[11], as *siꜣ mḥ*[12], and, more frequently than either, as *siꜣ šmꜥ*[13], but this is not itself significant, in as much as several substances, both mineral and otherwise, are similarly qualified. However, the greater number of references to *siꜣ šmꜥ* may suggest so-

[1] P. Dem. Mag. 3, 25.
[2] Elder Pliny II p. 176.
[3] Campbell Thompson, Dictionary p. 113.
[4] Dend. Mar. II, 28 = Düm. G. I. I, 69; Düm. G. I. III, 63.
[5] Med. Kent. p. 98.
[6] Düm. G. I. III, 63. In lacuna in I, 69.
[7] Eg. Eb. 31, 19; 56, 2; 59, 2; 59, 5; 60, 2; 61, 5; 61, 16; 63, 5; 78, 9; 104, 4; P. Ram. V 41 upper.
[8] Med. Kent. p. 98.
[9] JEA. 19 p. 135.
[10] Campbell Thompson, Dictionary p. 43.
[11] Eb. 31, 19; 104, 4.
[12] Eb. 56, 2; 63, 5; 78, 9; P. Ram. V 41 upper
[13] Eb. 59, 2; 59, 5; 60, 2; 61, 5; 61, 16; 63,5; 78,9.

mething coming primarily from Upper Egypt, which might be against natron, of which the prime source was the Wadi Natrun. But *si3* may well be a saline deposit of some description, or perhaps an obscure species of astringent earth or the like, though there is no certain evidence that it was even a mineral, and identification is impossible.

sʿḳw (no Wb. ref.) occurs only in a broken context in the Chester Beatty list[1], from which nothing can be deduced. It is not altogether certain that the reading is in fact *sʿḳw*, since it is preceded by *m*, which may be the preposition or, alternatively, part of the name, other signs being perhaps lost in the lacuna.

sbẖt (no Wb. ref.) also occurs in a broken context in the Chester Beatty list[2], followed by *ḫsbḏ m3ʿ*, and preceded by *ẖry*, which may perhaps belong to it, though what the expression then means it is not easy to see. A literal interpretation as 'that which is under the door (or palace)' conveys nothing.

snt (Wb. III. 458.5) seems to be known from three examples, through which the general nature of the substance can be ascertained. It occurs among the products of the Wadi Natrun[3] in the Edfu mineral list, together with *sty*[4], and in an offering-list[5], where it is glossed by *sty pw*. There can be little doubt that all refer to the same substance, and the description *sty pw* shows that this was a species of ochreous earth, probably somewhat uncommon since it required explanation. Apart from the knowledge that *snt* was found in the Wadi Natrun, and was apparently of some commercial value, no more precise identification is possible.

snm (no Wb. ref.) occurs only in the Edfu mineral list[6], in the phrase *w3y nfr m snm*, the significance of which cannot be appreciated, since *w3y* is also unknown. It has been suggested above (p. 166) that the latter may be a mistake for *ḫt-ʿw3*, which would, if correct, suggest that *snm* was the name of a kind of silicified wood, but this is mere hypothesis.

snn (Wb. IV. 166.9–13) is clearly a word with more than one meaning, referring both to a mineral substance and to a species of *ʿntyw*. This duality, recognised by the Wörterbuch, has led to considerable confusion in the past, and it is not, indeed, easy to determine to which category some examples of the word belong.

Brugsch[7] takes *snn* to mean stone or precious stone in general, though adding[8] that it was a mineral used for eye ailments, and that a connection between it and *ʿntyw* is apparent. Lüring[9], on the other hand, quoting an apparent parallel between *snn* and *ʿntyw*[10], regards the former as a type of incense or resin, and explains the description of *nšmt* as *snn pr m wsir* as a comparison of opal with a resinous excrescence. He points out that *snn* and *ʿntyw* cannot be identical, because they occur together in Ebers, though they are expressly related in a text at Edfu[11]. This latter passage is discussed by Ebbell[12], who identifies the variety of *ʿntyw* called *inḥ3sʿʿ-snn* and said to be of a golden colour, as Mecca-balsam, and concludes that the *snn* of Ebers and elsewhere is the same.

[1] P. Ch. B. IV vs. 8, 2. [2] P. Ch. B. IV vs. 8, 4. [3] Peas. R. 23. [4] Edfou VI, 203, 6. [5] Ann. Serv. 16, 225. [6] Edfou VI, 203, 2. [7] Wb. p. 1250. [8] Wb. Erg. p. 1074. [9] Med. Kent. p. 106. [10] Düm. G. I. II, 63, 2 & 75,4. [11] Düm. G. I. II, 86, 9. [12] Act. Or. 17 p. 105; cf. also Lefebvre in Fest. Grap. p. 211.

Now while it is clear that in several instances[1] *snn* represents a kind of *ꜥntyw*, probably Mecca-balsam as Ebbell suggests, it is equally certain that in others it has the appearance of being a mineral product, possibly a kind of eye-paint. Thus *snn* is found in the lists of the 24 *ꜥꜣt n mꜣꜥt* at Dendera[2], and twice in the Edfu mineral list[3], and is once closely associated with *wꜣḏ* and *msdmt*[4]. Moreover, although the usual determinative is 𓈒𓏥, in at least two examples[5] it is 𓐍𓏥, which strongly suggests a mineral and would be inappropriate to a species of resin.

The connection of *snn* with *wꜣḏ* and *msdmt*, and with the earthy minerals in the Dendera lists, is borne out in the medical contexts, where *snn*, regarded as the resin by Ebbell, occurs together with *wꜣḏ*[6], *msdmt*[7], and *ḥtm*[8], chiefly in prescriptions for the eyes. This would seem to suggest that it was an analogous material, possibly a species of eye-paint, and there is little reason for accepting Ebbell's view.

Various sources of *snn* are mentioned, including *tꜣ-nṯr*[9], *tꜣ-ỉꜣt*[10], and *nhr*[11], of which the last two were probably regions in Nubia, though nothing can be deduced from these. The most significant passage in which the word occurs is that in the Edfu mineral list[12], where *wꜣḏ m bꜣẖ*, *ḥb m ḏt.f* and *nšmt* are described as *snn pr m wsỉr*. Lüring's view of this has already been mentioned, and although not altogether impossible seems unlikely, since the only evidence for the colour of the resin *snn* suggests that it is golden, which none of the stones mentioned is. On the contrary, they are all green, which would perhaps point to a meaning 'green stuff' or the like for *snn* in this instance, though such a general interpretation is not suitable elsewhere. The occurrence of *snn* in the medical texts must presuppose a specific substance, as also its inclusion among the 24 *ꜥꜣt*. But the colour may none the less be of importance, and it is perhaps possible to see in *snn* a species of green mineral.

No very satisfactory conclusion can be reached as to the meaning of *snn*, beyond the fact that in some instances it represents a resin, possibly Mecca-balsam. In others it appears to be a mineral, found in Nubia, probably of an earthy nature, and possibly green in colour. Certain evidence seems in favour of an eye-paint, and it could perhaps be a specific name for one of the green substances used, though in this case it might be expected to occur as a variant of *wꜣḏ* in the Dendera tables[13].

ssmỉt (Wb. IV. 276.17) is altogether obscure, and cannot be discussed with profit. It forms, with gold, the material for an astronomical figure[14], and appears to have a variant *ṯsmd*[15] By analogy with the other materials cited in this connection, it must be some kind of mineral, but since it is otherwise unknown, some corruption seems possible.

šꜥšt (no Wb. ref.) is known only from the Edfu mineral list[16], where it occurs in the phrase *šꜥšt m tꜣ-sty r sḏsr ꜥꜣwy.k*, a context which might suggest some mineral used in decoration,

[1] Eg. Düm. G. I. II, 86, 9; 86, 10; possibly also Ostr. Cairo 25677 rt. 29.
[2] Dend. Mar. IV, 36, 49–50; 39, 140–2.
[3] Edfou VI, 202, 4; 202, 9.
[4] Dend. Chass. IV, 154.
[5] Philae ⟨2002⟩ Phot. 289; Junker, Große Pylon p. 277.
[6] Eb. 48, 18.
[7] Eb. 55, 21; 59, 4; 62, 14; 62, 17; Hearst 11, 7.
[8] Eb. 62, 14.
[9] Edfou VI, 202, 9; cf. also Dend. Mar. II, 12.
[10] Philae ⟨2002⟩ Phot. 289.
[11] Junker, Große Pylon p. 277 (*snn n tꜣ sty*). Düm. G. I. IV, 184 and Brugsch, Dict. géog. 1034 read *nhꜣw*.
[12] Edfou VI, 202, 4.
[13] Dend. Mar. I, 70; I, 71; II, 12 etc.
[14] Düm. H. I. II, 57b–d; Thes. 20 A', B'.
[15] Dend. Chass. IV, 178 – not *ṯsmk* as Dend. Mar. II, 10.
[16] Edfou VI, 203, 5.

possibly an ochreous earth, since *sty* and *mnšt* follow in somewhat parallel expressions. On the other hand, since the material is mentioned in connection with doors, it may perhaps be a metal with which they are to be overlaid. The source, *tꜣ-sty*, could apply equally to either.

šby (no Wb. ref.) occurs in an obscure context in the Chester Beatty mineral list[1], where it is apparently a material from which seals are made. By analogy with a number of other items in the same list, it may be a loan-word, and a very tentative connection with Akkadian *šubû*[2] might perhaps be suggested.

škr (Wb. IV. 550.6–7) is in all probability a word of Semitic origin, as BURCHARDT assumes[3], though no Akkadian or other equivalent suggests itself. It is found three times, once in the Chester Beatty mineral list[4], once in a list of Asiatic tribute[5], and once in the London medical papyrus, in a prescription to prevent blanching of a burn[6]. A passage in Ebers parallel to the last[7] has *kšw*, known only once elsewhere[8], which may point to a corruption, though the presence of both *škr* and *kšw* on the same page of the London papyrus might be against this.

The nature of *škr* is not clear from the few examples known, though the proximity of *ḫrr*[9], and of *ḏiḏi* and *msdmt*[10], would perhaps suggest an earthy mineral, possibly even a rare pigment, and this could account for its use to prevent blanching in Ebers. In default, however, of further evidence, such an interpretation cannot be pressed, and *škr* may possibly be a species of semi-precious stone as BURCHARDT supposes.

gsfn (Wb. V. 206.9–11) has long been recognised as a substance closely related to *msdmt* and *ḥtm*, with which, however, it is clearly not identical. This is pointed out by BRUGSCH[11] and WIEDEMANN[12], and discussed by LÜRING[13], who suggests a translation 'Kupfervitriol' (χάλκανθος), used by the Greeks as an eye-paint and in remedies for afflictions of the eye.

The textual evidence for *gsfn* is somewhat restricted, and the word does not occur as frequently as *ḥtm*, the other species of *msdmt*. In three instance[14], *gsfn* is associated with *msdmt*, and is evidently, therefore, a variety of black eye-paint, though a quite distinct substance, as is shown by the appearance of both *gsfn* and *msdmt* in the same prescriptions in Ebers[15]. In the medical papyri it is used both for eye ailments[16] and other diseases[17], and is found but rarely elsewhere, though once described as *gsfn nfr m Pwnt*[18].

From these few references, the only significant fact that emerges is that *gsfn* is a black eye-paint, which is perhaps inconsistent with an identification as χάλκανθος which seems to have been principally a blue or green copper compound. Of the black eye-paints mentioned by LUCAS[19], the one which might most readily be equated with *gsfn* is manganese oxide,

---

[1] P. Ch. B. IV vs. 7, 12.
[2] Campbell Thompson, Dictionary p. 91.
[3] Fremdworte II p. 45 n. 886.
[4] P. Ch. B. IV vs. 7, 11.
[5] Urk. IV, 715.
[6] P. Med. Lond. 15, 11.
[7] Eb. 69, 12.
[8] P. Med. Lond. 15, 7.
[9] P. Ch. B. IV vs. 7, 11.
[10] Urk. IV, 715.
[11] Wb. Erg. p. 1281.
[12] Aeg. Stud. p. 27.
[13] Med. Kent. p. 91; p. 155.
[14] Dend. Mar. I, 70; I, 71; II, 12 etc.
[15] Eb. 56, 4; 57, 16; 58, 4.
[16] Eb. 57, 16; 58, 4; 60, 7.
[17] Eb. 48, 13; 68, 14; P. Med. Berl. 11, 8.
[18] Dend. Mar. II, 12 = Dend. Chass. IV, 158; cf. Düm. G. I. II, 85B.
[19] Materials p. 100.

though this may, as has been suggested above, be represented by *ḥtm*, if the latter does not in fact mean calamine and cadmia. None of the other substances used, except apparently brown ochre, is of sufficiently regular occurrence to suggest *gsfn*, and no satisfactory interpretation is therefore possible. What does seem clear is that the Egyptians are likely to have had a term for the well known and widely used manganese compounds, and that this was in all probability either *ḥtm* or *gsfn*. Without more precise evidence it is, however, impossible to determine which, and the meaning of *gsfn* must remain uncertain.

In conclusion, mention must be made of a substance which is not mineral in origin, but was apparently regarded as such by the Egyptians and is therefore relevant to the present discussion.

ỉnr n r mw, ỉnr n spt mw (Wb. I. 97.14), used in treating burns and growths[1], would at first sight appear to be a type of stone, as JOACHIM regards it, and might be taken to refer to pebbles or the like. But EBBELL[2], translating literally as '(lime)stone from the shore', shows that the expression probably refers to shells and other marine organisms cast up by the sea. He points out that DIOSCORIDES[3] calls fragments of shell λίϑων κοχλάκων (ἄσβεστος) and mentions them together with τῶν ϑαλαττίων κηρύκων τὰ ὄστρακα (shells of sea-snails) as the material from which lime, ἄσβεστος, was made. *ỉnr n spt mw* and ἄσβεστος ἐκ τῶν λίϑων κοχλάκων are both used against the same kind of ailments, namely burns[4], and growths such as cancer penis[5], and identification of the two seems certain. There can be little doubt that EBBELL's argument is correct, and that this is a case where the Egyptian conception of a material is at variance with our own[6]. As such it is particularly important, and demonstrates the danger of assuming that a substance is mineral merely because it has a characteristic determinative or is compounded with *ỉnr* or *ʿꜣt*.

As in chapter 4, many of the substances discussed above have not been identified with any degree of certainty, and this is no doubt largely due to lack of archaeological evidence concerning the use of certain minerals in ancient Egypt, and to the imperfect scientific knowledge of the Egyptians themselves. Among the terms collected here are several which seem, like *bḳs-ʿnḫ* and *ḥtm*, to represent materials the use of which cannot be confirmed, while, on the other hand, there are minerals known to have been employed, such as graphite, mica and sulphur, for which no name can be distinguished. But it must be borne in mind that although a particular substance has not as yet been revealed by chemical analysis, it does not necessarily imply that it was unknown, and, conversely, that some minerals which were known may have been confused with others and not recognised as such. Moreover, some materials which were not clearly defined may have been given more than one name, and several of the terms occurring in the Chester Beatty mineral list and elsewhere may be borrowed Semitic designations of substances more commonly referred to by some other word. Inconsistency is therefore to some extent inevitable, and no very clear conception of Egyptian mineralogy is possible.

---

1 Eb. 47, 2–3; 68, 10; 68, 19; 95, 17; Hearst 5, 13; P. Med. Lond. 6, 2.

2 ZÄS. 64 p. 49.

3 5, 115 (W).

4 Dioscorides περὶ ἁπλῶν φαρμάκων 1, 170 = 161 (W).

5 Celsus 6, 18, 3.

6 Cf. the *gʿrʿb n yʿm* of P. Dem. Mag. 3, 26, which appears to be a mineral, but may well be some kind of shell.

CHAPTER 8

ALUM, SALT AND NATRON

The minerals to be discussed in the present chapter, alum and similar astringents, salt, and the various kinds of natron, do not, perhaps, form a very satisfactory group according to the criteria of modern science. There is, however, some justification for treating them together here, in that the Egyptians may well have regarded them as in some way related and seem to have confused alum and other analogous substances with natron, at least before the New Kingdom. In order, however, to avoid any unnecessary confusion, each class of substance will be treated separately within the chapter itself.

Alum and other astringent minerals

ỉbnw (Wb. I. 63.8), Coptic [S]**OBN̄**, [B]**ⲰⲂⲈⲚ**, is clearly the στυπτηρία of the Greeks and alumen of the Romans, loosely translated as alum. The evidence for the equation is stated more or less fully by Loret[1], who points out that *ỉbnw* corresponds to Coptic [B]**ⲠⲒ-ⲰⲂⲈⲚ**, the equivalent of the Arabic term translated alumen by Kircher, and used by Íbn-Baithar as a heading for extracts from Greek medical authors concerning στυπτηρία. He notes that Dioscorides[2] and Pliny[3] mention the alum of Egypt before all other, and also refers to Herodotus' story[4] that Amasis offered 1000 talents of στυπτηρία to Delphi. The Ebers papyrus[5] prescribes an eye salve containing *ỉbnw*, which agrees with passages in Pliny[6] and Dioscorides[7], and the use of *ỉbnw* for a tumour in the Berlin medical papyrus[8] can also be parallelled[9].

The cogency of Loret's argument cannot be gainsaid, and the identification of *ỉbnw* with Demotic *ȝbn* and Coptic [B]**ⲰⲂⲈⲚ**, proposed simultaneously by Müller[10], who suggested a Semitic origin[11], has been accepted by Jéquier[12], Chassinat[13] and the majority of scholars. Jéquier, however, suggests a somewhat wider meaning than alum, pointing out that to the ancients it signified a whole group of salts, particularly sulphates of alumina and potassium, which were not strictly distinguished, and which were used more or less indifferently. *ỉbnw* therefore included all alum compounds, and perhaps even natron, a confusion for which he is able to cite a modern parallel. Conversely, he suggests that before *ỉbnw* was introduced into the language, alum compounds were included in the terms for natron, eg. *bd*, which is used to clarify water, and perhaps *ḥsmn*, a point which will be discussed more fully below.

The question of the breadth of meaning of ancient terms for alum and the like is of peculiar interest, and requires some consideration here. Beckmann[14] tried to show that the

---

[1] Rec. Trav. 15 p. 199.
[2] 5, 122 = 5, 106 (W).
[3] HN. 35, 52, 184.
[4] II, 180.
[5] Eb. 63, 10.
[6] HN. 35, 52, 189.
[7] 5, 122 = 5, 106 (W).
[8] P. Med. Berl. 5, 3.
[9] Dioscorides loc. cit.
[10] Asien und Europa p. 188 n. 3.
[11] Cf. Hebrew *ȝbn*, Akkadian [a]*aban*, 'stone'.
[12] BIFAO. 19 p. 106.
[13] P. Méd. Copte p. 69.
[14] History of Inventions 1 p. 180.

classical peoples did not know our alum, and that alumen represents sulphate of iron. Moore[1] disputed this, pointing out that 'some of the properties ascribed to alumen belong to our alum, but not in the least to sulphate of iron . . . Thus Dioscorides[2] says of the fibrous variety, which he terms σχιστή, that it is exceedingly white (λευκὴ ἄγαν), and from Pliny's description of this kind[3] we may infer that it was such an efflorescence as that called plume, or fibrous alum . . . Pliny, too, speaks of alum as white, – as used for whitening wool, – as being prepared for certain medicinal uses by burning it in pans until it cease to flow liquid, all which is quite inapplicable to sulphate of iron'. On the other hand, 'the στυπτηρία of the Greeks, called alumen by the Romans, may in some instances have been sulphate of iron, or some other vitriolic salt beside what we call alum'. Liddell & Scott accordingly give στυπτηρία a wide range of meaning, referring it to 'any of a group of astringent substances containing (a) alum or (b) ferrous sulphate (χαλκῖτις)', and Bailey[4] says that 'it is clear that alumen represents a group of substances, rather than a single substance, the only excuse for grouping them together being their characteristic astringency'. He further points out that the species of alum proper and halotrichites are not easily distinguishable by taste and external characteristics, and early authors probably included under alumen such substances as alunogen and all minerals containing sulphate of iron.

Although, however, several apparently similar astringents were commonly referred to as στυπτηρία and alumen, certain distinct species were recognised. Dioscorides[5] states that στυπτηρία has many varieties, and Pliny[6] cites several types of alumen, including a white liquid kind used for dyeing with bright colours, and a darker sort for darker colours. He also makes specific mention of alumen schiston, otherwise called trichitis, from its habit of dividing into white hair-like threads[7], and of the variety called strongyle, of two kinds, one fungus-like, the other akin to pumice in appearance and occurring in globular forms. Coptic also recognises different types of 'alum'. In addition to simple **ⲟⲃⲛ̄**, the terms **ⲟⲃⲛ̄ ⲛ̄ⲗⲁⲥ**, **ⲟⲃⲛ̄ ⲛ̄ⲥⲓϣⲡ** and **ⲟⲃⲛ̄ ⲛ̄ⲥⲏϥⲉ** are also used, and **ⲥⲧⲉⲡⲧⲉⲣⲓⲁⲥ**, **ⲥⲧⲩⲯⲓⲛ** and **ⲥⲁⲣⲁⲕⲟⲡⲡⲁⲕⲟⲥ** are borrowed from Greek[8]. **ⲥⲧⲉⲡⲧⲉⲣⲓⲁⲥ** is of course στυπτηρία, **ⲥⲧⲩⲯⲓⲛ** is from στύψις, 'astringent matter', and **ⲥⲁⲣⲁⲕⲟⲡⲡⲁⲕⲟⲥ** apparently from σαρκοφάγος (λίθος)[9]. Of the three Coptic compounds, **ⲟⲃⲛ̄ ⲛ̄ⲗⲁⲥ** is schistous alum[10], **ⲗⲁⲥ** probably not the word for tongue, but connected with **ⲗⲁⲥ**, στυπτίον. **ⲟⲃⲛ̄ ⲛ̄ⲥⲓϣⲡ** is translated 'flaked alum', by Crum, but Chassinat renders 'l'alun rond', comparing Latin strongyle from στρογγύλος 'round'[11]. The third species, **ⲟⲃⲛ̄ ⲛ̄ⲥⲏϥⲉ** is connected by Chassinat with *sfy*, and translated 'l'alun liquide, mou, humide', perhaps correctly. It is also worthy of note that the Assyrians seem to have distinguished stony alum and white and black (liquid) alum[12], and that they probably had a word for alunite[13].

The alum of Egypt was famous in antiquity, and is first mentioned in Akkadian texts[14]. Dioscorides[15] says that nearly every type of στυπτηρία is found in Egypt, and Pliny[16] states that the best quality alumen came from there. The Egyptian alum monopoly is referred to in a papyrus from Oxyrhynchus[17], and Lucas[18] mentions extensive ancient workings of aluminium sulphate, and quotes Arab authors for more recent alum trading.

[1] Ancient Mineralogy p. 124.
[2] 5, 123 = 5, 106 (W).
[3] HN. 35, 52, 186; cf. Celsus 5, 2.
[4] Elder Pliny II p. 233.
[5] 5, 106 (W).
[6] HN. 35, 52; cf. also Isidorus 16, 2, 2.
[7] Cf. Dioscorides 5, 106 (W).
[8] Cf. Chassinat, P. Méd. Copte p. 134f.
[9] Dioscorides 5, 124 (W); Pliny, HN. 36, 27, 131.
[10] Crum 144b; 254a; cf. Dioscorides 5, 106 (W); Pliny, HN. 35, 52, 186; Celsus 5, 2 etc.
[11] Cf. also Bailey, Elder Pliny II p. 187, 'burnt alum'.
[12] Campbell Thompson, Dictionary p. 33.
[13] ib. p. 176.
[14] ib. p. 33, *ᵃgabû*.
[15] 5, 106 (W).
[16] HN. 35, 52, 184.
[17] P. Oxy. 2116; cf. P. Oxy. 1429.
[18] Materials p. 291.

The textual references to *ỉbnw* are somewhat disappointing, though the general meaning 'alum' is to some extent confirmed. The word occurs in the medical texts[1], in contexts which can roughly be parallelled in DIOSCORIDES and PLINY, as has been shown by LORET. It is also found in various lists of mineral and other substances[2] from which little or nothing can be deduced, though the use of the block determinative in two instances[3] might suggest that alum stone, alunite, is meant. More interesting is the appearance of *ỉbnw*, together with *ỉp*, madder[4], in a fragmentary context evidently referring to some process applied to skins[5]. The obvious inference is that the passage refers to the dyeing of skins, using alum as a mordant[6], though it is possible that the tawing of skins is meant[7]. But it is perhaps unnecessary to attempt to draw too fine a distinction, since any dyeing process involving the use of alum would also result in a degree of tawing.

The texts in which *ỉbnw* occurs are all of New Kingdom date, and there can be little doubt that *ỉbnw* is a loan-word, perhaps related to Hebrew *ꜣbn*, Akkadian *ᵃaban*, though both have only a general meaning. Before that time, the Egyptians must have used some other word in reference to alum, and JÉQUIER is perhaps right in suggesting that it was originally regarded as a species of natron. Either *bd* or *ḥsmn* might possibly refer to alum as well as natron, the former through its general significance of 'purification material', the latter because it was perhaps the most impure form of natron and is more likely to have embraced other analogous substances. But the problem is incapable of satisfactory solution, and it may even be that some such word as *bsn* was used of alum in earlier periods.

To sum up, it is clear from the evidence collected above that *ỉbnw*, though perhaps primarily our modern alum, had, like στυπτηρία and alumen, a fairly wide range of meaning, including various sulphates of alumina, potassium and iron, and possibly also other vitriolic salts.

ỉmrw (Wb. I. 87.19) is apparently confined to the Smith papyrus, where, however, it occurs on ten occasions, and with two different writings[8]. In eight of the ten instances, it is used alone, in the remaining two together with honey, and is bound onto a fracture or sprain. BREASTED[9] is of the opinion that *ỉmrw* was a particular substance employed for its disinfectant properties, and this seems to be borne out by the accidents which it is used to treat. Four are fractures or the like having an open wound[10], two are fractures without an open wound[11], and four are sprains[12]. In each case the use of an antiseptic would be appropriate, though in those where the wound is not open, *ỉmrw* may have been used merely as a soothing agent. The determinative suggests that *ỉmrw* was a mineral, and a disinfectant with astringent and styptic properties would perhaps be apposite. FORBES[13] regards *ỉmrw* as an alum compound, and this may well be correct. The characteristic astringency of alum is mentioned by PLINY[14], and it is known to have been used as a styptic by

[1] Eb. 24, 8; 63, 10; P. Med. Berl. 4, 5; 5, 3; Hearst 2, 17.

[2] Harr. 64c, 15; 73, 16; Med. Habu 160; P. Ch. B. IV vs. 7, 10; 7, 14.

[3] P. Ch. B. IV vs. 7, 14; Med. Habu 160.

[4] Coptic ⲁⲛⲉⲓ; cf. Spiegelberg, OLZ. 27 p. 187; Loret, Kemi 3 p. 29; Caminos, LEM. p. 385.

[5] Ostr. Cairo 25596, 7; cf. Ostr. Petrie 37 rt. 4 = Hier. Ostr. 29, 3, where *ỉbnw* may be lost.

[6] Cf. Lucas, Materials p. 293; Jéquier, BIFAO. 19 p. 108–9; Pliny, HN. 35, 52, 183.

[7] Cf. Campbell Thompson, Dictionary p. 33.

[8] Seven examples have [hieroglyph] as the second sign, three [hieroglyph].

[9] Pap. Smith p. 264.

[10] Smith 6, 17; 12, 17–18; 13, 2; 15, 2. 6, 17 is possibly a case of gangrene.

[11] Smith 12, 8; 12, 14.

[12] Smith 11, 6; 14, 20; 9, 6; 10, 11.

[13] Studies III p. 159.

[14] HN. 35, 52, 189.

the Assyrians, eg. in menorrhagia[1]. But the evidence is by no means conclusive, and acceptance of *ỉmrw* as a species of alum or halotrichite is open to question. GRAPOW, whose knowledge of Egyptian medical texts is unequalled, favours translation of *ỉmrw* by '*ỉmrw*-Verband', regarding it as 'ein Stofflappen von länglich viereckiger Gestalt'[2], which does not conflict with the rectangular determinative. On the other hand, one would rather expect a more precise sign such as , and in default of this it is perhaps preferable to interpret *ỉmrw* as a mineral.

ỉnr n šn (Wb. IV. 501.3) is known from only one example, in a context which is of no assistance in determining the meaning[3]. *šn* is regarded by the WÖRTERBUCH, no doubt correctly, as the common word for hair, so that the term is evidently to be translated 'hair-stone' or the like, presumably in reference to its texture or appearance. It is suggestive that among the various species of alum recognised in classical times was one, alumen schiston, known as trichitis owing to its habit of dividing into white hair-like threads[4], and it seems probable that this is the substance called hair-stone in this instance. No other mineral, except perhaps fibrous gypsum, would fit the name so well, and *ỉnr n šn* may therefore be identified provisionally as schistous alum.

wšbt (Wb. I. 373.7) was known to the WÖRTERBUCH and to BREASTED from only three examples, all in medical contexts[5], to which, however, more may now be added. On the basis of these references, where *wšbt* is used in treating wounds, and as a styptic, BREASTED[6] concluded that 'as a medicinal agent it was doubtless an astringent', and this seems essentially correct. A further medical reference has since become known[7], which adds little to our knowledge, and it seems probable that *wꜣšbt* (Wb. I. 262.15), applied externally to a fracture in an ointment[8], is also to be regarded as an example of the word. The uses of *wšbt* in the medical texts are thus strikingly similar to those of *ỉmrw*, and it seems probable that it is likewise to be identified as a species of alum or other analogous astringent. But *wšbt* also occurs in two other contexts, in company with *ḳnỉt*[9], where the meaning may require some modification. In both instances it is written *wšb*, and MACADAM[10] strangely compares the Kawa example with Wb. I. 373.8, ie. *wšbyt*. But the passage in question is a list of raw materials, including gold, orpiment and blue pigment, and the intrusion of 500 *dbn* of beads would be most extraordinary, the more so since beads would be numbered and not weighed. Unless, therefore, *wšb* is a new and otherwise unknown word, it is likely to be a writing of *wšbt*. The mention of *wšbt* together with *ḳnỉt* and *ḫsbḏ n sš* perhaps suggests that it was some kind of pigment, and this may be confirmed by its occurrence with *mnšt* and *ryt* in another context[11]. Other pigments were also used medicinally, and indeed *šs(y)t nt wꜣḏw* accompanies *wšbt* in the Smith passage. In spite, therefore, of the difficulty of identification, it seems at least possible that *wšbt* represents some astringent earth or the like, which may also have been used as a pigment. In conclusion it must be noted that what is in all pro-

---

[1] Campbell Thompson, Dictionary p. 33.
[2] Grundriß III p. 129; but cf. Grundriß IV (1) p. 183 etc., '*ỉmrw* – Mineral'.
[3] Anast. VIII, 1, 4.
[4] Pliny, HN. 35, 52, 186; Dioscorides 5, 106 (W).
[5] Eb. 71, 2; 80, 10; Smith 14, 5.
[6] Pap. Smith p. 382.
[7] P. Ram. V 70 upper.
[8] Eb. 80, 3.
[9] Kawa VI, 12; P. Tur. unpublished [25] (sv. *wšbyt* in the Wb. Zetteln).
[10] Kawa I p. 39.
[11] Ostr. DM. 609 vs. 1, probably a list of painting or writing materials.

bability *wšbt* occurs between *sty* and *ibnw* in a badly damaged part of the Chester Beatty mineral list[1]. This juxtaposition with both a pigment and an astringent may well serve to support the interpretation here proposed.

Salt

ḥmꜣt (Wb. III. 93.14f.), Coptic ϩⲘⲞⲨ, salt. The meaning seems never to have been doubted[2], though earlier authors like CHABAS and LIEBLEIN took it to be sea salt, which is specifically rejected by LÜRING[3] on the ground that the Egyptians thought of the sea as impure[4]. LUCAS[5], however, mentions present day sources of salt in Egypt, including Lake Mareotis and salines at Port Said, and quotes PLINY[6] who says that in addition to various inland sources, there were artificial salines on the coast for the extraction of salt from sea-water. This would appear to invalidate LÜRING's objection, though it is undoubtedly true that a great quantity, and perhaps the larger part, of ancient Egyptian salt was obtained from inland lakes and other similar deposits.

References to *ḥmꜣt* are very frequent, particularly in medical contexts, and it is often found associated with *ḥsmn*[7], to which it was no doubt thought to be related. The Wadi Natrun[8] and another locality *š-Ḥr*[9] are mentioned as sources of *ḥmꜣt*, and *ḥmꜣt mḥt*[10] and *ḥmꜣt nt ḫꜣst*[11] are also referred to. The former, which occurs with some regularity, is regarded by BREASTED[12] as 'evidently one of the salt minerals from the northern desert along the west side of the Delta, chiefly chloride of sodium', and may well refer to salt from Lake Mareotis. The latter, literally 'desert salt', is perhaps that spoken of by PLINY[13] as occurring beneath the sand in the desert between Egypt and Arabia, and also in the western desert. Bricks of salt appear in lists of donations[14], and such bricks have in fact been found[15], and salt is also measured in *ḫꜣr*[16].

The characteristic taste of salt was noted by the Egyptians and commented upon[17], and both salt[18] and salt water[19] were used medicinally, fresh salt being occasionally specified[20]. ARRIAN[21] states that salt was used in temple observances, which is disputed by FORBES[22], who maintains that natron was preferred. But ARRIAN was doubtless right, as is shown by a reference to *ḥmꜣt r wpš* at Edfu[23], which must point to the scattering of salt for purification.

ḥmꜣt occurs in a number of somewhat obscure compounds in the medical texts, *šspt ḥmꜣt*[24], *imy n ḥmꜣt*[25] and *bdt ḥmꜣt*[26], the precise meaning of which is by no means clear. It is, however, possible that they refer to salt crystals or efflorescences, to be used in their naturally occurring state.

---

[1] P. Ch. B. IV vs. 7, 14. [2] Gardiner, Onom. 1 p. 15. [3] Med. Kent. p. 153.
[4] Plutarch, Is. et Os. 7c; 32b. [5] Materials p. 304. [6] HN. 31, 39–42.
[7] Eg. Peas. 48; Peas. Bt. 4; Harr. 16b, 3; 19b, 5; 65b, 4; 71b, 12 etc.
[8] Peas. 48; Bt. 4; cf. also *sḫt ḥmꜣt* – Peas. 87; P. Kah. 18, 44; Piehl Inscr. II, 116 etc.
[9] Anast. III, 2, 9; cf. Gauthier, Dict. géog. V p. 124–5.
[10] Smith 14, 5; Eb. 22, 3; 70, 11; 72, 6; 77, 10; 80, 20; 83, 16; 104, 19; Hearst 7, 1; P. Med. Berl. 3, 3; 7, 9 etc.
[11] Eb. 83, 7. [12] Pap. Smith p. 383. [13] HN. 31, 39, 78.
[14] Harr. 16b, 3; 19b, 5; 71b, 13; Med. Hab. 152, list 22, 625; 160, list 47, 1069.
[15] Lucas, Materials p. 305; Forbes, Studies III p. 167.
[16] Harr. 65b,14; 71b,12.
[17] Lansing 10, 1; Meir IV, 13; Love Harris 5, 1; P. Boul. 18, 24, 18.
[18] Eb. 32, 2; 63, 10; 69, 16; Hearst 5, 13; 7, 3 etc.
[19] P. Med. Berl. 2, 11. [20] Eb. 94, 15; 94, 17. [21] Anab. III, 4, 3.
[22] Studies III p. 168. [23] Edfou VI, 203, 8. [24] Eb. 41, 3; Hearst 1, 3; 13, 16.
[25] Hearst 4, 7. [26] P. Med. Berl. 2, 4.

The orthography of *ḥmꜣt* varies considerably, and it seems likely that three words accorded separate existence by the Wörterbuch are to be regarded as nothing more than aberrant writings. *ḥmꜣrt* (Wb. III. 94.7), known only from a passage in Hearst[1], corresponds to what is undoubtedly a variant of *ḥmꜣt* at Meir[2], and is probably therefore a similar garbling of the word for salt. *ḥmyt* (Wb. III. 82.4), occurring in Ebers in the compound *šspt ḥmyt*[3], is in one instance parallel to *šspt ḥmꜣt* in Hearst[4], suggesting that it too is merely an irregular writing of *ḥmꜣt*, though it may possibly be a distinct but closely related substance. *ḥmwt* (Wb. III. 81.22), also from Ebers[5], cannot be identified, and, like *ḥmyt*, may well be a mistake for *ḥmꜣt*.

In conclusion, it may be remarked that *ḥmꜣyt* (Wb. III. 94.6) is not identical with *ḥmꜣt wꜣḏt* as the Wörterbuch suggests, but is evidently a plant[6], probably the same as *ḥmꜣw* (Wb. III. 94.4), and perhaps *trigonella foenum graecum*[7].

Natron

bḫḫ (Wb. I. 471.15) is regarded as granular natron by the Wörterbuch, and by Forbes[8], and this seems to be the meaning in at least two contexts. In one, *bḫḫ* occurs closely connected with *nṯry*, and parallel to *dr-ḳd*[9], and in the other, written *bšš*, it is more explicitly described as *bšš nw bd*[10]. On the other hand, *bšš* is known from other examples[11] to be a species of *ʿntyw*, which gives rise to some difficulty. Either one must assume two separate words, or else give *bḫḫ* or *bšš* a primary meaning 'pellet' or 'granule', which could be applied indifferently to natron, incense, resin, and other such substances.

bsn (Wb. I. 475.11–14) presents a somewhat complex problem, and has been subjected to widely divergent interpretations. The word occurs in the Pyramid texts[12], and in offering lists[13], apparently as a substance used in purification, and is found parallel to *bd*[14], *ḥsmn*[15], and *nṯry*[16]. It is also found in the phrases *wpš bsn*[17] and *(r)dit bsn*[18], in ceremonies connected with building, and in the medical texts, either alone[19], or in the compounds *bsn n iḳdw*[20], *bsn n tꜣ*[21], and *bsn n ṯḥnt*[22].

In view of the use of *bsn* for purification, and the parallelism with natron, Brugsch[23] regards it as an alkaline salt, similar to *ḥsmn*, *bd* and *šrp*, and a like interpretation is adopted by Lüring[24], who reconciles the references to *bsn n iḳdw* by taking *bsn* there to be a saltpetre exudation from walls, or a species of natron employed in making mortar. Sethe[25], however, while admitting *bsn* as a typical purification material, insists that because of the connection with building in the compound *bsn n iḳdw*, it must be lime or gypsum used as

[1] Hearst 3, 15. [2] Meir IV, 13. [3] Eb. 65, 21; 69, 21; 89, 22.
[4] Eb. 89, 22 = Hearst 13, 16. [5] Eb. 55, 3; 55, 11.
[6] Gardiner, Onom. 1 p. 19; p. 21; cf. in particular Smith 21, 9f.
[7] Loret, Mél. Masp. 1 p. 854 n. 1; p. 866f; cf., however, Keimer, BIFAO. 28 p. 84.
[8] Studies III p. 159. [9] Piehl Inscr. II, 106. [10] Ombos II 48, 592, 7.
[11] Wb. I. 478.15.; cf. Steuer, Myrrhe und Stakte p. 41. [12] Pyr. 774b; 101g.
[13] Cairo 1392; Giza J. V pl. 26, 4; Mar. Mast. B. 16. [14] Edfou II, 32; Giza J. V pl. 26, 4.
[15] Düm. Baug. 47. [16] Piehl Inscr. II, 75.
[17] Piehl Inscr. II, 75; Edfou II, 32; II, 62; Düm. Baug. 47; Abydos Sethos Saal C ⟨15⟩ etc.
[18] Young, Hier. 42, iii. [19] Eb. 84, 5.
[20] Hearst 7, 11; Eb. 73, 21; 74, 1; P. Med. Lond. 4, 4.
[21] Eb. 72, 21; 74, 7; Hearst 9, 18. Wb. V. 228.17.
[22] Eb. 105, 15. [23] Wb. Erg. p. 448. [24] Med. Kent. p. 152. [25] Dram. Texte p. 139.

whitewash on wooden or brick buildings, noting the relationship between *bsn* and 'die weiße Kapelle'[1]. He interprets *wpš bsn* and *(r)dit bsn* as the scattering of some granular substance playing the same part symbolically in relation to a stone building as whitewash did to one of wood or brick. EBBELL[2] also translates gypsum, presumably because it seems more appropriate to the compounds occurring in the medical texts.

The evidence is, indeed, somewhat equivocal, and neither of the interpretations proposed is suitable in every instance. The references in the Pyramid texts, and the apparently close connection with *bd*, *ḥsmn* and *nṯry*, suggest an alkaline salt, and this would also be appropriate to the examples of *wpš bsn*, where the granular substance scattered could be natron crystals or soda[3]. Alternatively, this might be gypsum, as SETHE suggests, and such a translation would be particularly apposite for *bsn n iḳdw*, gypsum being the regular mortar of the ancient Egyptians, while natron was not so employed[4]. If *ṯḥnt* in the phrase *bsn n ṯḥnt* is the common word for faience or glass, as seems likely, either lime or natron would be applicable, since both were used, the one in making glass[5], and glazing faience[6], the other for glass[7], and as a binding medium for faience[8]. *bsn n tꜣ*, *bsn* of the kiln, is not easily understood, and neither natron nor gypsum was used in connection with pottery. It may, however, be a reference to the slaking of gypsum before use as mortar, and if so would support the latter interpretation.

In considering *bsn* as gypsum, it must be borne in mind that a quite different word, *ḳḏ*, was used in the New Kingdom. From other similar instances of the introduction of new words, it seems probable that *ḳḏ* was adopted to replace a rather less precise term, the meaning of which was not altogether clear. Now it is significant that gypsum sometimes occurs in scattered masses of loosely aggregated crystals[9], and that it is found in the Wadi Natrun[10], whence came natron and salt, both in similar crystalline forms. Further, a form of granular soda or salt was found in the oasis of Ammon, not modern sal-ammoniac, but real rock salt associated with gypsum, and having a salt content of 10 or 20 per cent[11] It seems possible, therefore, that *bsn* was the name of various similar crystalline substances which the Egyptians found associated in nature and failed at first to distinguish, and that it covered both granular soda, ie. natron, certain forms of gypsum, and perhaps rock salt[12].

bd (Wb. I. 486.5–9) is quite clearly a species of natron, though the exact definition is uncertain, and a somewhat wider significance is perhaps possible. NAVILLE[13], noting the existence of the cognate verb *bd*, and the fact that *bd* follows *mw* in offering lists, thinks that it was a composite substance including *snṯr*. He translates conventionally as perfume, explaining that 'c'était une substance en forme de boules ou de grains, qui n'était pas destinée à être brûlée comme de l'encens, mais qui avait sans doute un effet analogue au bétel, celui d'agir sur la bouche et de rendre l'haleine odorante'. Though grasping many essential points,

[1] Cf. also Kees, Farbensymbolik p. 440; Kulturgeschichte p. 146.
[2] Pap. Ebers. p. 132.
[3] Cf. the parallel *ḥmꜣt r wpš* of Edfou VI, 203, 8.
[4] Lüring's salt-petre efflorescence is ruled out, since there is no evidence for salt-petre (nitre, potassium nitrate) in Egypt; cf. Lucas, Materials p. 303.
[5] Lucas, Materials p. 213. [6] ib. p. 181. [7] ib. p. 214. [8] ib. p. 204.
[9] ib. p. 470; p. 97. [10] Hume, Geology 1 p. 167; p. 170.
[11] Forbes, Studies III p. 167; Pliny, HN. 31, 39, 78–80; Strabo I, 3, 4, 49; Arrian, Anab. III, 4, 3; Herodotus IV, 181–5.
[12] Lucas, Materials p. 305, notes that sulphur also occurs associated with gypsum.
[13] ZÄS. 11 p. 28.

Naville is unfortunate in his interpretation, and Brugsch[1] is nearer the truth in translating *bd* as 'das helle, reine, zum Abwaschen und Abspülen benutzte (alkalische) Salz, eine besonders reine Natronart', a definition which has been generally accepted. Jéquier[2], on the other hand, discussing *ỉbnw*, notes that the earlier word for alum and similar substances is not known, and suggests that they were originally covered by a more general term, 'peut-être un de ceux qu'on traduit en général par natron, comme *bd*, qui désigne une substance servant à clarifier l'eau, ou encore *ḥsmn*', which would conflict with Brugsch's view that *bd* was a particularly pure form of natron.

It is evident that the idea of purification was uppermost in the use of *bd*, and its appearance in the offering lists, where, however, it is not found before the 4th dynasty[3]. The cognate verb *bd* (Wb. I. 486.10–13) means simply to purify, and includes cleansing with water and incense as well as *bd*. The substance was generally used in the form of pellets or crystals[4], to be chewed or dissolved in water, and Jéquier[5] suggests that , the normal determinative of *bd*, represents a packet of such pellets on the end of a stick, which could be dipped into water and then withdrawn[6]. Water so clarified seems to have been referred to as *mw nw bd*[7], and this would account for the juxtaposition of *bd* and *mw* in offering lists.

The use of *bd* in the form of pellets or crystals might suggest that this was the state in which it was found, and that it was, therefore, a species of granular soda. As early as the Pyramid texts[8], it appears in parallelism with *nṯry* and *ḥsmn*, and this relationship is repeated elsewhere[9]. Of particular significance is the interchange of *bd* and *ḥsmn* in the Dendera tribute lists. In one[10], a figure representing El Kab or Hierakonpolis brings a casket labelled *bd*, while in the accompanying text reference is made to both *bd* and *ḥsmn*. In the other[11], which is similar, the casket is labelled *ḥsmn*, and the text refers to *ḥsmn*, *bd* and *nṯry*. From this it is clear that there was a very close affinity between the three words, and that in the Greek period at least, *bd* was simply a term for natron.

Various sources of *bd* are mentioned in the texts, the most important being *šrp*[12], the Wadi Natrun, the location of the largest natron deposits in Lower Egypt[13]. *Nḫb*, El Kab, the chief source in Upper Egypt[14], is also referred to[15], and *ỉnt ḥḏ nḫn*[16] is cited as a third region from which *bd* was obtained[17]. The latter may be identical with *ỉnt ḥsmn* as Junker[18] suggests, and would seem to refer to Hierakonpolis, though it may also be a more general designation of the region round El Kab, where five distinct deposits are known[19].

An exact appraisal of the meaning of *bd* is hardly possible from the evidence, and it seems likely that the word had a somewhat different significance at various periods. In Graeco-Roman times, and perhaps even from the New Kingdom, it was evidently a term for natron,

---

[1] Wb. Erg. p. 462. [2] BIFAO. 19 p. 107. [3] Selim Hassan, Giza VI (2) p. 315.

[4] *ṯ3*, D. el B. 10; Karn. ⟨744⟩; Abyd. Rit. Tabl. 34–5 etc. The colour of the pellets varies, cf. Cairo 28024, 28; 28036, 10; 28037, 9.

[5] Frises p. 318.

[6] For the use of as det. cf. Gardiner, Onom. 1 p. 12, who thinks that there may have been some mythological nexus between natron and bone, or that hieratic confusion had arisen as early as the M. K.

[7] Dend. Mar. IV, 22a. [8] Pyr. 849c.

[9] LD. II, 25; Hier. Ostr. 113, 1 rt. 1 (*nṯry*); Totb. Nav. 169, 7; Destr. 76 (*ḥsmn*); Totb. Nav. 79, 8 (both).

[10] Dend. Mar. I, 70. [11] ib. I, 71. [12] Düm. G. I. II, 85A; Mammisi 97.

[13] Lucas, Materials p. 297. [14] ib. p. 299.

[15] Düm. G. I. II, 85A; Dend. Mar. I, 70; II, 28.

[16] Gauthier, Dict. géog. I p. 89. [17] Dend. Mar. I, 70; I, 71.

[18] Onurislegende p. 78. [19] Lucas, Materials p. 299.

and as such is not easily distinguished from *ḥsmn* and *nṯry*. In earlier periods, however, the notion of purification was probably uppermost, and *bd* may have embraced both crystalline natron, and certain other minerals of approximately similar nature[1]. Among these were perhaps alum and kindred astringent salts, for which no name can be traced before the introduction of *ỉbnw*, though these occur chiefly at Dakhla and Kharga, which are not mentioned as sources of *bd*. The Egyptians may even have differentiated between liquid natron, and granular soda, and in this case *bd* would probably refer to the latter, but such a distinction cannot be proved.

nṯry (Wb. II. 366.8–11), Akkadian *nitiru*, *nitru*, Hebrew *neter*[2], Syriac *nethrâ*, Aramaic *nitra*, Greek νίτρον, Latin nitrum. That *nṯry* means natron, a naturally occurring compound of sodium carbonate and sodium bicarbonate, has never been in doubt. Brugsch[3] translates *nṯry* as 'das mineralische Laugensalz, Reinigungssalz, Natron', deriving the Hebrew *neter* from the Egyptian, and this is also the view of Lutz[4], who renders 'carbonate of soda'. Steuer[5] goes somewhat further, and suggests that 'mit großer Wahrscheinlichkeit ist *nṯr* eine allgemeine Bezeichnung für Natron, während *bd* und *ḥsmn* bestimmte Sorten verschiedener Mischungsverhältnisse zwischen Natriumkarbonat, Natriumbikarbonat, Natriumsulfat und Kochsalz bezeichnen', comparing Lucas' analyses[6].

The fact that *nṯry* is ultimately the origin of the English word nitre has been the cause of some confusion, though the position has been clarified by Lucas[7]. *nṯry*, νίτρον and nitrum all refer to natron[8], a natural soda consisting of sodium carbonate and sodium bicarbonate, with sodium chloride and sodium sulphate as impurities. Natron is not sodium nitrate[9], which is employed as a fertilizer but cannot be shown to have been used anciently, nor is it potassium nitrate, nitre or saltpetre, which was likewise unknown in Egypt.

nṯry occurs with great frequency in Egyptian texts, and two varieties, *nṯry mḥ*[10] and *nṯry šmᶜ*[11] are mentioned. The former would most naturally refer to the natron of the Wadi Natrun, the chief Lower Egyptian source, and this is in fact implied in the Pyramid texts[12], where *nṯry mḥ* is qualified by *št-pt*, ie. *šrp*, the name of the Wadi. Similarly, *nṯry šmᶜ* is elsewhere qualified as *nḫb*[13], the name of El Kab, the most important locality from which natron was obtained in Upper Egypt. But El Kab is also specifically cited as the source of *nṯry*[14], and *Ḥr m nḫn*[15], evidently a region near Hierakonpolis, and *ỉrw*[16], may refer to one or other of the El Kab deposits. In the Dendera tribute lists[17], a casket of *nṯry* is brought by a figure representing the oasis of Baharia, and although this is not a known source of natron[18], it is significant that it is connected with the Wadi Natrun by a road, so that it may have been a stage on a desert trade route between the Wadi and Upper Egypt.

---

1 Not, however, kali-saltpetre, as Lutz, Textiles p. 76, suggests, since the substance was not used in Egypt.

2 Cf. JAOS. 73 p. 152.

3 Wb. Erg. p. 708.

4 Textiles p. 75.

5 Wohlriechende Natron p. 28.

6 Materials p. 548–9; cf. ib. p. 303.

7 ib. p. 303.

8 Bailey, Elder Pliny I p. 169, however, thinks it possible that in one or two instances Pliny's nitrum refers to nitre not natron; cf. HN. 31, 46, 109.

9 Sloley, in reviewing Lucas, JEA. 34 p. 125, wrongly states that it is.

10 Pyr. 27; D. el B. 110; Cairo 1486; Mem. Miss. V pl. 8 etc.

11 Pyr. 26; 27; D. el B. 110; Cairo 1486; Mem. Miss. V pl. 8 etc.

12 Pyr. 27e.

13 Mem. Miss. V pl. 8; Pyr. 26.

14 Edfou I, 471; II, 33 etc.

15 Edfou II, 75.

16 Pyr. 864 c, d; Sethe, Kommentar IV p. 128.

17 Dend. Mar. I, 70.

18 Cf. however, Ball & Beadnell, Baharia p. 37; p. 41 etc.

Like *bd*, *nṯry* was sometimes used in the form of pellets or the like (*ṯꜣ*)[1], and was employed for purifying the mouth[2], while the expression *nṯry m ḏt.f*[3] may refer to natron in its natural state. *nṯry* occurs in conjunction with both *bd* and *ḥsmn*[4] and parallel to *bsn*[5] as a granular substance to be scattered. It is also related to *snṯr* in the similar phrases *ḳbḥ nṯry*[6] and *ḳbḥ snṯr*[7] and Steuer[8] thinks that *snṯr* originally had a dual meaning.

It seems clear that *nṯry* is the most general word for natron, from whatever source, and it is interesting to note that *ḥsmn* is described as *nṯry šmꜥ*[9], while *bd* is also referred to in a way which would suggest that it was regarded as a species of *nṯry*[10]. Such a view would be consistent with the borrowing of *nṯry* by Hebrew, Greek, Latin and other languages, though it is somewhat surprising that the regular Coptic word for natron should be derived from *ḥsmn*.

In conclusion, a word must be said concerning the various forms of the ideogram *nṯry*, a question discussed in detail by Steuer[11], who is of the opinion that 𓊹 was originally for *snṯr*, and 𓊹 for natron, though in the Pyramid texts, 𓊹, 𓊹 and 𓊹 are interchangeable. Forms such as 𓊹𓂋, 𓊹𓂋 and 𓊹𓂋, where *r* is written, he would prefer to read *snṯry*, which he concludes, probably incorrectly, had a double meaning, primarily 'Gottesduft', more specifically 'duftendes Natron'.

He also makes the following points[12].

1. 𓊹 alone, without phonetic complement, usually refers to natron, and the reading is *nṯr(y)*. Both *bd* and *ḥsmn* require a fuller writing[13].
2. 𓊹 in the Pyramid texts is *nṯr(y)*.
3. 𓊹 with complement *r* or *n* (only once) is also *nṯr(y)*.

These conclusions seem to meet the facts, and further discussion here would be out of place.

nṯryt (Wb. II. 366.14) is known largely from medical contexts[14], and was formerly regarded as a compound *nṯr-ryt* or *nṯr-tit*, translated by Brugsch[15] as 'Gotteskoth', by Stern as 'metallum quoddam seu color', and by Joachim as 'netr-tit Metal'. Breasted[16], more reasonably, sees *nṯr(y)t* as an unidentifiable drug, 'possibly a preparation of natron, the native sesquicarbonate of sodium . . . known to the Egyptians at a very early date under the name *nṯry*'. This seems a likely suggestion, and is to some extent confirmed by a recently published text[17] referring to 'those who are anointed in the mouth with *nṯryt*'. Caminos[18] here accepts *nṯryt* as a variety or by-product of *nṯr(y)*, the use of which for the ritual purification of the mouth is well known[19]. More precise definition is not possible, and *nṯryt* may be accepted as a species of natron or something prepared from it, though it is just feasible that it is nothing more than an aberrant feminine form of the more common *nṯry*.

---

[1] Pyr. 26; LD. II, 28; Mem. Miss. V pl. 8; D. el B. 110 etc.
[2] Pyr. 2015; 1368f.
[3] Edfou VI, 202, 2.
[4] LD. II, 25; Hier. Ostr. 113, 1 rt. 1 (*bd*); Pyr. 2015; 1368a; Abyd. Mar. I, 28b, 7 (*ḥsmn*); Pyr. 849; Totb. Nav. 79, 8; Dend. Mar. I, 71 (both).
[5] Piehl Inscr. II, 75; cf. Edfou II, 33.
[6] Amamu 17; Cairo 1422; D. el B. 110 etc.
[7] Cairo 1422.
[8] Wohlriechende Natron p. 28.
[9] Pyr. 26f.
[10] Abyd. Rit. Tabl. 34–5, *nṯry . k bd*.
[11] Wohlriechende Natron p. 22f.; cf. also FuF. 14 p. 54f.
[12] Wohlriechende Natron p. 28f.
[13] This distinction is maintained in the Wörterbuch.
[14] Eb. 48, 16; 48, 18; 79, 1; Smith 16, 4.
[15] Wb. Erg. p. 1311.
[16] Pap. Smith p. 412.
[17] Lit. Frag. 10, 3, 4–5.
[18] Lit. Frag. p. 32.
[19] Cf. Lucas, Materials p. 319; Steuer op. cit. p. 36; p. 92; JEA. 5 p. 161.

ḥsmn (Wb. III. 162.11f.), Coptic ϩⲟⲥⲙ̄, is undoubtedly natron, to be distinguished sharply from the two similar words for bronze and amethyst[1]. Concerning the more precise identification of *ḥsmn*, varying views have been advanced. BRUGSCH[2] regards it as a particular species of natron, connecting it with the verb *ḥsmn* (Wb. III. 163.3), and the noun meaning menstruation (Wb. III. 163.8), and translating 'das Reinigungssalz'. LÜRING[3], on the other hand, takes *ḥsmn* as a more general term, noting two varieties, one from *šrp*, the other from *nḫb*, and a third, *ḥsmn dšr*, which he regards as natron made impure by iron ochres and the like, such as occurs in the natron lakes northwest of Cairo. A similar view is taken by LUTZ[4], and JÉQUIER[5] goes even further in suggesting that *ḥsmn* may at one time have included alum compounds.

ḥsmn occurs with great frequency as a substance for purification, and in conjunction with salt and other terms for natron. As a purifying agent[6], particularly of the mouth[7], *ḥsmn* is parallel to *bd* and *nṯry*, with both of which it is associated in other contexts[8]. It is also, like *nṯry*, found together with *snṯr*[9], where a general meaning of natron as opposed to incense seems necessary. More often, *ḥsmn* appears parallel to *ḥmꜣt*, salt[10], and it seems too, like salt, to have been compressed into bricks[11], which suggests a crystalline form. A solid rather than a liquid can also be inferred from the depiction of a heap of *ḥsmn*[12], and its transport in boxes[13]. Unlike *bd* and *nṯry*, *ḥsmn* does not appear to have been used in pellets (*ṯꜣ*), though the expression *dḳw nw ḥsmn* is known[14], and may refer to natron crystals. *ḥsmn* occurs very frequently in the medical texts[15], where *mw nw ḥsmn* is also found[16], and both are used in embalming[17], which is surely significant.

A specific variety of *ḥsmn*, *ḥsmn dšr*, which is known from medical contexts[18], is of particular interest. It is clearly the same as Coptic ϩⲟⲥⲙ̄ ⲉϥⲧⲣⲉϣⲣⲟϣ, the Greek βερενικάριον[19] (a diminutive of βερενίκιον[20]), and refers to red natron, so coloured by some impurity, presumably an iron compound. PLINY[21] mentions a species of nitrum coloured red by the surrounding earth, and also refers[22] to a lake near Memphis from which salt of a red colour was obtained. It is perhaps possible, therefore, that *ḥsmn dšr* represents not only red natron, but also other similarly tinted deposits of an analogous nature, including certain astringent minerals, as suggested by JÉQUIER. Apparently opposed to this impure species is *ḥsmn wꜥb*[23], the existence of which may even imply that most *ḥsmn* was crude.

---

[1] Erman, ZÄS. 30 p. 31. [2] Wb. p. 996. [3] Med. Kent. p. 154.

[4] Textiles p. 75. [5] BIFAO. 19 p. 107.

[6] Totb. Nav. 169, 7; Nauri 13; Urk. IV, 874; Pianchi 97 etc.; cf. also *ḥwt ḥsmn*, Gauthier, Dict. géog. IV p. 114.

[7] Pyr. 1368a; P. Tur. PR. 58, 11; Totb. Nav. 172, 1; Admon. 11, 2.

[8] Totb. Nav. 79, 8; 169, 7; Destr. 76; B. v. Atmen 1, 10 etc. (*bd*); Pyr. 1368a; 2015; Coff. Texts I, 62d; Abyd. Mar. I, 28b, 7 etc. (*nṯry*); Pyr. 849; 864; Dend. Mar. I, 71; Totb. Nav. 79, 8 etc. (both).

[9] Pyr. 864d; Totb. Nav. 133, 19; 172, 1 etc.

[10] Peas. 48; Bt. 4; Harr. 16b, 2–3; 19b, 5; 65b, 4; 71b, 12–3; Anast. III, 2, 9 etc.

[11] Harr. 65b, 1; 71b, 13; Abyd. III, 54, 10.

[12] Urk. IV, 1130.

[13] Kagemni 4; Bersh. II, 44; Dend. Mar. I, 71.

[14] Eb. 87, 5 = Hearst 10, 16; cf. Appendix I.

[15] Eb. 66, 2; 71, 14; 71, 21; 73, 20; 75, 9; 78, 13; 78, 18; 85, 7; 88, 2 etc.; Hearst 12, 8; Smith vs. 21, 4; P. Leid. 343 vs. 4, 7–8 etc.; cf. also Louvre Apisstele 5 (Rec. Trav. 21, 72).

[16] Eb. 97, 16; cf. P. Dem. Mag. 16, 23; 28, 4.

[17] Emb. Rit. 7, 8; 7, 10; etc.; P. Dem. Mag. 12, 12; 12, 31.

[18] Eb. 57, 1; 57, 5; 72, 6; 75, 17; Smith vs. 21, 4 etc. [19] Cod. Paris 44 fol. 88.

[20] Cf. Galen 13, 568. [21] HN. 31, 46, 111. [22] HN. 31, 41, 86.

[23] P. Med. Berl. 5, 8; Gurnah Sethostempel ⟨100⟩.

Several sources of *ḥsmn* are known from the texts. Like other varieties of natron, it was obtained from the Wadi Natrun[1], the Coptic ⲠⲦⲞⲞⲨ ⲘⲠⲒϨⲞⲤⲈⲘ, and from El Kab[2], where at least five separate deposits are known. One or other of these may be the location *tꜣ ỉnt*[3], and *ỉnt ḥḏ nḫn*[4], evidently a region near Hierakonpolis, perhaps refers to another of the same group. Baharia, from which *nṯry* was brought, is also mentioned as a source of *ḥsmn*[5], and may represent a trade station on the route from the Wadi Natrun, though there are dried up salines at the oasis itself[6]. Another area from which *ḥsmn* is said to have been obtained is the Oxyrhynchite nome[7], which may also have been a transit point[8], though Al Kalkashandi describes natron deposits near Behnesah, now no longer extant, which may be the ones referred to[9]. Lastly, we read of *ḥsmn* gatherers at Elephantine[10], which corresponds to no known source, and Coptic specifies ϨⲞⲤⲘ ⲚⲀⲢⲀⲂⲒⲔⲞⲚ[11].

To arrive at a satisfactory interpretation of *ḥsmn* is by no means easy, and in the Greek period at least it seems to have been a general term, parallel to *bd* and *nṯry*. As a substance used in purification (cf. the related verb) it may, like *bd*, have included natron and other minerals not clearly distinguished from it. The variety referred to as *ḥsmn dšr* also suggests that *ḥsmn* embraced less pure species of natron, and perhaps even other salts as well. On the other hand, it may be significant that the majority of the sources from which *ḥsmn* is stated to have been obtained are in Upper Egypt, and that *ḥsmn* is once described as *nṯry šmᶜ*[12], which might imply that the distinction of *ḥsmn* was partly geographical. This is to some extent borne out by the fact that, according to Lucas' analyses[13], natron from El Kab contains a higher percentage of impurities than that from the Wadi Natrun. Altogether, therefore, it seems reasonable to suppose that *ḥsmn* refers primarily to the less pure forms of natron containing large quantities of sodium chloride and sodium sulphate, which might explain the apparent relationship between *ḥsmn* and *ḥmꜣt*.

swᶜb (Wb. IV. 67.4) is apparently ἅπαξ in a passage where *swᶜb pr m nḫb* is balanced by *ḥsmn pr m tꜣ ỉnt*[14]. The Wörterbuch is therefore right in regarding *swᶜb* as a term for natron, though it is clearly an artificial word derived from the verb *swᶜb* on the analogy of other pairs of noun and verb[15], and cannot have any real meaning. There is thus no justification for interpreting *swᶜb* as a specific variety of natron from El Kab.

šrp (Wb. IV. 528.4), occurring largely in later texts, is clearly a word for natron derived from the name of the Wadi Natrun, originally *št-pt* and later *šrp*[16]. Brugsch[17] rightly regards it as the name of a particular type of natron, though his translation 'Himmels-Salz' is based

---

[1] Peas. 48; Bt. 4; also *ḥsmn n sḫt*, Hearst 15, 14; Q. Saqq. III, 82f. 1.9 (cf. ZÄS. 56, 51), and *š n ḥsmn*, Totb. Nav. 17, 22.

[2] Q. Saqq. III, 82f. 1.9; Emb. Rit. 14, 16 (personified as Nekhebet).

[3] Emb. Rit. 20, 8; 25, 12; cf. Gauthier, Dict. géog. I p. 89.

[4] Dend. Mar. I, 70; I, 71.

[5] Q. Saqq. III, 82f. 1.9.

[6] Ball & Beadnell, Baharia p. 37; p. 41 etc.

[7] Pyr. 864.

[8] Cf. Sethe, Kommentar IV p. 128; Mercer, Pyr. Texts II p. 440.

[9] Lucas, Materials p. 299.

[10] Cat. des Mon. I, 118c. *ḥsmn* parallel to *ḥmꜣt* and certainly, therefore, natron.

[11] Chassinat, P. Méd. Copte p. 176. Lucas' natron from Retenu (Materials p. 300) is a mistake, *ḥsmn* being the stone of the same name.

[12] Pyr. 26f.

[13] Materials p. 548–9.

[14] Emb. Rit. 27, 1.

[15] Cf. *bd* and *bd*, *ḥsmn* and *ḥsmn*.

[16] Cf. Fakhry, Ann. Serv. 40 p. 840f.

[17] Wb. Erg. p. 1200.

on a false etymology and may be disregarded. The Wadi Natrun was the source of natron par excellence, and *šrp* must refer primarily, if not exclusively, to natron from that source. In some cases, it is difficult to determine whether *šrp* is the place-name or a designation of natron. Thus *ṯ3 nw šrp*[1] and *mw nt šrp*[2] may be either crystals and water from the Wadi Natrun, or of natron, and *ṯ3 mḥ nw šrp*[3] is similarly equivocal, though the writing suggests that *šrp* is here natron. Likewise, *bd nw šrp*[4] may simply be natron from the Wadi, though again the determinative of *šrp* in this instance suggests a substance, and a translation 'natron of the type *šrp*' is possible, particularly since *bd* and *šrp* are elsewhere distinguished[5].

Why the Wadi Natrun was called *št-pt* is not known, and Selim Hassan[6] may be right in suggesting that the Wadi with its natron lakes was thought of as exemplifying the early conception of heaven with its celestial pools.

dni (Wb. V. 467.11) occurs only once[7], and is evidently a material used in embalming, though the context is too obscure to allow of identification. Items of *ḥsmn* have previously been mentioned, and honey follows, and the large amount of *dni*, 2000 *dbn*, is perhaps significant. In all probability *dni* represents some kind of salt, perhaps a particular species of natron.

dr-ḳd (Wb. V. 474.18) is apparently confined to a single context, where it is approximately parallel to *bẖẖ*[8]. The Wörterbuch regards it as a term for natron, and this seems likely to be correct. The word is, however, a fanciful creation, and can have no real significance, unless it is an aberrant writing of *dḳw*, a possibility which cannot altogether be ruled out.

An attempt must now be made to summarise the various conclusions reached in the present chapter, and to see how far they present a coherent picture, though no very clear definition will be possible, since it is evident that the Egyptians were unable satisfactorily to distinguish between a number of analogous saline deposits.

Of the words for alum and the like, little need be said. *ibnw* is the most general term, and no doubt covers most sulphates of alumina, potassium or iron, while the other words, *imrw*, *inr n šn*, and *wšbt*, represent particular species of alum or other similar astringents. The chief point of interest is that none of these words occurs before the New Kingdom, which seems to imply either that alum was not used before that date, which is extremely unlikely, or that one of the terms for natron originally embraced the alum compounds also. Jéquier suggests that either *bd* or *ḥsmn* may have had such breadth of meaning, particularly the former, and while he may be right in this, evidence would rather point to *ḥsmn*.

Salt, *ḥm3t*, was also closely associated with natron, especially *ḥsmn*, and it is significant that sodium chloride is present as an impurity in all natron, some samples from El Kab containing more than 50 per cent. It is hardly surprising, therefore, that some confusion should have existed, and in particular *bsn* seems to have included certain forms of gypsum, granular soda, and perhaps rock salt.

šrp, natron from the Wadi Natrun, the two 'artificial' terms *swᶜb* and *dr-ḳd*, and *dni*, in all probability a species of natron or other saline deposit, present no problem. *bẖẖ* is evidently

[1] Mammisi 97. [2] Edfou II, 62. [3] Lib. Fun. I, 39.
[4] Düm. G. I. II, 85A. [5] LD. III, 255i, 8; Philae ⟨2979⟩ Phot. 859/60.
[6] Giza VI (2) p. 184. [7] Q. Saqq. III, 89f. [8] Piehl Inscr. II, 106.

granular soda, corresponding to one meaning of *bsn*, and *nṯryt*, if indeed a distinct word, must be a natron preparation related to *nṯry*. The three principal terms for natron, however, are not so readily distinguished.

Egyptian natron is basically similar, a natural compound of sodium carbonate and sodium bicarbonate[1] containing various impurities, but the proportion of these impurities, chiefly sodium chloride and sodium sulphate, varies considerably in samples from different localities. This seems to have been recognised by PLINY, and may have been realised by the Egyptians themselves, though a distinction of *nṯry*, *ḥsmn* and *bd* according to quality or provenance alone does not appear altogether satisfactory.

nṯry is undoubtedly the most general designation of natron proper, and is qualified as *nṯry mḥ* or *nṯry šmᶜ*, corresponding to the two principal sources of natron at Wadi Natrun and El Kab, which are also mentioned specifically. There is, however, nothing to suggest that any other substance than natron was referred to as *nṯry*.

ḥsmn too is obtained from both Wadi Natrun and the region around El Kab, though more frequently from the latter, and since it is described in the Pyramid texts as *nṯry šmᶜ*, it is perhaps reasonable to suppose that it was primarily the less pure southern variety of natron. Further, it is closely associated with salt, and may well have a somewhat wider meaning, including other saline deposits not clearly distinguishable from natron. This would also seem to be suggested by the term *ḥsmn dšr*, which implies considerable impurity.

bd might then be regarded as principally the purer northern variety of natron, but this does not seem to fit the evidence. Nowhere is *bd* directly opposed to *ḥsmn*, nor is it exclusively connected with the Wadi Natrun. Admittedly it was widely used for ritual purification, but this does not necessarily imply that the substance was itself pure, and in this context *bd* may include not only natron, but other crystalline salts. In fact, no very exact definition of *bd* is possible.

Finally, a word is necessary concerning the sources from which natron was obtained, in comparison with those mentioned in the texts. Of the three chief sources at the present day, the Wadi Natrun, the El Kab deposits, and those at Barnugi near Naucratis, only the first two occur in connection with *bd*, *nṯry* and *ḥsmn*. The latter, the Coptic Pernoudj and Nitria, does not seem to correspond to any of the ancient Egyptian localities cited, which is certainly strange, though of course *nṯry mḥ* may refer to Lower Egyptian natron in general, including that from both Wadi Natrun and Barnugi. On the other hand, various other regions are said to have produced natron, principally *ḥsmn*, among them Baharia, the Oxyrhynchus nome, and Elephantine, where no deposits are now extant. There are, however, dried up salines at Baharia, and extensive natron beds at Tarabiya near Behnesah in the Oxyrhynchus nome are mentioned by AL KALKASHANDI, so that either may have been exploited anciently. But Elephantine is most unlikely as a source of natron, and the reference to natron gatherers there is puzzling. Of the alum deposits, chiefly at the oases of Dakhla and Kharga, there is no apparent trace in the texts.

In short, there are still many problems to be solved concerning the occurrence of natron in Egyptian texts, and further evidence is necessary before any definite conclusions can be reached.

[1] The proportion of these two salts in ancient specimens analysed by Lucas, Materials p. 549, varies from 94 per cent. to as low as 15.5.

CHAPTER 9

EARTHS

One further group of minerals now remains to be considered, namely the common earthy materials widely used in the making of bricks and pottery, and for various other purposes. Certain ochreous and astringent earths have already been treated above, and the present discussion will be confined to words representing soil, mud, clay, dust, sand, rubble, and similar substances. Of these, several have more frequent primary meanings such as 'land', 'earth's surface', or 'field', from which their use 'als Stoff' is derived, but it is only the secondary, material, designation that is here relevant. As might be expected, reference to earth in its different forms is neither precise nor consistent, and while certain divisions of meaning are quite clear, it must be conceded that there is a considerable overlap between words, and that exact definition is not always possible.

ꜣḥt (Wb. I. 12.18), the usual word for field or arable land, is also found as a term for the Nile alluvium of which cultivated fields in Egypt were largely composed, and from which bricks and the coarser kinds of pottery were commonly made[1]. Textual evidence for both these uses can be cited, *ꜣḥt* being mentioned as the material from which bricks are struck[2], and also as the stock in trade of the potter[3]. In the former case only Nile mud is applicable, while in the latter *ꜣḥt* must refer to the coarser alluvial pottery clay, since the finer variety seems to have been called *sin*. A further passage appears to connect *ꜣḥt* and *ḳꜣḥ* with the plastering of tomb walls[4], and this is confirmed by the use of a coarse plaster of Nile alluvium, usually faced with a clay or gypsum plaster, as a base for wall paintings[5]. From these few examples it is abundantly clear that *ꜣḥt* as a material is simply clay made from Nile alluvium, for which it is evidently the most precise, though not the only, term.

iwtn (Wb. I. 58.5–10), a fairly common word for ground or ground-level, particularly in Late Egyptian, is relevant here through its extended meaning of dirt or dust, cf. Coptic ⲉⲓⲧⲛ[6]. This meaning rests on three passages, two of which suggest dirt or mud as a translation, the other requiring dust or the like. In Lansing[7], the potter is said to be anointed with *iwtn*, which must surely in this case be mud or clay, though a more general meaning dirt is possible. The same expression also occurs in Orbiney[8], in connection with mourning, and can there be parallelled by a Coptic example quoted by Crum[9], 'each put ⲉⲓⲧⲛ on head and threw ϣⲟⲉⲓϣ into air'. This reference to the outward signs of grief enables the meaning of *iwtn* to be fixed as mud rather than dirt or dust, and thus supports the Lansing passage. Herodotus[10]

---

1 Lucas, Materials p. 62; p. 420.
2 Urk. IV, 837; 1152.
3 Sall. II, 5, 7; 5, 8; cf. Hier. Ostr. 13, 1, 1.
4 Urk. IV, 57.
5 Lucas, Materials p. 95.
6 Crum. 87b.
7 Lansing 4, 4.
8 Orb. 8, 7.
9 87b.
10 II, 35; cf. Loret, Kemi 9 p. 103.

states that when a man of consequence died, the ladies of the house daubed their faces and heads with mud (πηλῷ), and the same is related by DIODORUS[1], who also speaks of mud. LANE[2] and BLACKMAN[3] cite modern parallels. From this it seems evident that the best translation in Orbiney is mud, and this also suits the reference to the potter. But mud is not the only meaning of *iwtn*, as is shown by a text[4] concerning pigments, in which the ochres are said to have become reduced to *iwtn*, which must be dust or powder, and the same meaning seems likely in a medical context[5], where the *iwtn* of two bricks is mentioned. It appears, therefore, that *iwtn* refers both to mud and dust, and 'dirt' is perhaps the most satisfactory translation.

im (Wb. I. 78.2–3) is evidently some plastic material, probably a plaster of some kind. BRUGSCH[6] compares Coptic ⲞⲘⲈ, and translates accordingly 'lutum, pulvis, argilla', and LÜRING[7], commenting on the phrase *im n twt*, also compares with ⲞⲘⲈ and suggests 'Formthon' or 'Staub von Statuen'. The identification with Coptic ⲞⲘⲈ, and the consequent translation clay is perpetuated in the WÖRTERBUCH, and has been generally accepted until comparatively recently. GARDINER[8], however, points out that ⲞⲘⲈ is in fact derived from *ʿmʿt*, referring to an observation of GRIFFITH[9], for only in this way can the occasional broken vowel of ⲞⲞⲘⲈ be explained. With the etymological connection destroyed, the meaning of *im* is at once called in question, and a translation such as 'frit' or 'faience' is not impossible, cf. *imyt* discussed above (p. 97–8).

The various examples of *im* are not in themselves conclusive, and all that can be said is that it was a material from which statues were made[10], which was used in wall decoration[11], and which could be glazed[12]. It was therefore some plastic substance, and at first sight clay seems a suitable translation. But although statuettes could be made of clay, which could also be used as a wall plaster, it does not seem that the Egyptians ever glazed earthenware or pottery, so that the reference to glazing is difficult to reconcile unless one assumes a magical deviation from normal practice. The most common material for glazing was faience, from which small statues were also made, but whether it was ever employed as wall decoration seems doubtful. There is clearly some inconsistency here, and it may be that in *im* we have a more or less generic term for plastic material, covering clays and frits of any kind. Possible support for this is afforded by the expression *m im nb*, which occurs parallel to *m ḫt nb* and *m ʿꜣt nb* in reference to statues of gods[13]. But the question cannot finally be decided, and more evidence is required.

ʿmʿt (Wb. I. 185.17; I. 186.12; I. 185.12), separated under three heads in the WÖRTERBUCH, is clearly a unity[14], and all discussion must be based on GARDINER's observations in Onomastica[15]. He there translates 'mud', 'clay', 'mud-flat', comparing Coptic ^{S}ⲞⲘⲈ, ⲞⲞⲘⲈ, ⲀⲘⲈ, ^{B}ⲞⲘⲒ (Greek πηλός), and remarks that 'the entries in Wb. relating to this and some

[1] I, 72, 2; 91, 1. [2] Modern Egyptians (Everyman) p. 533.
[3] Fellahin p. 294. [4] Ostr. Len. 2973.
[5] P. Med. Lond. 3, 7. [6] Wb. Erg. p. 66.
[7] Med. Kent. p. 100. [8] Onom. 1 p. 10. [9] Kah. Pap. p. 100; p. 107.
[10] Eb. 85, 13; 69, 19; 59, 22; Hearst 10, 1; Salt 825, 19, 3; Denk. Memph. 60 etc.
[11] Louvre C 12 (Lesest. 17a).
[12] Salt 825, 19, 3. [13] Denk. Memph. 60.
[14] Eg. *ʿmʿ(t)* of Onom. 7, 12, parallel to *ʿmʿm* of Hood 11, cited by Gardiner.
[15] Onom. 1 p. 10.

cognate words need radical revision'. The writings with *ꜥmꜥ* and *ꜥmꜥm* ought not to have been separated, and Wb. I. 78.2 wrongly connects with *im*, as also does Spiegelberg's Handwörterbuch[1], though quoting a Demotic *ꜥmꜣ*. Such an equation leaves unexplained the occasional broken vowel of ⲞⲞⲘⲈ, and the true etymology is given by Griffith[2], who also realised that *ꜥmꜥt kmt* of Ebers[3] was the same word. Gardiner maintains that *ꜥmꜥt*, 'Salbe o. ä.' is to be eliminated. That *ꜥmꜥt* means mud or clay was recognised by Erman and Lange in editing Lansing[4]. Gardiner distinguishes three meanings: 1. 'mud' as substance in medical texts; 2. 'muddy ground'; 3. 'mud-flat' as agricultural land. The Egyptian word is feminine, the Coptic mostly masculine, but at least once feminine.

There is little to be added to this exhaustive treatment. Parker[5] rightly equates Demotic *ꜥmꜣ* with Coptic ⲞⲘⲈ, and refers to Lange's note on Amenemope 24,13, pointing out that mud and straw represent the constituents of brick, which is of interest. To some extent, therefore, *ꜥmꜥt* seems to coincide with both *ꜣḥt* and *ḳꜣḥ*, though it is perhaps less precise than either, and 'mud' is the most satisfactory translation where the substance is concerned. There does not appear to be any Egyptian original for the Coptic ⲀⲘⲈ ⲚⲢⲀϨⲦ[6], 'fuller's earth'.

pns (Wb. I. 510.9) is rightly recognised by the Wörterbuch as earth, though greater precision is impossible from the two examples known[7]. Stern mistranslates 'cibus quidam', and Joachim follows with 'penes-Brod', but the determinatives are more suited to a word for earth or the like, and Ebbell follows the Wörterbuch, rendering *ṯꜣ n pns*[8] as 'a lump of earth'. It is possible that some particular quality of earth is meant, but nothing can be deduced.

nš (Wb. II. 338.6–9) and *nšꜣ* (Wb. II. 338.16) are clearly the same word, the meaning of which can be demonstated from very few examples. The most significant is a passage in Ebers[9] where the phrase *nšꜣ n šꜥy* occurs, from which it is evident that *nš* is the form in which *šꜥy* is found. A similar connection is implied by a text[10] describing *ḥwt nš* as *pr Mnw nb šꜥy*, and the two words are also brought together elsewhere[11]. *šꜥy* being sand, the obvious translation of *nš* is grains, which is consistent with the comparison *ꜥšꜣ r nš*[12], 'more numerous than grains (of sand'). It may be noted that whereas in the Ebers passage *nš* means simply 'grains' and is qualified by *n šꜥy*, in the majority of examples 'grains of sand' seems to be understood.

ḫmꜥ (Wb. III. 282.6) appears to be connected with certain other words of similar orthography, and the question is discussed at some length by Gardiner[13]. The most significant point is that Smither, in studying a papyrus of Middle Kingdom date, came across a word *ḫmꜥ*, parallel to stone, sand, bricks and *srft*, in a calculation concerning the amount of manpower required to move each class of material. It seems possible that rubble may be meant, and that the word is related to *ḫmꜥ* (Wb. III. 282.7) meaning to 'demolish' or 'break through'

---

[1] Hwb. 88. [2] Kah. Pap. p. 100; p. 107. [3] Eb. 67, 17.
[4] Lansing 4, 5; cf. Lansing 6, 6; P. Tur. A vs. 2, 7 = LEM. 122, 16.
[5] JEA. 26 p. 109. [6] Crum. 255a. [7] Eb. 73, 3; Düm. G. I. III, 75.
[8] Eb. 73, 3. [9] Eb. 41, 19. [10] Ann. Serv. 16, 226.
[11] Thes. 1267 = Piehl Inscr. II, 76.
[12] Edfou I, 466; Dend. Mar. I, 48a; Düm. T. I. I, 79, 32; Düm. Res. 30, 2 etc.
[13] Onom. 2 p. 217.

walls, perhaps to be read *ḫm* and identified with *ḫm* (Wb. III. 281.1)[1]. The similar *ḫmꜥ* of the Kahun text[2] was rendered 'allotments' by Griffith, who thought that the recovery or reclamation of land was referred to. Gardiner connects it with the present group of words, and also suggests 'demolition' or 'rubble' for the *ḫmꜥ* of On. Am. In fact, there does seem to be a certain community of meaning between the various words, and 'rubble' is perhaps the most satisfactory translation of *ḫmꜥ*.

ḫmw (Wb. III. 277.15) may conceivably be related to *ḫmꜥ* discussed above. That it is a term for earth or the like seems evident, since the phrase *wḥꜣ n.k ḫmw.k*[3] is alternated with *wḥꜣ n.k tꜣ.k*, so that *ḫmw* appears to be regarded almost as a variant for *tꜣ*. Sethe[4] renders 'Staub', noting that the two determinatives ∘∘∘ and ⌒ suggest a granular substance and a liquid, perhaps the perspiration or mildew of a corpse. If this is so, then *ḫmw* has the nuance of decay, and the Biblical 'dust' is perhaps the most appropriate English equivalent for the passages in the Pyramid texts[5]. In other contexts, the word has a wider application, referring to dust created by human activity[6]. The Wörterbuch translation of *ḫmw* as 'der Staub, der bei der Arbeit, beim Gehen aufwirbelt' and 'Staub, den der auferstehende Tote von sich abschüttelt' is quite adequate, though in the latter case the meaning of Staub may require extension. That *ḫmw* is necessarily an earthy, mineral, dust cannot be maintained, though in most instances it seems probable.

sꜣṯw (Wb. III. 424.1–3), is a common word for ground, ground-plan etc., the meaning of which is extended to refer to earth or the like as a material. Most interesting in this connection are two references to buildings silted up or concealed by heaps of rubble[7], where *sꜣṯw* would seem to cover earth and debris alike. In one of these examples[8] a brick boundary wall is obscured by *sꜣṯw*, which suggests the collapsed material of the wall itself. The other contexts[9] refer to the digging and heaping up of *sꜣṯw*, pointing to its being simply the ground surface, though here also surface debris may be included. Any precise translation of the word is impossible, though it seems clear that something more than just earth is implied.

sin (Wb. IV. 37.11f.; IV. 64.11) is clearly a word for clay, and this basic meaning seems never to have been questioned. Brugsch[10] has 'die Siegelerde, eine zum Versiegeln von Schlössern, Briefen, Töpfen usw. verwendete Tonart, terra sigillata', the Wörterbuch simply 'der Ton', both of which are somewhat misleading. Careful reconsideration of the evidence will show that *sin* is not in fact the common term for clay, but represents a particular variety, the precise nature of which can be suggested.

The various uses of *sin* are significant. *sin wꜣḏ* (fresh *sin*) is the material from which the four magical bricks are made[11], and from *sin* is also made a similar brick on which an inscrip-

[1] Cf. Hier. Texts I p. 19 n. 1; JEA. 22 p. 178; Brunner, Lehre p. 31 (on Sall. II, 5, 7).

[2] P. Kah. 23, 16.

[3] Pyr. 735c; 747b; 1363a; 2008.

[4] Kommentar III p. 366.

[5] The earth (*tꜣ*) is then identified with the decay (*ḫmw*), which might suggest that the Egyptians regarded corruption as the ground eating into the body.

[6] Eg. Urk. IV, 502; 837 etc.; cf. also Sehel 3, where 'debris' is possible.

[7] Urk. IV, 834; Tell el Yah. p. 70.

[8] Urk. IV, 834.

[9] Pianchi 91; 92; Urk. II, 67, 1.

[10] Wb. p. 1165.

[11] Marseille 40; 41; 42; 43; cf. P. Dem. Mag. 4, 5.

tion is carved[1]. A figure of Anubis[2] and of an ibis[3] are of *sin*, and another Anubis is of *sin* mixed with incense[4]. *sin* is the raw material of vessels[5], and the phrase *wd sin* occurs over representations of potters at work[6], as also does *ḥw sin*[7]. Lastly, mention is made of a writing-tablet of *sin*[8], and sealings of *sin* are frequently referred to[9].

The employment of *sin* for the four magical bricks, also on occasion made of *ṯḥnt*[10], is of peculiar importance, since the actual composition of such bricks is known. Mlle. MONNET, in a detailed study[11], quotes an analysis as follows: Identité: brique crue. Aspect: petits grains. Coloration: brun marron, avec petits grains blanc de chaux. Composition: sable (SiO^2) 55.5; oxyde de fer (FeO^3) 13.8; alumine (Al^2O^3) 8.7; chaux (CaO) 4.7; magnésie (MgO) 2.8; perte au feu 12.2; alcalins, non dosés. With this she compares an analysis of pottery clay from Ballas given by LUCAS[12], and it is indeed clear that the material of the magical bricks was not the ordinary clay of Nile alluvium, but a special clay akin to, if not identical with, that used for certain types of pottery[13].

The fact that *sin* is actually used in the making of pots tends to confirm this view. It has already been pointed out above that the common pottery clay was *ꜣḥt*, so that *sin* must be something quite distinct, the obvious conclusion being that it is 'Ballas' clay. *ḥw sin r ḳḥt* then refers simply to the kneading of the clay to give it the correct consistency for sealing jars, but the phrase *wd sin* requires further consideration. The action which the phrase describes is evidently connected with the interior of the pot, but is otherwise somewhat obscure. MONTET[14] is of the opinion that 'tout d'abord l'ouvrier plonge la main dans des pots vides, non pour les nettoyer, mais plutôt, d'après les légendes, pour y introduire un produit, *sin*, avec lequel on frottait l'intérieure', and in his index gives *sin* as 'produit pour frotter l'intérieure des jarres'. JUNKER[15], on the other hand, declines to recognise a new word, and regards *sin* as clay. *wd sin* is literally 'Ton (in den Krug) stecken', though it cannot mean that, since the jars were to be filled with beer. *sin* must therefore be 'ein sehr harter, also gebrannten Ton zum Abreiben der Innenfläche des Kruges'. Such an interpretation is not altogether impossible, though the inside of a pot is far more likely to have been smoothed by scraping or polishing with a pebble. An alternative explanation of the action *wd sin* is that it represents the smearing of the interior of a vessel with a slip of special clay both to smooth it and to make it more watertight[16]. In certain cases it could also be used to change the colour of a pot made of common clay.

The use of *sin* for statuettes of deities, for a writing tablet, and for sealing the doors of shrines etc. requires little comment, though it is likely that in the last instance, a specially pure clay would be used rather than common Nile mud. It may perhaps be noted that in the expressions *ḫḫ sin*, *sd sin* and *swꜣḏ sin*, *sin* appears to have the extended meaning of clay seal, though the point cannot be pressed. *ḫḫ sin*, 'ein Siegel auflegen'[17] and *sd sin* 'das Siegel

---

[1] Totb. Nav. 151d Nachschr.
[2] Totb. Nav. 151g Nachschr.
[3] P. Med. Lond. 14, 5.
[4] Marseille 43.
[5] Totb. 144, Nu 26, 25–6; P. Salt 825, 18, 6; Meir IV, 13; Amonrit. Berl. 1, 9; Smith 22, 5 etc.
[6] LD. II, 74; Karlsruhe taf. 5.
[7] Meir IV, 13.
[8] Urk. IV, 1314.
[9] Pianchi 104; Karn. ⟨252⟩; Abyd. Rit. Tabl. 2 & 21; Amonrit. Berl. 3, 5; Dend. Mar. III, 37h; Bersh. II, 13, 10; Cairo 20712, 2; Berl. 8815, 6; Hatnub Gr. 17, 2; Theb. Grab. 69 ⟨509⟩.
[10] Louvre E 3239 1.10–11 (Rec. Trav. 14 p. 10f.).
[11] Rev. d'Ég. 8 p. 151. Analysis p. 153 n. 2.
[12] Materials p. 550.
[13] ib. p. 437; Cem. Armant p. 50; cf. also ZÄS. 42 p. 104 for *ḏꜣḏꜣyw n Gbtyw*.
[14] Scènes p. 251.
[15] Giza VII p. 112.
[16] Lucas, Materials p. 422. Whether slip was applied both outside and in is not stated.
[17] Pianchi 104.

zerbrechen'[1] are self-explanatory, and need not be commented upon. The parallel *sw3ḏ sin*[2], connected with the opening of the doors of a shrine, is taken by the Wörterbuch to refer to the loosening of the seal, presumably by moistening it until it became plastic again, as there is evidently some distinction between *sw3ḏ* and *sd*. Alternatively, however, it may simply indicate the renewal of the seal after the breaking.

Worthy of note, finally, are two references to *sin* as coming from the sea, both in magical contexts[3]. If any cognisance is to be taken of such examples, the inference must be that *sin* was found not only at the mouth of certain small valleys[4], but also on the shore of the sea and of lakes, and that it had therefore a slightly wider meaning than 'Ballas' clay.

To sum up the evidence for the meaning of *sin*, it is quite clear that a particular species of clay is meant, in contradistinction to the common Nile mud, *3ht* or *ḳ3ḥ*. Analysis seems to suggest that this clay is the desert product recognised by Lucas and others as the material of certain classes of pottery. Whether *sin* was applied solely to this kind of clay, or was a more or less general term for high quality 'special' clay, it is impossible to determine, and any precise rendering such as 'Ballas' clay is to be avoided. But something more than 'clay' is required, and 'special clay' or 'marl' may be suggested as translations.

srft (Wb. IV. 197.16) requires careful revision. Of the four examples cited by the Wörterbuch, two[5] are in fact *sdf* 'foundation'[6], as has been pointed out by Gardiner, and the third[7] seems to be a corrupt writing of this. There remains therefore only one certain reference[8], to which may be added the *srft* or *sdft*[9] found by Smither in the Middle Kingdom papyrus mentioned above (p. 201). Gardiner[10] regards this as 'possibly a kind of earth or building material', which seems the only possible conclusion from the parallelism with stone, sand, bricks and *ḥmʿ*. No more precise translation seems possible. If there is any confusion with *sdf* 'foundation', *r* being perhaps incorrectly written for *d*, then some extension of meaning similar to that of *s3ṯw* may be suggested, *srft* (or *sdft*) being foundation-material or the earth removed from foundation trenches. The *srft* of Onomastica would then be ground suitable for foundations, perhaps even bed-rock, though the determinative does not suggest this. More examples would be required to justify a translation on these lines, and 'earth' seems the most satisfactory rendering on the present evidence.

š3w (Wb. IV. 405.5) occurs in late texts in a number of set expressions, eg. *š3š3 š3w*, with the meaning 'soil' or 'domain'. In common with other words of similar connotation, it appears to have an extended significance as soil or the like as material. This meaning depends on two passages, of which one is, however, susceptible of alternative explanation. In the Sallier papyrus[11], the unfortunate potter is said to be more bespattered with *š3w(t)* than a pig, which Brunner[12] thinks must be a rare word for clay or mud, comparing *š3*[13] in the same text, and explaining the peculiar orthography by the close proximity of a plant *š3wt*[14]. One is bound to agree with Brunner's contention that *š3wt* is here a word for mud or dirt, and in this case a connection with *š3w* seems likely.

---

[1] Karn. ⟨252⟩; Abyd. Rit. Tabl. 2 & 21; Amonrit. Berl. 3, 5; Dend. Mar. III, 37h.
[2] Bersh. II, 13, 10; Cairo 20712; Berl. 8815, 6; Hatnub Gr. 17, 2; Theb. Grab. 69 ⟨509⟩.
[3] P. Salt 825, 18, 5 (*w3ḏ-wr*); P. Salt 825, 18, 6 (*ḥ3t ḏf3t*).
[4] Lucas, Materials p. 440.
[5] Harr. 10, 12; 11, 2.
[6] Wb. IV. 370.1; Onom. 2 p. 219; Wilbour II p. 116.
[7] Harr. 52a, 7.
[8] Onom. 12, 4.
[9] Wilbour II p. 116.
[10] Onom. 2. p. 218.
[11] Sall. II, 5, 8.
[12] Lehre p. 31.
[13] Sall. II, 7, 5.
[14] Sall. II, 6, 7.

In the Bremner-Rhind papyrus[1], what may be either *šꜣw* or *ḥsb* (above p. 144) occurs in a context concerning writing on the ground which is to be erased with the foot. Faulkner[2] questions the reading, and translates non-committally 'pigment'. Fairman[3] refers to this, and notes that Piehl established *šꜣ* as the reading of □ in *šꜣw* and certain other words, so that *šꜣw* is to be read here. He doubts the translation 'pigment', since it would hardly be suitable for writing on the ground, and suggests 'clay' or 'soil', which would square with the normal meaning of *šꜣw*. This certainly seems a more plausible translation, since writing could quite well be traced in the soil on the ground, and then erased with the foot, whereas to write on the ground with pigment seems impracticable, although some latitude must be allowed in the case of magical requirements. A possible third example of *šꜣw* as a material[4] has been shown above (p. 144) to be a writing of *ḥsb*, and although this does not imply that the reading *šꜣw* should be abandoned in all cases, it is just possible that in the Bremner-Rhind passage *ḥsb* is intended. In this case *šꜣw* as a material would rest on the Sallier example alone, and the exact meaning and the connection with *šꜣw* of the late texts might be doubted. In default, therefore, of further evidence, no definite conclusion can be reached.

šꜥy (Wb. IV. 419. 23f.), Coptic ϣⲱ, 'sand'. The meaning is certain from Coptic, but could equally well be deduced from Egyptian examples. Such phrases as *šꜥy n wꜣḏ-wr*[5], *šꜥy n wḏb*[6] and *wꜣbt n šꜥy*[7] point unmistakably to sand, and the comparison *ꜥšꜣ r šꜥy*[8] is likewise significant. Interesting also is the expression *šꜥy n ḥ(ꜣy)r(w)*[9], translated 'Flugsand' by Brugsch, the second element of which occurs in the Wörterbuch[10] as *ḥꜣjrw*, where the reader is referred to *ḥr*, only to be disappointed. Uses of *šꜥy* as a material are not very numerous, though the phrase *wšꜣ šꜥy* 'Sand ausschütten' als Zeremonie – bei der Gründung des Tempels'[11] is fairly common. But sand is mentioned with other building materials in the Middle Kingdom papyrus referred to above (p. 201), and other instances of its use are known elsewhere[12].

ḳꜣḥ (Wb. V. 12.9–12), Coptic ⲕⲁϩ, 'earth' or 'clay'. The true meaning of the word is apparent from two passages in Ebers[13], where the phrase *ḳꜣḥ n iṯ mw* occurs, which can only refer to Nile alluvium. This sense is confirmed by an El Bersheh text[14], in which a man asks that seed may reach the fertile soil (*ḳꜣḥ*) at the right moment, and is further supported by a reference to *ꜣḥt* and *ḳꜣḥ* in connection with the plastering of tomb walls[15]. The parallelism with *ꜣḥt* suggests that common clay plaster[16] is meant, though there is presumably some distinction between the two. Peet[17] suggests that the cognate verb *sḳ(ꜣ)ḥ* has the meaning 'plaster with stucco', because it is applied to wood[18], where clay plaster would not be used. But there is nothing to show that *ḳꜣḥ* means stucco, and the verb must have a more general

---

[1] Apoph. 29, 15. [2] JEA. 24 p. 49; p. 42. [3] Ann. Serv. 44 p. 274.
[4] Edfou VI, 203, 3. [5] P. Salt 825, 18, 6.
[6] Anast. I, 16, 7; LD. III, 153; Rec. Trav. 20, 40 1.8. etc.
[7] Anast. IV, 10, 12. [8] LD. III, 187c d; LD. III, 153 etc.
[9] Thes. 600. [10] Wb. III. 16. [11] Wb. I. 369.9–10.
[12] Anast. I, 16, 7. Note also the reference to the transporting of *šꜥy* in the shawabti formula, which may indicate manuring with *ṯafl* as Keimer suggests – cf. OLZ. 29 p. 98f.
[13] Eb. 68, 16; 96, 18. [14] Bersh. II, 18 no. 7, 6.
[15] Urk. IV, 57; PSBA. 22 p. 90; cf. Admon. 11, 3 and Gardiner's note.
[16] Lucas, Materials p. 95. [17] Tomb. Robberies p. 101.
[18] P. B. M. 10068 rt. 6, 11; Harr. 15b, 10; 71a, 4; 71a, 6. Hor. & Seth 13, 5–6.

primary sense[1]. Of possible relevance to the use of clay plaster is the mention of a *nḏw ḳ(ꜣ)ḥ* among a list of workmen[2], whose function was presumably to prepare the mud for use as plaster or the like[3].

One apparent reference to *ḳꜣḥ*[4] has been the cause of much confusion in the past, and thus requires some consideration here. Misled by it, BRUGSCH[5] regarded *ḳꜣḥ* as 'schwer zu bestimmendes Mineral, sicherlich eine besondere Steinart', and MASPERO[6] took it to be red jasper. The passage in question concerns a lady, some part of whose anatomy, unfortunately in lacuna, is favourably compared to *ḳꜣḥ* or *dḳw n ḳꜣḥ*. MASPERO saw a *dšr* bird in the lacuna, regarded the ⌐ as *dḳw*, and thought that the comparison referred to the colour of the lady's skin, maintaining that 'la teinte rose de la peau' was greatly admired by the Egyptians. MÜLLER[7], on the other hand, took ⌐ as the determinative of *ibḥ*, and restored *ḥḏ* where MASPERO had seen *dšr*, regarding *ḳ(ꜣ)ḥ* as flint. PAGE HURD[8] sees the line as a reference to the whiteness of the girl's teeth, but rejects 'flint' as a translation because the Egyptian for flint was *ds*, and *ḳḥ* was in any case a white stone.

The passage is certainly obscure, but MASPERO's 'grains du jaspe' can be discounted, since it is highly unlikely that any Egyptian ever thought of a girl's skin as the colour of jasper, the name of which is, moreover, *ḫnmt*. Equally unlikely is a comparison of the skin with mud, and it seems possible that interpretation of ⌐ as *ibḥ*, *nḏḥt* or *ṯs(t)* is correct. In this case, *ḳḥ* cannot be equated with *ḳꜣḥ*, and must be regarded as a white material, which cannot be identified, and may perhaps be the *ḳḥ* discussed above (p. 90). At all events, this disputed and unsatisfactory passage can be set aside in a discussion of *ḳꜣḥ*, which is clearly to be identified as Nile alluvium.

tꜣ (Wb. V. 214.3–10) cannot be fully discussed here, the entries under *tꜣ*, *tꜣwy*, *tꜣw* etc. covering several pages in the WÖRTERBUCH. The only aspect relevant to the present discussion is the use of *tꜣ* as a material[9], a not unnatural development of meaning, which can be parallelled for several other words, eg. *iwtn* and *sꜣṯw* discussed above. Of particular interest are three cases in which *tꜣ* is evidently the material from which bricks are made[10], one stating precisely *snn.n.i tꜣ ḥr mw*[11], and another passage[12] where *tꜣ* is evidently thought of as a material to be handled. In the Pyramid texts[13], *tꜣ* is found parallel to *ḥmw*, and seems to refer to the earth surrounding the body, with perhaps the same nuance as the Biblical 'dust', while in medical contexts it appears to have the meaning 'dirt'[14]. This also seems to be the significance of *tꜣ* in the phrase *ikw n tꜣ*[15], 'dirt gatherer', an alternative name for the scarab[16]. Lastly, *tꜣ* appears to be mentioned as the material from which a hill[17] and an island[18] are constituted.

From the diversity of the references to *tꜣ* as a substance, it is clear that the general translation 'earth' or 'dirt' is the most satisfactory, and the meaning of *tꜣ* does not therefore conflict with *ꜣḥt*, *ḳꜣḥ* or *ꜥmꜥt*, which refer primarily to mud or clay.

1 Cf. eg. P. Ch. B. III rt. 9, 13, where the material used is *sty*.
2 Ostr. D. el B. ⟨var. 198⟩. 3 But cf. above p. 90. 4 Louvre C 100.
5 Wb. Erg. p. 1263. 6 ZÄS. 17 p. 53.
7 Liebespoesie pl. 16 and p. 44; cf. however, ib. pl. 18.
8 World's Oldest Love Poem p. 10–19, in particular p. 19.
9 Cf. also *tꜣwy* 'als Stoff', Dend. Mar. IV, 36, 33 (Wb. V. 219.5).
10 Dend. Mar. I, 21a; Düm. Baug. 46; Dend. Mar. text. p. 133.
11 Dend. Mar. text p. 133 (from Edfu – Brugsch Wb. p. 1362).
12 Dend. Mar. IV, 38, 117. 13 Pyr. 654; 1067; 1732; 747; cf. Totb. Nav. 170, 1 Pb.
14 Eb. 92, 12; 93, 20. 15 Pyr. 806; 959; Eb. 74, 15.
16 Pyr. 959. 17 Pyr. 542. 18 Pyr. 1188.

dbn (Wb. V. 438.11–14), 'etwas das Maurer und Töpfer gebrauchen, auch in offizineller Verwendung' according to the Wörterbuch. Stern has 'argilla quaedam', Joachim 'deben-Körner', Wreszinski 'Material (des Töpfers)', Reisner 'pottersclay', and Breasted 'mortar', while Ebbell does not offer a translation. The references to *dbn* in the medical texts are not in themselves very significant, occurring in remedies to be bound on externally. *dbn* is to be ground or kneaded[1], which suggests something more or less solid, and the compound *dbn n iḳdw*[2] shows it to be a material used by potters or builders, more probably the former[3]. The determinative of *dbn* varies, including ▭[4], probably by confusion with *dbn* weight, ı°ı[5], which suggests an earthy substance, and ⁰ııı[6], which might point to a paste or liquid. The meaning of *dbn* is, however, shown by the Sallier passage[7], where one of the miseries of the potter's calling is said to be that his clothes are stiff (*nḫt*) with *dbn*. This appears to suggest that some semi-liquid substance has been spattered on the potter's clothes and has dried, and only clay or slip seems possible. But there is evidently some distinction between *dbn* and the raw materials *ꜣḫt* and *ḳꜣḥ*, the former being perhaps clay as prepared for potting, in which case Reisner's 'pottersclay' is the best translation.

ḏꜥꜥt (Wb. V. 535.12) deserves brief mention as the name of a type of earth or rock, though not strictly a material. Gardiner[8] notes that apart from the reference in Onomastica[9] the word is found only in the phrase *nbw ḥr ḏꜥt.f* (*ḏꜥm ḥr ḏꜥt.f*), discussed above (p. 34), and suggests that it refers to the vein of rock from which gold is obtainable. This seems a very plausible suggestion, and there can be little doubt that the Wörterbuch entries[10] should be combined into one word with the meaning gold-bearing vein.

ḏbt (Wb. V. 553.7f.), Coptic ⲦⲰⲂⲈ, Arabic *ṭûb*, and hence, through Spanish, English adobe[11]. The word occurs in Egyptian texts with considerable frequency, and has three principal divisions of meaning, 1. brick, 2. block or bar, and 3. pillow or the like (of material), of which only the first will here be discussed. The sense of *ḏbt* has never been in doubt, but it is interesting to compare the textual references with information gathered from archaeological sources concerning the fabrication and use of brick[12].

Lucas points out that bricks were commonly made of Nile alluvium, essentially a mixture of clay and sand containing small amounts of impurities, and this can be parallelled by references to *ꜣḫt*[13] and *tꜣ*[14] as the material of *ḏbt*. *ḏbt* were also made of *sin*, which has been shown to correspond to the calcareous clay or marl from which magical bricks and drab pottery were fabricated[15]. The common alluvium was also mixed with sand and chopped straw, and there is textual evidence for the latter practice[16].

---

[1] Eb. 75, 1.
[2] Hearst 14, 17; Eb. 48, 17; Smith 16, 5.
[3] Cf. the det. in Hearst, and the Sallier passage discussed below.
[4] Sall. II, 5, 8; Hearst 14, 17.
[5] Smith 16, 5; Eb. 75, 1.
[6] Eb. 48, 17.
[7] Sall. II, 5, 8.
[8] Onom. 2 p. 218.
[9] Onom. 12, 4.
[10] V. 535.12; V. 535.1–2.
[11] Wiesmann, Adobe, in ZÄS. 52 p. 130.
[12] Cf. Lucas, Materials p. 62f.; Clarke & Engelbach, Ancient Egyptian Masonry p. 207f.
[13] Urk. IV, 837; 1152.
[14] Dend. Mar. I, 21a; Düm. Baug. 46; Dend. Mar. text p. 133.
[15] Cf. above p. 202.
[16] Anast. V, 3, 1; cf. Nims, Biblical Archaeologist 13 (1950) p. 22f.

Bricks were, and still are, struck from wooden moulds, called *ḫt ḏbt* by the Egyptians[1], the process of striking being known as *sẖt ḏbt*[2]. Ancient Egyptian bricks were made of different sizes for specific purposes, and although actual references to size are somewhat rare[3], bricks for particular uses are more frequently mentioned, eg. for walls[4], kilns[5] etc.

Information concerning brick production is unfortunately somewhat meagre. A passage in Anastasi III[6] suggests that a daily quota of bricks was required from workers, which is what one would expect to find. The Louvre parchment[7], however, seems to reveal a rather different system in operation, though an exact interpretation of the document is not possible. It appears that the workers whose production is listed in the parchment were required to furnish 2000 bricks per month, perhaps as a kind of tax, and any number required daily is nowhere mentioned. Total receipts for a day vary between 1800 and 6290, so that the bricks were apparently delivered as and when the worker pleased. The account is made at the half month, when all those whose deliveries of bricks were on schedule, ie. who had produced 1000 or more, were marked with ⲁ, the sign also attached to each separate delivery. These individual deliveries vary from 65 to 660, five entries for one worker totalling only 620 (150, 150, 65, 85, 170), while another reached 1630 in three entries (660, 410, 560), leaving only 370 to be produced in the latter half of the month. Now 2000 bricks per month is only 67 per day, which is a ridiculously small output, and even 660, the highest entry, represents less than an average day's work; 1000 a day is easily accomplished, and between 2000 and 3000 is not unknown[8]. The obvious conclusion seems to be that the individual entries represent deliveries of part of a monthly impost of 2000 bricks, which workers or their overlords were required to furnish in their own time, and perhaps with their own materials. Such a system can only have been supplementary to normal production, and may even have been an ad hoc measure to provide bricks for some particular work.

The uses to which brick was put were many and varied, and this is reflected in the textual references. Temples[9], pylons[10] and girdle walls[11] of brick are recorded, the disturbance of the brickwork of a tomb is forbidden[12], a window[13] and a ramp[14] of bricks are mentioned, and bricks are used as supports in childbirth[15] and in other operations[16]. Brick is frequently opposed to stone[17], and there are several instances of buildings of *ḏbt* reconstructed *m inr*[18].

In all these examples, where the nature of the brick or brickwork is not precisely stated, *ḏbt* refers to the common brick of sun-dried Nile mud, the familiarity of which led to the use of *ḏbt* as a term for a block or brick of any specified material, as noted above (p. 30).

Before passing to a summary of the Egyptian terms for earth, mud, clay and the like, some reference must be made to a word which is very closely connected with clay and potting, but the nature of which has not satisfactorily been established.

---

1 Düm. Baug. 46; Edfou II, 61.

2 LD. III, 40 = Rekhmirēʿ pl. 58, beside Nubians making bricks; Urk. IV, 837; 1153; Dend. Mar. I, 21a; Edfou II, 61 etc.; cf. *pip ḏbt*, Anast. V, 3, 1.

3 P. Kah. 23, 38–9; cf. Ostr. UCL. ⟨2⟩.

4 Eb. 47, 2; cf. also Reisner, Kerma II p. 509.

5 P. Med. Berl. 8, 2; 8, 10.

6 Anast. III vs. 2, 2.

7 Mem. Miss. I, 2 p. 481f.

8 Cf. Perrot & Chipiez, Histoire de l'Art, l'Égypte p. 506; Nims, Biblical Archaeologist 13 (1950) p. 27.

9 Urk. IV, 879; 765.

10 Urk. II, 68, 13.

11 Urk. IV, 765; 834; Amada 12; LD. III, 152b.

12 Urk. I, 260; Leipzig ⟨Gizeh⟩.

13 Harr. 78, 4.

14 Urk. IV, 1152; cf. also Anast. I, 14, 7.

15 Rec. Trav. 2, 109; cf. Westc. 10, 12.

16 Smith 2, 7; 3, 15.

17 LD. III, 152a; Petos. 61, 34; cf. P. Berl. 11292 etc.

18 Urk. IV. 765; Mem. Miss. XV, 10, 1 etc.

iḏꜣ (Wb. I. 152.16) and *ꜣḏ* (Wb. I. 25.1) are almost certainly to be identified, as the Wörterbuch suggests, since both occur with reference to the texture of clay, and appear to have a similar significance. *iḏꜣ* is found in two related expressions, *sẖt iḏꜣ*[1] and *?ḫw(i) iḏꜣ*[2], both connected with scenes of pottery fabrication, and each apparently ἅπαξ. Montet[3] takes *iḏꜣ* as a word for clay, translating the two phrases as 'pétrir l'argile' and 'piétiner l'argile', and referring the latter to 'des gens qui foulent aux pieds une masse grise évidemment destinée à fabriquer les pots'. The Wörterbuch, more cautiously, has 'Ausdruck bei der Töpferei: vom Glattstreichen? Ausschmieren? eines Kruges', which is probably nearer the truth. There are serious objections to accepting Montet's interpretation, and regarding *iḏꜣ* as clay. As has already been shown, there are several distinct words for mud or clay, the material of pots, so that yet another is to be regarded with suspicion. *iḏꜣ* does not occur in any other context, which would be strange, nor has it a determinative suggesting an earthy nature. Although the Beni Hasan legend might refer to the trampling of clay, the other example occurs, together with *wd sin*, above a man apparently smoothing a pot, a similar action being elsewhere[4] described as *ꜥbb ḏwiw*.

Some light is thrown on the problem by the Smith passage[5], which Breasted[6] translates 'put into a hin jar after it is [of the consistency] of clay; rub and make thick its [consistency]', the point being that *ꜣḏ* is here of *sin*, and so cannot be a word for clay. On the contrary, in this context *ꜣḏ* must mean 'body fabric', 'constituents', 'texture' or the like, and the same range of meaning is applicable to the other two examples of *iḏꜣ*. *wd sin* and *sẖt iḏꜣ* together may perhaps refer to application of slip and smoothing of the fabric, and the parallel *ꜥbb ḏwiw* could be 'proofing' the pot. But whether or not so precise an interpretation be accepted, it is clear from the evidence that *iḏꜣ* is not a material.

In summarising the conclusions reached concerning the terms for earth and the like, some attempt must be made to relate them one with another, and to see how far apparent inconsistencies can be resolved.

The most general word for earth as a material is clearly *tꜣ*, which has a fairly wide range of application, and does not seem in any instance to demand a more precise translation. Nile alluvium, the common raw material of bricks and coarser pottery, is apparently referred to both as *ꜣḫt* and *ḳꜣḥ*, though there was perhaps some slight difference in meaning between these two words which it is now more or less impossible to appreciate. As distinct from the crude raw material, *dbn* is probably potter's clay prepared ready for use, while *sin* represents a quite separate calcareous clay or marl with certain specific uses. Bricks and brickwork are *ḏbt*.

Several words evidently apply to mud or dirt, with no very exact connotation. Such are *ꜥmꜥt*, *pns* and *šꜣw*, the last, however, being in some doubt, and mud is also one of the meanings of *iwtn*, which in other instances appears to refer rather to earthy dust. Dust, in a wider sense, seems to be the most apposite translation of *ḫmw*, the precise nuance of which may vary according to the context[7].

sꜣṯw, *srft* and *ḫmꜥ* have somewhat similar significance, and the meaning of each cannot be too nearly defined. *srft* may perhaps refer to surface earth and rubbish removed in making foundations, while *sꜣṯw* represents silt or debris. *ḫmꜥ*, on the other hand, is fairly clearly a term for rubble or the like resulting from demolition, and as such coincides to some extent with *ḫšḫš* and *sšnyt*[8].

[1] LD. II, 74. [2] Beni Has. II, 7. [3] Scènes p. 254.
[4] Ti 83 reg. 7 (Wb. I. 178.7). [5] Smith 22, 5. [6] Pap. Smith p. 495.
[7] For other words meaning dust or the like, cf. *ibw*, *ssw*, below p. 211; p. 214.
[8] Cf. above p. 28; p. 29.

nš and *šꜤy* are the words for grains (of sand) and sand respectively, and *ḏꜤꜤt* is gold bearing land, possibly with extension of meaning as 'nugget'.

There remains *im*, the most problematic of all the words discussed above, the meaning of which cannot be established with any degree of certainty. It may possibly be, as was originally thought, a term for clay, though the equation with Coptic ⲞⲘⲈ cannot now be maintained. Alternatively, there is some evidence to suggest that *im* is a word of more general connotation, perhaps a generic name for most plastic materials including frits.

From the above brief summary, it is quite clear that there is considerable inconsistency and lack of precision in the naming of earth in its various forms. Such vagueness is perhaps only to be expected, since the kind of criteria by which the Egyptians classified their stones and other minerals could not be applied to common earths. But the real cause of the difficulty is to be found in the extension of meaning accorded to words for earth and earth's surface, groundplan etc., quite distinct in their primary significance, but much less so when used of earth or rubble as material. Nevertheless, certain divisions of meaning can readily be established, and the probable translation of most terms suggested, from which it appears that the Egyptians did in fact differentiate between earth, mud and clay, and also perhaps recognised certain distinct types of soil.

CHAPTER 10

TERMS POSSIBLY REPRESENTING MINERAL SUBSTANCES

The foregoing chapters have been very largely concerned with the more precise identification of substances which are clearly established as minerals from the contexts in which they occur, or from the very determinatives of the words themselves. In addition to these, however, there are numerous unidentified terms not even the basic nature of which can be deduced with any degree of certainty. Some are apparent ἅπαξ λεγόμενα, others occur in isolation or in unhelpful contexts, and the majority are determined with $_{\mid}\overset{\circ}{\mid}_{\mid}$, which is altogether imprecise, being equally applicable to vegetable products, fruits, seeds and the like, or to minerals in granular, earthy or powder form. Mention of all the unknown words which might conceivably represent mineral substances would be of doubtful value, and beyond the scope of the present work. Accordingly, all terms which for any reason may tentatively be regarded as of vegetable origin have been excluded, and only those for which the evidence is equivocal or suggestive of a possible mineral connection are here discussed.

iꜣw (Wb. I. 28.7) is apparently ἅπαξ in Ebers[1], where it is included in a remedy against humour exuding from a blow in the eye. The accompanying medicaments are both mineral and vegetable, and nothing can be deduced concerning it.

iꜣgwt (Wb. I. 34.19), also apparently ἅπαξ in Ebers[2], is used in treating an illness of the knee. The other substances in the prescription are of various kinds, and afford no clue as to its nature.

ibw (Wb. I. 61.3) is found once in the Pyramid texts[3], in a somewhat obscure context. Sethe[4] regards *im ibw* as 'prädikative Verbalform nach *tm*', and translates 'Speichel der nicht eintrocknet (zu Staub wird?)'. Mercer[5] follows this, rendering 'spittle which is not dried up (in dust?)'. Though 'dust' seems an apposite translation, cf. *ssw* below, it is not impossible that some other meaning is here concealed.

ipšnn (Wb. I. 69.16) is virtually ἅπαξ, occurring in Ebers[6], and the parallel prescription in Hearst[7]. The remedy is to expel exudation and still itching, and the other constituents are varied. Stern suggests a possible connection with *pšnt* (Wb. I. 560.10), which is also unknown, though the contexts rather suggest a vegetable product. In fact, such a connection is mere hypothesis, and there is nothing to establish whether *ipšnn* is mineral or otherwise.

ir-pt (Wb. I. 114.12) occurs only once in Ebers[8], in a prescription against spotted baldness. The other ingredients include two mineral (*mnšt* and *šs*), one vegetable (*ḏꜣrt*), and honey.

[1] Eb. 56, 9. [2] Eb. 77, 11. [3] Pyr. 241a. [4] Kommentar I p. 215.
[5] Pyr. Texts I p. 72; II p. 111. [6] Eb. 72, 21. [7] Hearst 9, 18. [8] Eb. 92, 15.

STERN's 'coelo factum—species metalli vel mannae' is hardly convincing, and FORBES[1] specifically rejects heaven-made iron as a translation. There can be little doubt that he is right in this, but an alternative does not immediately suggest itself, and there is nothing to show whether *ỉr-pt* is mineral or not.

bbỉ (Wb. I. 419.9) is apparently ἅπαξ in Harris[2] in the phrase *bbỉ ḳrt tỉsti m nbw*. The WÖRTERBUCH takes both *bbỉ* and *ḳrt* as semi-precious stones, though either or both might equally well be designations of beads or the like. PIEHL renders each by 'anneau', though the determinative is perhaps against this. The word may possibly be a vox peregrina, but no Semitic original suggests itself.

bḥ (Wb. I. 468.5), occurring twice[3] in connection with leather working, is evidently some foul smelling substance used in one of the tanning processes. ERMAN and LANGE[4], BLACKMAN and PEET[5] and CAMINOS[6] all translate non-committally by 'tan', which begs the question. The Egyptians used three tanning methods, tanning proper with vegetable tannins, tawing with aluminium salts, and the oil process known as chamoising, and *bḥ* may refer to a material used in any one of these. The determinative of the Lansing example might suggest an oily secretion or a pod, and in the latter case the pods of *Acacia arabica*, probably used in the true tanning process[7], might be meant. On the other hand, tawing with alum yields a white, stiff leather, which is sometimes 'fed' by treading in a shallow tub containing various substances, maladorous and otherwise, and this operation seems to be depicted in scenes of Egyptian leather working[8]. Such a composition would accord well with what is known of *bḥ*, which was evidently mixed by the leather worker[9], and over or through which leather was passed[10]. In this case, *bḥ* would not be a material in the strictest sense, but a man-made mixture. In conclusion, it must be pointed out that the characteristic smell of a modern tannery, mentioned by LORET[11], is due largely to the fat scraped off the hides and skins, but since *bḥ* is quite clearly a substance used in the tanning, it cannot refer to this.

nṯ (Wb. II. 357.10) is apparently confined to one occurrence[12], where it appears parallel to *br*[13] as a material for filling the sound eye. The context is of little assistance in determining the meaning of the word, since although many precious stones and other minerals are mentioned in this connection, not all the parallel terms are of this nature, and *br*, for example, may be punningly derived from two similar words connected with sight[14]. In fact, nothing can be deduced concerning the meaning of *nṯ*, though it is possible that it may represent a mineral substance, or be an artificial designation of one for which a commoner term exists.

ḥꜤt (Wb. III. 40.1) is found only once, in the phrase *ḥꜤt nt ỉny wnnt m ḫt n dpt*[15], tentatively rendered by JOACHIM[16] as 'Überzug von Feuchtigkeit der sich in dem Schiffsholz findet', while EBBELL[17] has 'oakum-tar (?) that is on the wood of a ship (zopissa?)'. It is clear that

[1] Metallurgy p. 430. [2] Harr. 64a, 9. [3] Lansing 4, 5; P. Tur. 1887 vs. 3, 11 = RAD. 82. [4] Lansing p. 53. [5] JEA. 11 p. 288. [6] LEM. p. 385. [7] Lucas, Materials p. 47. [8] Klebs, Reliefs AR. p. 95; MR. p. 121; NR. p. 166; Waterer in History of Technology 2 p. 151; p. 155. [9] Lansing 4, 5. [10] RAD. 82. [11] Kemi 3 p. 29. [12] Dend. Thes. 41. [13] Wb. I. 465.7. [14] Wb. I. 465.5 and 6. [15] Eb. 93, 21. [16] Pap. Ebers p. 171. [17] Pap. Ebers p. 132.

*ḥ*ʿ*t nt iny* is something to be found in the wood of a ship, and either oakum or zopissa would be equally applicable. The former is loose fibres obtained by picking old rope to pieces, and is used in caulking the planks of ships. The latter, Greek ζώπισσα, Latin zopissa, is pitch mixed with wax, which is scraped off ships, and is mentioned by both DIOSCORIDES and PLINY. The former, after a discussion of πίσσα and its medicinal uses, comments[1] ζώπισσαν δὲ οἱ μὲν εἶπον εἶναι τὴν τῶν πλοίων ξυομένην ῥητίνην μετὰ τοῦ κηροῦ, καλουμένην ὑπ' ἐνίων ἀπόχυμα, οὖσαν διαχυτικὴν διὰ τὸ ἐν τῇ θαλάσσῃ βρέχεσθαι· οἱ δὲ τὴν πιτυίνην ῥητίνην οὕτως ὠνόμασαν. Now the determinative of *ḥ*ʿ*t* would suggest rope, and that of *iny* would be applicable to an oily or resinous material. Bearing in mind that oakum is by preference made from old tarry ropes and the cordage of ships, we may take the literal meaning of *ḥ*ʿ*t nt iny* to be something like 'tarry rope', and accept EBBELL's 'oakum-tar' as a convenient English translation. The passage from DIOSCORIDES just quoted provides an explanation of the virtue of the tar, namely its having been in contact with sea-water, and since the Egyptian remedy was to be taken internally, it may be assumed that only the tar or pitch extracted from the rope was of medicinal value. Whether this pitch was mineral or wood-pitch is uncertain, butthe latter seems more probable.

ḥḏt (Wb. III. 211.20–21) is altogether dubious. Of the two examples cited, the first[2], from the Begig 'obelisk' 'in unklarem Zusammenhang', is too fragmentary to admit of any conclusion. The second, from the Hearst papyrus[3], is regarded as corrupt by WRESZINSKI[4], who would emend to *wꜣḏ* or *ḥḏw*. Either is possible, and it is equally likely that *wḏ*[5] was intended. At all events, no conclusion can be drawn, but connection between the two examples seems unlikely.

ḥḏwy (Wb. III. 213.24) is apparently ἅπαξ in a context which is of no assistance in defining the meaning[6]. The substance, measured in *mn*[7], was evidently something loose and perhaps granular, and a mineral cannot be excluded.

sꜣkswt (Wb. IV. 27.1), mentioned once among the products of the Wadi Natrun[8], may be either mineral or vegetable. The former might seem more probable in as much as the word follows *sꜣhwt*, which is almost certainly the semi-precious stone *shrt*[9], but no definite conclusion is possible.

sꜣtt (Wb. III. 422.10) occurs once, 'hinter Weihrauch genannt'[10], which might indicate a general classification, though the context is by no means clear.

sẖ (Wb. III. 468.6) seems to be found only once[11], together with *dšr* wood and ivory as the material for part of a chariot, the precise nature of which is unfortunately obscure. JÉQUIER[12] connects with the verb *sẖt*, 'weave', and translates 'en (forme de) métier de tisserand', which, as CAMINOS[13] points out, is hardly consistent with the determinative, quite apart from being meaningless. There may be an etymological connection with *sẖ*, 'strike', ie. per-

---

[1] 1, 72.
[2] LD. II, 119.
[3] Hearst 11, 8.
[4] P. Med. Lond. & P. Hearst p. 109.
[5] Cf. above p. 104.
[6] P. Tur. PR. 9 right 6.
[7] Wb. II. 66.4.
[8] Peas. R. 21.
[9] Cf. above p. 130.
[10] P. Boul. 18, 3, 23.
[11] Anast. IV, 16, 10.
[12] BIFAO. 19 p. 153f.
[13] LEM. p. 214.

haps something beaten or pulverised, in which case *sḫ* may be the covering of the *im* of the chariot, the determinative of which suggests a wooden pole or the like. Altogether, the context is too obscure to warrant any more precise suggestion.

ssw (Wb. III. 474.13) is somewhat obscure. The Pyramid text utterance[1] refers to the ascent of the Pharaoh to heaven on the wind, so that a dust or sand storm seems appropriate. Mercer[2] is perhaps justified in regarding Sethe's[3] corn storm as over ingenious, though the ∴ det. does look rather like grains of corn. The Wörterbuch comparison with *ss(f)*, 'ash', seems unlikely. Any precise rendering of *ssw* is impossible, and 'dust' seems most satisfactory.

sti (Wb. IV. 334.2) is apparently ἅπαξ as a material for collars, parallel to *ḥḏ* and *ḏʿm*[4]. It may possibly be a mineral substance, otherwise unknown, though a translation 'beads' or the like is perhaps more likely[5]. The Wörterbuch comparison with *str* is hardly significant, and no very satisfactory conclusion can be reached.

sṯt (Wb. IV. 349.2), is apparently confined to two similar contexts, neither of which is clear. In one[6] *sṯt* and [hieroglyphs] occur as labels to three vases, and in the other[7] they describe two altars. If the latter word is a writing of *ḥsmn*, *sṯt* would by inference be a material also, either the material of the vase and altar, or the contents thereof. On the other hand, the fact that *ḥsmn* is correctly written in the same inscriptions, and that in the former [hieroglyphs] occurs as the content of a chest, would seem to suggest that *sḥtit* and *sṯt* are both unknown and represent some kind of offering. In this case, *sṯt* might tentatively be identified with *sṯt* a kind of bread or cake[8], found in some Old Kingdom offering lists. But nothing is certain, and it is not impossible that *sṯt* and *sḥtit* represent the material of the vessels and altars.

š-py (Wb. IV. 399.2), apparently occurs only once, in a list at Medinet Habu[9], where the reading seems certain. Very little can be deduced concerning the nature of the substance represented, but the fact that it is measured in *dbn*, and its occurrence next to *ibnw*, might suggest a mineral. It is just possible that there is some connection with Akkadian *šîpu*[10] which seems to be an arsenic compound, but the point cannot be pressed, and no conclusion of any real value can be reached.

ḳ3(3)t (Wb. V. 6.1) seems to be found in only two prescriptions in Ebers, once to expel exudation[11], and once for the knee[12], on both occasions in the phrase *ḳ3w n ḳ3(3)t*. The contexts are unhelpful, and *ḳ3w* conveys nothing, since it is used indifferently of both vegetable and mineral substances[13]. The meagre evidence does not therefore admit of a conclusion, and the nature of *ḳ3(3)t* must remain obscure.

---

[1] Pyr. 308a. [2] Pyr. Texts II p. 144. [3] Kommentar I p. 376.
[4] Ann. Serv. 16, 196. [5] Cf. *swtit* (Wb. IV. 77.10). [6] Ann. Serv. 16, 207.
[7] Ann. Serv. 16, 211. [8] Wb. IV. 349.1. [9] Med. Hab. 160, list 47, 1092.
[10] Campbell Thompson, Dictionary p. 46f. [11] Eb. 73, 15.
[12] Eb. 77, 2, spelled *ḳ3t*. [13] Cf. Appendix I. In both examples [hieroglyph] is written.

ḳ3ȝt (Wb. V. 6.2) is found only once, in a legend above boatmen fighting[1], where it can scarcely have any connection with the foregoing word. The phrase *ḏbw ḳȝȝt* is quite incomprehensible, and its relevance to the scene below is not clear.

ḳrt (Wb. V. 66.6) is apparently ἅπαξ in Harris[2], in the expression *bbi ḳrt tisti m nbw*, where the WÖRTERBUCH takes both *bbi* and *ḳrt* to be semi-precious stones, though either or both might equally well refer to beads or the like. PIEHL renders each as 'anneau', though the determinative is against this, and BURCHARDT[3] has *ḳrt* as 'unklares Wort', while omitting *bbi*. No Semitic original seems to suggest itself, and no conclusion can be reached. Whether a somewhat similar word occurring in a fragment of the Chester Beatty mineral list[4] is in fact related, it is impossible to say.

ḳsnty (Wb. V. 71.5) is seemingly confined to the medical papyri, where it occurs in a number of prescriptions together with a variety of other medicaments[5], so that from the contexts alone little can be gathered. BREASTED[6] thinks the word 'may possibly refer to some form of preparing the preceding ingredients', though if the terms in Smith and Ebers are to be identified, it may be a herb, possibly the *ḳsntt* of Ebers[7]. The problem is complicated. From the Smith examples alone, *ḳsnty* might seem to be a mineral[8], and this is also suggested by one of the Ebers references[9], where *ḳsnty* replaces *sty* in the parallel passage in Hearst[10]. On the other hand, an identification of *ḳsnty* with *ḳsntt* is not unlikely, and in that case a vegetable product is certain, since the latter regularly has the plant determinative. On the whole it is perhaps better to separate the two, and to see in *ḳsnty* an ochreous or astringent earth, though no proof can be offered.

kmw (Wb. V. 130.13) occurs in two prescriptions in Ebers[11], one a remedy for tape worm, the other for discharging exanthema of the scalp, the former being taken internally. In both instances the accompanying ingredients are predominantly vegetable. STERN renders 'species granorum, fructus, ni fallor, herbae *gmy*', and JOACHIM translates 'qemu-samen', referring to BRUGSCH's 'Schwarzbaum'[12]. But although a seed or the like seems probable, a quasi-mineral product cannot altogether be excluded.

kmmnt (Wb. V. 131.1) is apparently ἅπαξ as the material of a royal headdress[13]. The determinative would seem to suggest that it refers not to a real headdress of linen, but to a symbolic or model one, perhaps of some mineral substance. In fact, nothing can be deduced from a single occurrence of this kind.

krk (Wb. V. 136.4) appears once in a remedy for coughing[14], the other ingredients of which are too varied to be of any assistance. Either a mineral or a plant product is equally possible.

---

[1] Cairo 1535; Montet, Scènes p. 82.
[2] Harr. 64a, 9.
[3] Fremdworte II p. 47 n. 928.
[4] P. Ch. B. IV vs. 8, 6.
[5] Smith 14, 4; 16, 4; Eb. 44, 10; 45, 4; 96, 4.
[6] Pap. Smith p. 380.
[7] Eb. 46, 18; 73, 2 etc. (Wb. V. 71.7).
[8] Cf. Smith 16, 4.
[9] Eb. 45, 4.
[10] Hearst 7, 3.
[11] Eb. 22, 16; 65, 6.
[12] Wb. Erg. p. 1245.
[13] Karn. ⟨825⟩.
[14] Hearst 4, 11–12.

ṯwtkn (Wb. V. 360.11) is apparently ἅπαξ in a remedy for the eyes[1], where it stands between three vegetable and five mineral substances. The prescription is said to have been imparted by an Asiatic from Byblos, and the word may therefore be a vox peregrina as Stern thinks. Joachim goes further and regards *ṯwtkn* as 'Pflanze aus Phönicien', while Grapow[2] describes it as 'ein offenbar fremdländisches Mineral'. But no possible Semitic loan-word suggests itself, and apart from the probability that the substance and its name were of foreign origin, nothing can be inferred.

dbyt (Wb. V. 434.14) is found once in a context which provides little clue as to the meaning[3]. The accompanying term *ꜥꜣd* seems to be the same as that meaning 'bewässertes Land'[4], which might suggest that *dbyt* was somehow connected with *dbw* 'bank'[5]. On the other hand, it could equally well be the seed of the plant *dbyt*[6], or related to the similarly written drink[7]. Other words of comparable form include *dbyt*[8], discussed above (p. 30), *db*, 'horn'[9], and *db*, *dbt*, 'hippopotamus'[10], any of which might be relevant. No conclusion is possible, though it seems most likely that *dbyt* represents some kind of sand or mud.

For want of a more appropriate place, reference may here be made to a substance, unidentified, but probably of animal origin, which may yet have been regarded as mineral by the Egyptians.

ns-š (Wb. II. 320.18) is known only from the medical texts, in contexts where its use is of little or no significance[11]. All early translators regard it as the seed of some plant, which may, indeed, be correct. Brugsch[12] has 'Seezunge' quoting Stern's 'nomen seminis herbae', while Joachim has 'Seezungekörner', and Wreszinski '*ns-š* – körner'. Breasted[13] also renders literally 'garden-tongue' on the basis of the Ebers writings, though without commenting on the determinative in Smith[14], and Sobhy[15] takes *ns-š* as the name for common algae. Ebbell, however[16], translates 'cuttle-bone', probably in reference to Chassinat's[17] interesting discussion of the Coptic term ⲗⲁⲥ ⲛ̄ⲉⲓⲟⲙ, σηπία, 'l'os de seiche', the use of which is mentioned by both Dioscorides and Pliny. Chassinat tentatively refers the Coptic expression to Egyptian *ns-š*, though admitting that 'il est possible que ce soit le fruit ou la graine d'un végétal aquatique'. The determinative is not conclusive, but it is perhaps of negative significance that no example is known with the plant det. . The chief difficulty in the identification is that there is no precedent for connecting *š* with ⲉⲓⲟⲙ (Eg. *ym*), one meaning, lake or pond, the other sea, and to regard  as a graphic variant of *ym* is too daring. The majority of the medicinal uses of *ns-š* are of no consequence, though two prescriptions against blisters[18], and one against a swollen gland[19] are to some extent parallel to Dioscorides'[20] prescription of cuttle-bone for skin diseases, and Pliny's[21] mention of it as a remedy for mumps. Thus, although a translation of *ns-š* as cuttle-bone is tempting, it cannot be pressed because the philological evidence is tenuous, and the therapeutic parallels

[1] Eb. 63, 9. [2] Unt. Med. II p. 90. [3] P. Kah. 19, 58.
[4] Hamm. 114; Wb. I. 168.12. [5] Wb. V. 434.8.
[6] Wb. V. 435.1. [7] Wb. V. 435.3. [8] Wb. V. 434.13.
[9] Wb. V. 434.3. [10] Wb. V. 433.14; 434.1.
[11] Eb. 31, 12; 65, 2; 68, 9; 83, 21; P. Med. Lond. 6, 3; 6, 6; Smith 14, 7 etc.
[12] Wb. Erg. p. 696. [13] Pap. Smith p. 383. [14] Smith 14, 7.
[15] Bull. Inst. Ég. 20 p. 14. [16] Pap. Ebers. p. 132.
[17] P. Méd. Copte p. 160; cf. Grapow, Grundriß II p. 79 n. 1.
[18] Eb. 72, 3; 72, 6. [19] Eb. 105, 16. [20] 2, 21. [21] HN. 32, 28, 2.

are not sufficiently cogent in themselves. One point remains to be made; if *ns-š* is in fact cuttle-bone, it may be that the Egyptians regarded it as mineral in the same way as they did *inr n spt mw*[1], but this is pure hypothesis.

Finally, something must be said concerning three related expressions for dirt or dung, the precise nature of which is open to question, and which are therefore considered here rather than in the preceding chapter.

itnw (Wb. I. 146.6–7), *mky* (Wb. II. 162.6) and *ẖnnw* (Wb. III. 384.4) are all found with the same qualification *n ḥ*, 'of the courtyard' (Wb. II. 470.3–5), and are clearly parallel expressions, as is recognised by LORET[2]. *itnw* is found once in Ebers[3], in a remedy to prevent *kkt*-animals from eating the corn in a barn, and twice in magical contexts[4], *mky* also in Ebers[5] as an ingredient of ointments, and *ẖnnw* in Ebers[6] and the parallel passage in Hearst[7] in a prescription for the toe. LORET regards *itnw* and *mky* as terms for κόπρος, stercus, fimus, connecting the former with *iwtn*, Coptic ЄІТN, and the latter with Coptic MHOЄIK, and pointing out that both are parallel to an Arabic medicament 'fiente de chauve-souris', cf. Coptic ϢϪⲒⲢⲀⲤⲒⲀ.

LORET's interpretation may well be correct, and 'dung' seems a likely translation of *itnw* in the Ebers passage. The *kkt*-animals are to be prevented from eating the corn in the barn by smearing the walls and floor *ḥr itnw.sn ḥr mw*. Supposing *.sn* to refer to the *kkt*-animals, and not to the walls and floor, 'dung' seems the most obvious rendering, though the *kkt*-animals are probably not bats[8]. On the other hand, *itnw* may be dirt from the walls and floor, which would better suit the qualifying *n ḥ* elsewhere, and the use of *itnw* to quench a fire[9].

In fact, no precise translation of these closely related expressions is possible, and the distinction between them cannot be determined. It is probable that each refers to some form of excrement, though a more general meaning such as filth or rubbish can equally be defended.

---

[1] Cf. above p. 184.
[2] Rec. Trav. 16 p. 92.
[3] Eb. 98, 8, without the qualification *n ḥ*.
[4] M. u. K. I, 7; I, 6.
[5] Eb. 25, 16; 26, 4.
[6] Eb. 78, 13.
[7] Hearst 12, 4.
[8] Bats do not eat corn, their teeth being unsuitable for grinding small grains. It is possible that *kkt* is onomatopoeic, and that the animals are geckoes, so called from their peculiar cry, and remarkable for their power of climbing up walls and across the ceiling.
[9] M. u. K. I, 7.

CHAPTER 11

TERMS MISTAKENLY IDENTIFIED AS MINERAL SUBSTANCES

No lexicographical discussion of Egyptian minerals would be complete without at least some allusion to the various words which have from time to time been thought to represent mineral products, but which can now be identified otherwise. To collect all the erroneous translations from Egyptological literature, from the older dictionaries and from vocabularies such as those of STERN and PIEHL to the Ebers and Harris papyri, would be of little value, and the present chapter includes only those few of particular interest or concerning which some ambiguity remains in the WÖRTERBUCH.

i͗ḳḥw (Wb. I. 138.18) is tentatively identified by the WÖRTERBUCH with *ꜣḳḥw* (Wb. I. 22.1), an equation which can now be substantiated. On the basis of the two examples originally known[1], BRUGSCH translated *ḫpš n i͗ḳḥw* as 'Schwerte in Flammenform'[2], but later preferred 'aus Silex'[3], which was accepted by BREASTED[4], but with some reservation. JÉQUIER[5], on the other hand, laid stress on the determinative 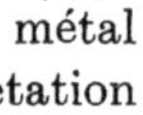, and saw in *i͗ḳḥw* 'un métal, sans doute un alliage spécial à base de cuivre, peut-être un métal particulièrement brillant', comparing the verb *i͗ḳḥ* (Wb. I. 138.15). The true interpretation has, however, been given by MONTET[6], who takes *ḫpš n i͗ḳḥw* as the name of a particular kind of weapon, 'la harpé'. It is clear that in either example, the designation of material is *ḥsmn*, with *i͗ḳḥw* and *ḫpš n i͗ḳḥw* as the names of the objects made from it, and this is put beyond any doubt by a recently published text[7] in which occurs *i͗ḳḥw m ḏꜥm, i͗stnw m ḥsmn*, where *i͗ḳḥw* is determined with an axe. Thus not only is the identification of *i͗ḳḥw* as a weapon made certain, but the identity of *i͗ḳḥw* and *ꜣḳḥw* as one term is established. The more precise definition of *i͗ḳḥw* is problematic. Three instances[8] suggest a kind of axe by their determinatives, and the remaining two examples with the copper sign are equivocal. On the other hand, the phrase *ḫpš n i͗ḳḥw* seems to require that *i͗ḳḥw* be a particular type of scimitar, hence the tentative 'Beilschwerter' of the WÖRTERBUCH[9]. Perhaps the inconsistency can be resolved by regarding *i͗ḳḥw* alone as an axe, and *ḫpš n i͗ḳḥw* as a scimitar with an axe-like cutting edge towards the point[10]. Finally, it must be noted that in three cases[11] the axes are made of precious metals, and are evidently ornamental rewards for valour.

[1] Urk. IV, 722; 726.
[2] Wb. p. 131.
[3] Wb. Erg. p. 21.
[4] AR. II 525.
[5] BIFAO. 19 p. 196.
[6] Reliques de l'Art Syrien p. 35f.
[7] Urk. IV, 1344.
[8] Urk. IV, 1344; Urk. IV, 39 (twice) – *ꜣḳḥw*.
[9] sv. *ḫpš* (Wb. III. 270.4).
[10] Cf. Wolf, Bewaffnung pl. 7, pp. 66–8 (also p. 64); Bonnet, Waffen figs. 33–7, pp. 85–96; Rutimeyer, Verh. Naturforsch. Gesellsch. Basel 40 p. 472f.; Paton, Records III, 2 p. 78 no. 188.
[11] Urk. IV, 1344; IV, 39 (twice).

ʿmꜣ (Wb. I. 185.3) is regarded as 'Art kostbarer Stein' by the WÖRTERBUCH, though almost certainly incorrectly. The reference cited[1] has no mention of the word, but it does occur on the same monument[2], though written with the determinative ⌓, which suggests a country rather than a material, and is in keeping with the context. No land thus written is known to GAUTHIER, though among the variants of *ʿmꜣmw* and *ʿmw*[3] are several sufficiently similar to suggest that *ʿmꜣ* should be identified with one or the other. It would thus be a gold producing region to the south of Egypt, and not a precious stone as the WÖRTERBUCH states

rwd (Wb. II. 410.1), which appears to be confined to a single occurrence in a damaged context at Philae[4], is regarded by the WÖRTERBUCH as 'ein mineralischer Stoff', presumably a variant of *mfkꜣt*. But the WÖRTERBUCH interpretation, based on an imperfect copy of the text in the Zetteln, is erroneous, and comparison with DÜMICHEN's earlier version[5] shows that *'rwd'* is in fact *ršd*, the name of the region from which *mfkꜣt* is frequently stated to come in Ptolemaic texts[6]. The word *'rwd'* does not therefore exist, and the WÖRTERBUCH entry is to be eliminated.

ḥpʿ (Wb. III. 70.9) occurs only once, in a medical text[7], where STERN has 'lapis quidam durus', JOACHIM 'ein (harter) Stein'. EBBELL translates the whole passage as follows: 'If thou examinest an oozing swelling of the male genitals, (an oozing) that (also) has made the swelling on his belly, and if thy finger examines, and it (ie. the swelling) is like *ḥpʿ* under thy fingers and it slips away (?), then thou shalt say concerning it . . . etc.'. The determinatives of *ḥpʿ* are somewhat contradictory, ♉ suggesting something purulent and oozing and ▭ pointing to a stone, which appears inconsistent unless one assumes some such rendering as gallstone. That an actual stone is referred to seems most unlikely, since an oozing swelling would scarcely be so hard. Of course, *ḥpʿ* may refer to the shape or size of the swelling, with a meaning 'pebble' or the like, or to its slippery nature. On the whole, some less definite translation than the WÖRTERBUCH's 'Art Stein' seems desirable.

ḥḳk (sic.), regarded as a mineral, a species of semi-precious stone, by BRUGSCH[8], is certainly not so. In fact, it is the *bḥḳk* of the WÖRTERBUCH (Wb. I. 470.2), which is a kind of fruit[9].

sšt ḥḏt (sic.), occurring between *mnt* and *ḫt* in an offering list[10], was given a somewhat fanciful translation by MASPERO[11]. Through the juxtaposition with *ḫt*, he saw in *sšt ḥḏt* the flint with which the fire was kindled. It is, however, clear that it represents nothing more than a variant writing of *sḫt ḥḏt*[12], which, with its parallel *sḫt wꜣḏt*, is of frequent occurrence in offering lists, and is evidently some kind of vegetable product.

---

[1] Ombos II, 234, 863. [2] Ombos II, 235, 864. [3] Dict. géog. I p. 143.

[4] Philae ⟨1654⟩ Phot. 210 

[5] Düm. G. I. IV, 162 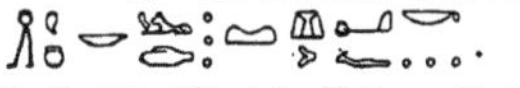

[6] Eg. Dend. Mar. I, 70; I, 71; II, 12; Düm. G. I. II, 63–4; II, 69; IV, 170.

[7] Eb. 106, 19. [8] ZÄS. 18 p. 7.

[9] Gardiner, Hier. Texts I p. 41 n. 6; Dévaud, Rec. Trav. 38 p. 195 n. 3; Caminos, LEM. p. 206.

[10] Ann. Serv. 9, 187. [11] ad loc. [12] Wb. IV. 267.9f.

Mention must also be made of confusion which has existed over certain writings of *ᶜntyw* the most interesting being from the tomb of Rekhmirēᶜ[1]. There, the *tyw* bird was misread as an *m*, and the word consequently misunderstood. Champollion[2] regarded it as the name of a precious stone, and translated 'cornaline', because the accomanying heap was red. Brugsch[3] noted this, but suggested that the writing was for *inr mꜣᶜ*, Coptic ⲀⲚⲀⲘⲎⲒ, 'echter Stein', later, however, preferring to read *ᶜnm* or *ᶜn m ib*[4]. Sethe reads the third sign as ꜣ, but the word is clearly *ᶜntyw*, as recognised by Brugsch elsewhere[5], and by Virey[6]. The red colour of *ᶜntyw* can be parallelled in other instances[7]. Another confusion seems to exist in the Wörterbuch itself, where there appears an entry *ᶜnḏw* (Wb. I. 208.1) 'Art Salbe oder Wohlgeruch', for which two references are given. Of these, the former[8] is somewhat equivocal, but the second[9] is evidently nothing more than a variant writing of *ᶜntyw*. It therefore seems likely that the entry is to be removed, and the two examples included under *ᶜntyw*.

---

1 Urk. IV, 1096.
2 Not. Descr. I p. 506.
3 Wb. p. 196.
4 Wb. Erg. p. 237.
5 Thes. p. 1111.
6 Mem. Miss. V, 1 p. 31.
7 Jéquier, BIFAO. 19 p. 150; Ebbell, Act. Or. 17 p. 89f.
8 Urk. I, 60.
9 Luxor ⟨561⟩.

APPENDICES

I.

In the medical texts, and occasionally elsewhere, certain mineral substances occur in the form [hieroglyph] *nw X*, an expression more frequently found with vegetable products, such as *ḏꜣrt*[1]. The reading of [hieroglyph] is problematic, since it is possible as an abbreviated writing of any of three terms having approximately similar meanings, namely *ḳꜣw*[2], *dḳw*[3], and *dḳr*[4]. But whereas in the case of plants any of these is equally applicable, it seems likely that with reference to minerals [hieroglyph] can only represent either *ḳꜣw* or *dḳw*, since the primary meaning of *dḳr* is fruit. This is confirmed by the existence of full writings of both *ḳꜣw* and *dḳw* with mineral substances, while no example of a full writing of *dḳr* in such a connection appears to be known.

To distinguish further between *ḳꜣw* and *dḳw* when only [hieroglyph] is written is by no means easy. *ḳꜣw n(w) šs* and *ḳꜣw nw ḥsmn* occur together, both written in full, in Hearst[5], and the reading *ḳꜣw* for the [hieroglyph] *n(w) šs* and [hieroglyph] *n(w) ḥsmn* of the parallel passage in Ebers[6] would therefore seem assured. In a similar context in Smith[7], however, a full writing reveals *dḳw šs* and *dḳw n(w) ḥsmn*, and *dḳw (nw) šs* is certain in two other examples[8]. An undoubted case of [hieroglyph] representing *dḳw* is found in a passage in the 'Book of the Dead'[9], where *dḳw n ḥmt wꜣḏt* of three variants is replaced by [hieroglyph] in two, and [hieroglyph] *nw wꜣḏw* in one.

Thus, although, where plants are concerned, *ḳꜣw* may seem the most satisfactory transcription of [hieroglyph], with minerals either *ḳꜣw* or *dḳw* is apparently admissable, and there is evidence for both. It is therefore impossible to settle the reading of [hieroglyph] *n dìdì*[10], for which no full orthography exists, though *dḳw* may be preferred from its use with another pigment[11].

In conclusion, it may be suggested that for mineral substances *dḳw* was in fact the only correct term, and that the *ḳꜣw* of Hearst[12] is a mistake, the apparent inconsistency being thereby resolved. Inaccurate expansion of summary writings is not infrequent in Egyptian texts, and the scribe of Hearst may have misinterpreted [hieroglyph] as *ḳꜣw* on the analogy of the vegetable examples. The converse, regarding *dḳw* as a wrong expansion of *ḳꜣw*, is much less likely, since it presupposes that the same error was made by five different copyists. But the view here put forward is a hypothesis at best, and it is perhaps preferable to regard *dḳw* and *ḳꜣw* as two similar and to some extent confused terms for minerals in granular or powder form.

II.

In discussing the semi-precious stones *nmḥf*, *nšmt*, *ḥsbḏ* and *shrt*, it was assumed that references in the 'Book of the Dead' to a heart, *ib*, of these materials indicated not the actual heart amulet, but the heart scarab, a view which requires some further justification.

---

[1] Wb. V. 526.5–13.
[2] Wb. V. 8.2f.
[3] Wb. V. 494.15f.
[4] Wb. V. 495.8f.
[5] Hearst 10, 16.
[6] Ebers 87, 5.
[7] Smith 21, 6.
[8] Smith 16, 4; Edfou II, 210, 4.
[9] Totb. Nav. 100, 7–8.
[10] Ebers 74, 15.
[11] Totb. Nav. 100, 7–8.
[12] Hearst 10, 16.

The context in which these hearts are mentioned is in each case similar, namely in the title to one of the spells ensuring that a man shall not be without his heart in the necropolis[1]. Thus spell 26 is entitled *r n ib n ḫsbḏ*, 27 *r n ib n nšmt*, 29 B *r n ib n shrt, and* 30 B *r n ib n nmḥf*, though the first two are found in only one papyrus of 19th dynasty date now in Berlin[2], and the third only there and in one other version[3].

Now whereas there is no evidence to show what is meant by the hearts of *ḫsbḏ*, *nšmt* and *shrt*, spell 30 B, that of the heart of *nmḥf*, is regularly inscribed upon the base of heart scarabs, generally of a green stone, and there can be little doubt that it is to this that the title refers[4], But if the 'heart' in one instance was in fact a heart scarab, it seems reasonable to suppose that the other spells also relate to the same amulet[5].

As far as the archaeological evidence is concerned, there is nothing to suggest one interpretation rather than the other. Neither *nmḥf* nor *shrt* can be identified with certainty, but *ḫsbḏ*, lapis lazuli, and *nšmt*, green felspar, occur among the stones used for both the heart scarab[6], and the heart itself[7]. Moreover, several other materials are common to both amulets, notably 'basalt', 'durite' and 'schist' (probably all in fact greywacke)[8], serpentine[9], steatite[10], green jasper[11], and porphyry[12], and only two, carnelian and haematite, occur frequently for one and not the other[13]. In fact, therefore, as regards the identification of *shrt*, evidently a green stone, the question of the meaning of *ib* is of no practical importance, since the same range of green stones was used for both scarabs and hearts.

In default of any more definite evidence, it is not possible to insist upon the interpretation here proposed for the 'hearts' of *ḫsbḏ*, *nšmt* and *shrt*, but the undoubted reference of the *ib n nmḥf* to a heart scarab of that material, must surely suggest that a similar meaning is likely in the other instances also.

III.

Among the many names and epithets of Hathor listed in the WÖRTERBUCH[14] are several which seem to connect the goddess more or less closely with the precious metals and stones or with other materials[15], and yet others which refer to her brilliance and beauty[16]. The frequency of metaphorical references to Hathor in these terms has been discussed by GRAPOW[17], who cites 'gold of the gods, *ḏʿm* of the goddesses, lapislazuli among the mighty ones'[18], 'gold

---

[1] Cf. Naville, Totb. Einl. p. 128. [2] Ba of Naville. [3] Ani.

[4] Cf. Totb. Nav. 30B Rubric; Totb. Leps. 64, 33; Totb. Budge p. 142; ZÄS. 18 p. 56; Yuya pl. 16; Neferrenpet pl. 11 etc.

[5] Petrie, Amulets p. 10 no. 7, divides the spells, taking 26, 27 and 29B as referring to the heart, and 30B to the heart scarab.

[6] Inscribed (Petrie no. 90): 2 lapis, 1 green felspar; Uninscribed (Petrie no. 89): 3 lapis, 1 felspar.

[7] Petrie no. 7: 8 lapis, 1 green felspar.

[8] Scarab inscribed: 'basalt' 18, 'durite' 9, 'schist' 4; uninscribed: 'basalt' 17, 'durite' 1, 'schist' 1; Heart: 'basalt' 19.

[9] Scarab inscribed: 5; uninscribed: 15; Heart: 6.

[10] Scarab inscribed: 5; uninscribed: 8; Heart: 6.

[11] Scarab inscribed: 4; uninscribed: 6; Heart: 8.

[12] Scarab inscribed: 6; uninscribed: 4; Heart: 8.

[13] 26 hearts are carnelian, 15 haematite. Petrie gives no heart scarabs of either. NB. *shrt* is certainly not carnelian as Naville, Budge and Petrie suppose.

[14] Wb. VI p. 74.

[15] *nbt insy* (I. 100), *nbt mfkȝt* (II. 56), *mfkȝt* (II. 57), *nbw* (II. 239), *ḫsbḏ* (III. 334), *sȝwy* (IV. 13), *ṯḥnt* (V. 394), *ḏʿm* (V. 539).

[16] *iȝḫt* (I. 33), *ʿnt* (I. 190), *wbnt* (I. 294), *rʿyt* (II. 402), *ḥȝytyt* (III. 15).

[17] Bild. Ausdr. p. 57. [18] Dend. Mar. III, 47.

of the gods, silver of the goddesses, lapislazuli among the ennead'[1], and 'the golden one ... the lady of silver, the lapislazuli one, the lady of lapislazuli, the faience one, the lady of faience[2], the turquoise one, the lady of turquoise[3]'[4]. He also notes[5] that she is called 'she of the turquoise skin'[6], 'she of the beautiful limbs, with turquoise hair, with lapislazuli head'[7], and[8], as Isis-Hathor, 'lady of faience in place of carnelian'[9], the significance of which is discussed below[10].

But the close association of Hathor with precious materials has never been examined or commented upon in any detail, and a study of Egyptian minerals would hardly be complete without some reference to it, particularly in view of the fresh light thrown by new material. The titles which connect Hathor with the precious stones and metals fall roughly into two groups, metaphorical names, and epithets of the form *nbt X*, though some designations do not conform to either. The first group includes *nbw*[11], *ḥḏ*[12], *ḏꜥm*[13], *sꜣwy*[14], *ḫsbḏ*[15], and *mfkꜣt*[16], four noble metals and the two most important semi-precious stones, while the second covers a greater variety of stones but no metals. At Sinai Hathor is *nbt ḫsbḏ*[17] and *nbt mfkꜣt*[18], at Wadi el Hudi she is *nbt ḥsmn*[19], at Tushka *nbt mḫn(m)t*[20], and elsewhere *nbt ṯhnt*[21] and *nbt shrt*[22]. The eye of Rēꜥ, probably Hathor, is *nbt šsmt*[23], and the only goddess other than Hathor to have an epithet of this form is Neith, who is *nbt wꜣḏ*[24]. Lastly, Hathor once bears the title *hnwt ꜥꜣt nbt*[25], which is perhaps the most significant of all.

It is surely suggestive that in the few precious stone mines that are known there are found inscriptions naming Hathor 'lady' of the stone in question, and it is worthy of note that she was also 'lady of the *mfkꜣt*-country'[26] and *nbt ḏw dšr*[27], both mining or quarry regions. It seems very probable that the discovery of other mines might reveal Hathor as lady of other stones such as *ḥnmt* and *nšmt*, though perhaps not *ḥrst* in view of its somewhat sinister connotation. Indeed, it rather looks as though she was, in the Middle Kingdom at least, the patron goddess of quarrymen and miners, and it is interesting that at Sinai she should be addressed as 'lady of good colour'[28], the thing most earnestly desired there in connection with the turquoise[29].

The concern of Hathor with precious stones is amply illustrated at Dendera, where the goddess is shown receiving minerals in various forms. Once[30] the Pharaoh offers her blocks of precious minerals, listed as *nbw*, *ḥḏ*, *ṯhnt*, *ḫsbḏ*, *mfkꜣt*, *ḥnmt*, *ḥrst*, *ꜥꜣt nb n mꜣꜥ*, and on three occasions[31] presents a casket containing stones of various kinds. In addition, Hathor is given an elaborate ornament of gold and precious stones[32], which, like the raw materials, is peculiarly appropriate to her. It may perhaps be in reference to the varied colour of the stones with which she was associated that Hathor has the title *ꜥšꜣ ỉwnw*[33].

---

1 Edfou I, 572.
2 Grapow has 'das Glas, die Herrin des Glases', but cf. Appendix IV.
3 Grapow has 'der Malachit, die Herrin des Malachits', but cf. above p. 106f.
4 Düm. Baug. 52. 5 p. 55. 6 Dend. Mar. III, 77c.
7 Dend. Mar. II, 81a. 8 p. 56. 9 Philae ⟨3013⟩ Phot. 866 (Wb. III. 150.15).
10 Appendix IV. 11 Wb. II. 239.3f. 12 Edfou I. 572.
13 Wb. V. 539.1; 539.4. 14 Wb. IV. 13.14. 15 Wb. III. 334.14.
16 Wb. II. 57.1. 17 Sinai 102E. 18 Sinai 102E and passim.
19 Wadi el Hudi 16, 3; 17, 3. 20 Ann. Serv. 33 pl. II, 4; pl. III, 3.
21 Dum. Baug. 52. 22 Ombos II, 130, 701. 23 Sinai 121.
24 Sinai 121; cf. Horus *nb wꜣḏ*, Pyr. 457c. 25 Dum. G. I. III, 68.
26 Wb. II. 57.5. 27 Rec. Trav. 30, 214. 28 Sinai 124(a); 126; 129A; 312.
29 Cf. Sinai 90. 30 Dend. Mar. I, 22.
31 Dend. Mar. I, 67a; II, 8a; IV, 23a.
32 Dend. Mar. I, 69; II, 80. 33 Lepsius, Métaux p. 19 n. 10.

The origin of Hathor's association with precious minerals is not easily established, and does not seem to be a logical development from any of her other attributes. It is possible that it arose from confusion and misinterpretation of her primitive title *Nbt*, which Jéquier[1] has shown to mean simply 'lady of *Nbt*'[2]. This title, mistakenly connected with *nbw*, 'gold', cf. χρυσέη Ἀφροδίτη, may have given rise to a general connection between Hathor and all precious metals and stones, which the Egyptians regarded as essentially the same. Alternatively, the association may date to the Middle Kingdom, and the first extensive exploitation of the mines at Sinai, Wadi el Hudi, Tushka and elsewhere. Hathor was perhaps regarded as the patron and protector of miners and quarrymen, and so came to be the lady of the stones mined[3]. On the other hand, being the goddess of Sinai and closely connected with the mining there, she may have become the guardian of other mines by association.

The problem is not easily solved, and neither of the above suggestions is wholly satisfactory. But whatever the origin of the attribute, there can be little doubt that Hathor was, from the Middle Kingdom at least, the peculiar deity of precious minerals, and that this aspect of her character gained considerable prominence in the Graeco-Roman period.

IV.

The Egyptian conception of green as the colour of joy and delight has been discussed in some detail by Erman[4], Grapow[5], Kees[6] and others, and is well enough known not to require elaboration. But some mention of the idea seems necessary in a work of this kind, because it is of particular relevance to the nomenclature of precious stones, especially in the Graeco-Roman period.

In a country where the contrast between arid desert and fertile river-valley was acutely felt, green, the colour of vegetation and growth, became, quite naturally, symbolic of fruitfulness and pleasant appearance. *mfk3t*, the hue of turquoise, to the Egyptians the green stone par excellence, was used of the green of the fields, and so acquired the verbal meaning 'to be joyful' or the like. In the same way, *ḥsbḏ* and *ṯḥnt* were used of the pleasing colour of fertile fields etc., and *mfk3t* and *ṯḥnt*, 'joy', were opposed to *ḥrst*, 'sadness'[7]. Hathor was *mfk inm*[8], and Isis-Hathor 'lady of *ṯḥnt* in place of *ḥrst*'[9].

The use of *mfk3t* as a word for joy was due to its green colour, but the similar employment of *ḥsbḏ* and *ṯḥnt* is less readily explained[10]. The former may have been used through its long and close connection with *mfk3t* rather than from any assimilation of green and blue as colours. The latter, if regarded as natural glass, is of no obvious significance, but as faience, which was predominantly green in colour, it is quite appropriate.

Among the green stones are some which seem to have their names formed upon related verbs with the meaning 'to be joyful', 'delight' etc., and which therefore reflect the same general feeling about greenness. *sḥrt*, from *sḥr*, cannot be identified precisely, but all the evidence points to its having been green in colour. *ḥb*, an artificial creation of the Greek period, is a synonym of *mfk3t*, and therefore green, and it is significant that it is formed upon a verb which is used in parallelism with the verb *mfk*, and with the same meaning.

[1] Considérations p. 193f.
[2] Gauthier, Dict, géog. III p. 84. Cf. Onom. 2 p. 28.
[3] It must, however, be admitted that there is little evidence of devotion to Hathor at eg. Hatnub or Hammamat.
[4] Wortforschung I p. 7f. [5] Bild. Ausdr. p. 54f.
[6] Farbensymbolik p. 425; p. 433; p. 465. [7] Grapow, loc. cit. for refs.
[8] Wb. II. 57.10. [9] Philae ⟨3013⟩ Phot. 866 (Wb. III. 150.15).
[10] Cf. Hornblower, Anc. Eg. 1932 p. 47; Daumas, Les Mammisis p. 231 n. 5.

Also relevant may be the use of *nmḥf* for heart scarabs, for although no verbal connection exists, it is possible that a typically green stone was chosen for good omen.

Mention must also be made of *wꜣḏ*, which had root meanings of greenness and freshness, and was for that reason an adjective of good connotation. The apparent substitution of *wꜣḏ* for *dšr* in some instances has given rise to serious misconceptions to be discussed elsewhere[1], but it may be that Lefebvre[2] is correct in maintaining that '*wꜣḏ*, nom d'une couleur gaie et de bon augure, a été par euphémisme substitué au mot *dšr*, lequel revêtait parfois la valeur d'une épithète péjorative'.

That *ḥrst* was the antithesis of *mfkꜣt* has already been noted, and it is, moreover, a commonplace that *dšr* was a colour of somewhat evil significance, owing to its connection with Seth. It might therefore be expected, on the analogy of the green stones, that the names of the red would largely be of unfavourable import, but such is not in fact the case. Although Seth is once described as a hippopotamus of *ḫnmt*[3], it is clear that the name of red jasper is derived from the verb *ḫnm* 'delight', and it seems therefore that the red colour of the stone was generally appreciated. Moreover, the same process that created *ḥb* as an artificial doublet of *mfkꜣt* also formed a synonym for *ḫnmt*, *ḥkn*, derived from the verb *ḥkn*, of similar meaning to *ḫnm*. Another red mineral, realgar, the red sulphide of arsenic, may, as Iversen has suggested[4], owe its name *ꜣwt-ib* to an appreciation of its bright colour.

Thus, only *ḥrst* of the red stones seems to have had an untoward significance, and the reason for this is not altogether clear. It may be that the full blood-red of jasper was distinguished from the orange-red flame-colour of some carnelian, and that only the latter was a Sethian colour and so ill-omened. Alternatively, *ḥrst* had a wider range of meaning in the Greek period, including chalcedony, and it is possible that it was paleness or lack of colour that was unfavourably contrasted with greenness.

In conclusion, it is worthy of note that of the semi-precious stones used by the Egyptians, the greater number were either green or red in colour, the green being particularly numerous. In view of the value attached to them and the evident esteem in which they were held, it is hardly surprising that the names of such stones should in some measure reflect the pleasure they gave. The green minerals were further associated with fertility and the fresh colour of vegetation, and it is from this that the metaphorical use of *mfkꜣt* and *ṯḥnt* arose.

V.

Colour was always of considerable significance for the Egyptians, and constituted one of their chief criteria for differentiating between semi-precious stones and other minerals. Some discussion of Egyptian colour terminology and conventions therefore seems appropriate here, but it will be confined as far as possible to those aspects which are of particular relevance.

The general word for colour was *iwn* (Wb. I. 52.10), Coptic ^{S}ⲗⲩⲁⲛ ^{B}ⲗⲟⲩⲁⲛ ^{A}ⲉⲟⲩⲉⲛ frequently confused with *inm* (Wb. I. 96.14), Coptic ⲁⲛⲟⲙ, 'skin', particularly when 'appearance' is meant. The reasons for this confusion have been discussed by Grapow[5] and Kees[6], and it is evident that colour was regarded as an essential element in the composition of an object. By means of the word *iwn*, *inm*, the Egyptians were able to supplement their

[1] Appendix V.
[2] JEA. 35 p. 74. Cf. however Kees, Farbensymbolik p. 435.
[3] Edfou VI, 123.
[4] Paints and Pigments p. 42.
[5] Bild. Ausdr. p. 106–7.
[6] Farbensymbolik p. 414.

somewhat restricted colour vocabulary, using phrases compounded with the names of metals and stones of distinctive hue. Such are *iwn m nbw*[1], *m iwn nbw*[2], *m iwn nbw nfr n ḫ3st*[3], *m iwn n nbw n sp 3*[4], *inm.s mi ḥḏ iʿ*[5], *inm m ḏʿm*[6], *inm m s3wy*[7], *iwn (inm) n ḫsbḏ*[8], *mfk3t iwn*[9], *iwn mi p3 ṯhnt*[10], *mi iwn nšmt*[11], and many others.

The simple colour designations which such phrases served to supplement were rather limited. Black was *km*, blue *irtyw*, *ḫsbḏ* or *tfrr*, green *w3ḏ* or *mfk3t*, red *mrš*, *ṯms*, *ṯr* or *dšr*, white *ḥḏ*, and yellow *ḳnit*. Of these, *km* and *ḥḏ* require no comment, save to note that the former may cover certain nuances of brown as well as black. *ḳnit*, the colour of orpiment, seems to be the only word for yellow, which would suggest that shades towards vermillion were not carefully distinguished. Of the other colours more must be said.

The terms for blue appear to distinguish two main nuances, pure blue, and a shade approximating to purple or black. The former is *ḫsbḏ*, the colour of lapis lazuli, and *tfrr*, the latter *irtyw*, the precise definition of which is uncertain. Breasted[12] points out that *irtyw*, like *ṯms*, was derived from a plant, or from its fruit or flower, and was a colour designation 'indicating a dark hue, verging into blue', occasionally having the meaning 'colour, hue, tint' or even 'surface'. Kees[13] notes the word as a variant for *dšr* in late ritual texts, and Jéquier[14] takes it as a variant for *idmi* 'le rouge foncé'. This seemingly conflicting evidence would suggest a purplish tint, and Iversen[15] may be correct in regarding *irtyw* as a derivative of *i3rrt*, with the root meaning of 'grape-coloured'.

Of the green adjectives, *mfk3t*, turquoise-coloured, requires no comment, but *w3ḏ* is of some interest. In addition to pure green shades, it was also used for certain blue-green nuances that we should consider primarily blue[16], and for some grey and yellow tints[17]. Whether *w3ḏ* originally referred to red as well as green seems extremely doubtful. Sethe was of the opinion that it did, and refers to the possibility on more than one occasion[18]. But Lefebvre[19] and Kees[20] have shown that, in the majority of instances, the difficulty can be resolved by translating *w3ḏ* as 'fresh' rather than 'green'[21]. In the few cases that cannot be explained in this way, *w3ḏt* replaces *dšrt* as the name of the crown of Lower Egypt, and some other explanation is required. Lefebvre[22] follows Erman[23] in regarding *w3ḏ* there as a euphemism, the substitution of the name of a colour of good omen for one which, through its association with Seth, was perhaps evil. Kees[24], more plausibly, connects *w3ḏt* with the goddess *W3ḏt*, and translates 'die Uto-Krone', thus disposing of the last remaining evidence for *w3ḏ* as red.

The red colour designations are perhaps the most interesting of all, and have been discussed by Lefebvre[25] with some profit. The most common term is *dšr*, the general word for red of whatever shade, including the red of flame, of quartzite sandstone, and of blood, and the tawny hue of the desert which we should not perhaps regard as red at all. *ṯms*, often a mere synonym of *dšr*, has in fact a quite distinct range, as is shown by a passage in the Pyramid texts[26] where a (?) baboon has ears which are *dšr*, and a behind which is

[1] Düm. G. I. II, 86–7.
[2] Harr. 6, 9.
[3] Harr. 47, 4.
[4] Harr. 78, 4.
[5] Ann. Serv. 38, 223, 13.
[6] LD. III, 140c, 3.
[7] Düm. T. I. I, 68.
[8] Dend. Mar. I, 72c; Nauri 11.
[9] Dend. Mar. I, 79.
[10] Love Tur. 2, 4.
[11] Love Tur. 2, 4.
[12] Pap. Smith p. 193f.
[13] Farbensymbolik p. 466.
[14] Frises p. 37.
[15] Cf. Paints and Pigments p. 10.
[16] Iversen, op. cit. p. 6.
[17] Iversen, op. cit. p. 7.
[18] Unt. III p. 127; Urgesch. p. 160 n. 2; Kommentar I p. 187; IV p. 284.
[19] JEA. 35 p. 74.
[20] Farbensymbolik p. 435.
[21] Cf. Gardiner, Onom. 1 p. 22.
[22] JEA. 35 p. 74.
[23] Religion p. 39.
[24] Farbensymbolik p. 435.
[25] loc. cit.
[26] Pyr. 1349a.

tms. This suggests a colour in which red is tinged with blue, and accords well with the Smith passages commented upon by BREASTED[1], where LUCKHARDT's 'cyanotic' seems a suitable translation. The word is derived from the flower or fruit of some tree, but cannot be identified more precisely. *tr*, the colour of blood, is not properly a colour word, and *mrš*, though surviving in Coptic ⲘⲞⲢϢ, is of only limited application in texts of the Greek period, as a term for light red. A possible red colour is *mkrr*[2], 'zwischen schwarz und weiß als Farbenbezeichnung einer Holzart genannt'. Assuming that an actual wood is referred to, possible colours seem to be limited to the yellow-brown of such woods as *ᶜš*[3], and the red of, for example, *mrw*[4]. In view of the range of meaning of *dšr*, either of these might perhaps be regarded as a shade of red, so that it does not seem unreasonable to see in *mkrr* a reddish colour.

Before leaving the question of Egyptian red colour vocabulary, it is worth mentioning that the Assyrians recognised three different reds: DIR (*sâmu*), the red of red ochre and blood to the brown of opium, alum etc., and also the tint of red gold; GUG (*sându*), the red of cinnabar, blood and litharge to a rose colour; ḪUŠ (*ruššû*), the russet of corn, gold etc.[5]. It is interesting to note that these terms cover much the same range of colour as *dšr*, from true reds to browns and tawny shades.

So much for colour distinctions in vocabulary; a word must now be said concerning the extent of the Egyptian palette, which was similarly limited. LUCAS[6] mentions black, blue, brown, green, pink, red, white, and yellow, and this is, broadly speaking, the complete range of colours available to the Egyptian artist. Of these, black, blue, green, red, white, and yellow were regarded as primary, and it is only for these that specific words existed. The pigments from which these 'primary' colours were derived have been treated at length in chapter 6, and are examined in detail by LUCAS[7]. From these discussions it will be seen that the actual pigments were confined to a small number of frits, earths and other simple substances. But although the reds, for example, are largely ochres or realgar, the Egyptians seem to have recognised several distinct reds, and many different nuances appear in their painting. This is explained by the variety of colour of red ochres (haematites), which are yet a scientific unity, and it must be remembered that different specimens of the same basic pigment might always provide the painter with varied shades of the same colour.

Nevertheless, the comparative poverty of the Egyptian palette, together with other considerations, led to the employment of certain colour conventions, and these must be borne in mind when attempting to deduce anything from the colour by which a material or an object is represented. Colour conventions have been studied at some length by RANSOM-WILLIAMS[8], and further detailed examination would be out of place here, though one or two points in connection with minerals may be mentioned.

Blue was undoubtedly a conventional colour for copper, the blades of tools and weapons etc., but RANSOM-WILLIAMS[9] doubts whether it was always so, since mirrors of polished copper are generally red, though such mirrors may, of course, be bronze. WOLF[10] thought blue was iron, and red bronze, but this is clearly incorrect. The inconsistency of the blue and red colouration of copper is shown by an example at Medum[11], where five objects occur

[1] Pap. Smith p. 193f.
[2] Wb. II. 163.5.
[3] Cairo 28024, 10.
[4] Cairo 28083, 12.
[5] Campbell Thompson, Dictionary p. xxxviii.
[6] Materials p. 391.
[7] ib. Ch. XIV, p. 391f.
[8] Per-Nēb p. 38f.; cf. also Smith, Sculpture and Painting p. 257f.
[9] ib. p. 51.
[10] Bewaffnung p. 65; p. 76.
[11] Medum 13; Junker, Mitt. Kairo 14 p. 92.

with the material designation O. The first is blue-grey, with the O red, the second red, with the O blue-grey. The third object is lost, but the O is red, the fourth is blue-grey and the O likewise, and the last is perhaps white, with the O red. Such inconsistency merely serves to show that nothing can be deduced from the colour of any particular base metal object. So far as can be ascertained, there is no recognisable example of green for copper with patina.

Bronze is generally chestnut-brown, gold yellow, and silver white. *ḏʿm* varies to some extent, but is usually yellow like gold. Precious stones are normally coloured as nearly as possible correctly, though there is some ambiguity over blue and black, and the various shades of deep red, brownish red, and brown are not carefully distinguished. Turquoise, by a curious convention, was regularly shown as green, though some at least must have been blue, which seems to suggest that the Egyptians preferred the greener shades, contrary to modern taste.

One point of particular interest is how the Egyptians conceived of and represented those colours for which they had neither a precise word nor a pigment. Amethyst, for example, has a very characteristic hue which does not approximate to any Egyptian pigment, and it is not known by what colour it would have been shown. Wendel[1] and Jéquier[2] both assumed that it would have been regarded as red, though on the false premise that *ḥmꜣgt* was amethyst, and it may be that this is correct. Some support can be adduced from the known colour of the other things to which the true name of amethyst, *ḥsmn*, was also applied, bronze, (red) natron, and possibly also menstruation[3]. On the other hand, either *ṯms* or *ỉrtyw* would more nearly describe the colour of amethyst, and while the former was certainly classed as a red, the latter was evidently regarded as a shade of blue. Similarly, the Egyptians seem to have regarded brown as a nuance of red, and it was no doubt sometimes referred to as *dšr*, although in other instances it is clear that *km* was used to describe brown substances.

Difficulties of this nature are frequently encountered in dealing with the problems of ancient Egyptian colouration, and are only to be expected from a people whose material and lexical resources were so restricted. It is only by keeping them in sight that unfortunate mistakes in identification can hope to be avoided.

VI.

In connection with the semi-precious stones discussed in chapter 5, some mention must here be made of the word *kꜣf*, occurring almost exclusively in the compound *psš-kꜣf* (Wb. I. 555.2), which can, with some certainty, be identified as the name of a material. Not only *psš-kꜣf*, but also *kꜣf-psš* and *psš n kꜣf* are known, and from the last it might seem reasonable to deduce that *kꜣf* was the substance from which the *psš* was made. But the point has passed unnoticed, save by Prof. Černý[4], who has, moreover, discovered a solitary example of *kꜣf* standing alone, in the unpublished Abusir papyrus[5]. There *kꜣf* is found parallel to *bỉꜣ*, *mnw*, *ḫtm* etc., as the name of a material, and below it *psš*, parallel to *sbꜣ*, *ḥꜣṯs*, *ḥnwt*, etc., as the

[1] Bau- und Edelsteine p. 108. [2] BIFAO. 19 p. 121.

[3] The name may, however, be derived from purification by natron, and be quite devoid of colour significance.

[4] I am indebted to Prof. Černý for permission to incorporate his information on this subject, which he himself hopes to present more fully in a separate article.

[5] P. Abusir rt. 21, written  [30].

name of the object made from it, the only possible interpretation being that *kꜣf* is the stone from which the *psš* implement was ideally to be made. What this stone was is open to question, but Černý tentatively suggests that *kꜣf* may be the ancient word for obsidian, some predynastic implements being of this material[1], and some *psš-kꜣf* on coffins being coloured black. It must be admitted that no specific name for obsidian is otherwise known, though *mnw km* almost certainly includes it, and it may be that *kꜣf* does in fact represent the stone. On the other hand, both *mnw* and *kꜣf* occur in the Abusir papyrus, and if the latter were a precise term for obsidian it might be expected to survive in other contexts, eg. as the material for *ḥnwt* and *hꜣṯs* vessels. Moreover, judging from the majority of *psš-kꜣf* implements, it would seem possible that *kꜣf* is merely an ancient name for flint or a particular variety of flint. In default of further definite examples of the word, no very satisfactory conclusion can be reached, and it is always possible that *kꜣf* as a material is nothing more than a misinterpretation of the element *kꜣf* in the term *psš-kꜣf*, which may originally have had some quite different meaning.

[1] Cf. eg. Rev. d'Ég. 2 p. 158, also ib. p. 157 for two *psš-kꜣf* amulets of obsidian.

ADDENDA

While correcting the proofs of the present work I have noted several points on which I am now able to adduce additional evidence, as well as a few statements that seem to require some modification. I have also had the opportunity of referring to the translation and commentary of the medical texts contained in Grundriß IV pts. 1 & 2, and of consulting the excellent Wörterbuch der ägyptischen Drogennamen (Grundriß VI), which appeared when my manuscript was already in the hands of the printers. In so doing I have observed a number of instances in which the proposed translations differ from my own, and I am grateful to Prof. GRAPOW and Fr. Dr. VON DEINES for drawing my attention to others. In addition, careful reconsideration of the text of the Dendera amulet table, Dend. Mar. IV, 87, has led me to the conclusion that the interpretation adopted by MARIETTE and PETRIE, and followed by me, is in fact incorrect. I now believe that the designations of material refer not to the amulets which precede, but to those which follow, and that it is the first group and not the last for which the material is lacking. As a consequence, the deductions based on the text are invalid, though only in the case of *ḳꜥ* and *bḳs-ꜥnḫ* is the proposed interpretation of a word materially affected. In the following notes the emendations resulting from this revision are indicated by an asterisk.

p. 29. *snt*. An apparent example of this word, Anast. I, 15,4, is almost certainly to be interpreted as *šnw* (Wb. IV. 491. 6f.), since the pedestal of the obelisk has already been mentioned as *dbyt*. The passage is then to be translated, '. . . its pedestal (*dbyt*) of 10 cubits, the circumference (*šnw*) of its base making 7 cubits on each side . . .'.

p. 33. *Gold*. Various sources of gold are treated in some detail by VERCOUTTER, Kush 7 p. 128f., who identifies a number of individual mines mentioned in a text of Ramses II, Müller Res. II, 84f., and shows that the 'gold of the desert of Coptos', the 'gold of Wawat (or *tꜣ-sty*)', and the 'gold of Kush' must correspond to the three main groups of mines, 1) near Wadi Hammamat and Wadi Abbad, 2) near Wadi Allaqi and Wadi Cabgaba, and 3) in the Nile Valley between Buhen and Kerma. He also discusses the output of the gold mines of Kush, and the organisation of gold production, with special reference to certain titles which may relate to gold working.

*p. 40. *wḥr* does not refer to the first group of seven (or nine) amulets, for which no material is given, but to the second group of three, – two scarabs and a *wḏꜣt* eye. Unless *nbw* has been omitted at the head of the list, the arguments for identifying *wḥr* as a term for gold are not affected.

p. 42. *Silver*. In noting Cyprus as a source of silver I overlooked WAINWRIGHT's article, Klio 14 p. 1f., showing that *isy* was in all probability a country in Western Asia. For other Asiatic sources of silver cf. LUCAS, Materials p. 280.

p. 56. *ḥmty*. For the expression *wdḥ ḥmty* cf. also Hor. & Seth 8, 12; Smith 2, 25; Ostr. Gard. 3 rt. 10 = Hier. Ostr. 22, 2 rt. 10 (Wb. I. 393.11–12).

p. 87. *mntt*. My reference to Tushkah as the source also of the coarser variety of diorite is due to a misunderstanding of ENGELBACH's note, Ann. Serv. 33 p. 65f. The 'white variety with black specks or flecks', which he distinguishes from the 'Chephren' diorite, is not in fact the coarser species, but merely another kind of anorthosite-gneiss, the same which LUCAS, Materials p. 466, includes under the general heading of 'Chephren' diorite. There is thus no evidence that the coarse-grained black and white speckled diorite was found at Tushkah, or that *mntt* ever referred to it, though no other name can be suggested.

p. 89. *nn*. There is reason to doubt the authenticity of the gold decoration on the B. M. figure (1482); cf. MÖLLER, Die Metallkunst der alten Ägypter p. 60 n. 100.

*p. 97. *ỉbhty* is in fact the material for three blocks, from which nothing can be deduced.

*p. 99. *ỉnr n ḏbꜣ* is not the material for two scarabs and a *wḏꜣt* eye, but for a single scarab.

p. 101. *ꜥꜣt ḥḏt*. It is perhaps possible that the name *ꜥꜣt ḥḏt* may also have been applied to particularly white specimens of alabaster (calcite), such as might now be termed 'onyx marble'.

*p. 105. *frtf* refers to only two amulets, – a *wḏꜣt* eye and part of a *wḏꜣt*, and not to eleven as stated. The suggestion that it may represent lapis lazuli cannot therefore be supported by the evidence of the amulets.

*p. 108. *mfkꜣt ḫꜣrw* was evidently of some importance, since it is the prescribed material for eleven amulets, while *mfkꜣt ỉmntt* is the material for a single white crown, and ordinary *mfkꜣt* (p. 109) that for a heart, a scarab, and (with *nšmt*) for two uraei.

*p. 116. *nšmt* is the material not for a scarab, but (with *mfkꜣt*) for two uraei.

p. 124. *ẖnmt*. My description of the items of *ẖnmt* listed in the Turin account, P. Tur. 1900 rt. col. II [20], as 'a form of payment' is perhaps misleading, since the material was evidently handed over by certain temple officials for some unspecified work, the nature of which was probably indicated in the lost portion of the papyrus. The exact interpretation of the text is not altogether clear, but it appears that separate lots of *ẖnmt* were contributed by different temples, perhaps as a kind of tax. If, as seems to be the case, the statues and parts of statues were finished pieces, it must be assumed that they were worked by craftsmen belonging to the temples, such large pieces of jasper possibly having been too precious to entrust to ordinary sculptors. Although the weight of the statues is lost, it is possible to make a rough estimate, since three together should account for the difference between the sum of the other weights (875 *dbn*) and the total figure given in 1.10 (1499 *dbn*). Three statues therefore weighed 624 *dbn* together, i e. about 208 *dbn* (42 lbs.) each. NB. In l.1 of the transcription, the last group should perhaps be understood as *tn* rather than *40*.

*p. 124. *ḫnmt* is the material for six amulets,—two *tit* amulets, two bulls, and two *wḏ3t* eyes, the *tit* amulets being of particular significance as has already been pointed out.

pp. 126, 127. *ḫsbḏ*. The reference to *ḫsbḏ* from Cyprus is incorrect, since *isy* has been shown by WAINWRIGHT, Klio 14 p. 1f., to be a country in Asia. *ḫsbḏ* from *isy*, like that from other regions in Western Asia, was probably lapis lazuli obtained from Badakhshan by trade, and passed on to Egypt as tribute.

*p. 127. *ḫsbḏ* refers not to the five amulets here discussed, but to the largest group of twenty-four, wrongly attributed to *bḳs-ʿnḫ* (p. 170), among which four of the five are by coincidence repeated. In fact, the direct confirmation of the meaning lapis lazuli is even more striking, as shown by the following table (from which the amulets of doubtful identity are omitted).

	lapis lazuli		blue 'glaze'	blue glass
Thoth	7	(0)	23	2
Isis	7	(11)	13	4
Nephthys	4	(7)	6	1
Neith	7	(7)	3	—
Selket	14	(4)	—	—
ḏd pillar	37	(2)	56	4
scarab	13	(8)	8	—
Maat	10	(3)	—	—
cartouche	10	(5)	—	—
wḏ3t eye	2	(7)	46	1
heart	8	(5)	16	3
hawk	26	(1)	20	3

The figures are those given in PETRIE's Amulets, with, for comparison, the number of specimens of lapis lazuli in the Ashmolean collection (in brackets). The divinities are particularly significant. In the case of Thoth, Isis, Nephthys, Neith, and Maat, no other genuine stone is cited, all others being faience, glass or metal, while in the case of Selket the only remaining example is of 'schist'. The total of blue 'glaze' *wḏ3t* eyes also includes those of green 'glaze'; for other *wḏ3t* eyes of lapis cf. p. 127 n. 23.

p. 130. *shrt*. Grundriß IV (1) translates tentatively 'Chalcedon', following DAWSON; cf. Wb. Drog. p. 458–9. My reasons for preferring a more typically green stone have already been given.

*p. 131. *shrt* does not refer to the ten amulets mentioned, but to three only,—two hearts, and a third, possibly also a heart, now lost. The suggestion that the various materials of the amulets seem to support the identification of *shrt* as a green stone is not therefore valid, though it should be noted that of the hearts cited by PETRIE 19 are 'basalt', 8 green jasper, 6 serpentine, 4 'beryl', 3 'prase', 3 obsidian, 1 green felspar, 1 green volcanic ash, and 15 green 'glaze'; cf. Appendix II.

*p. 133. *ḳʿ*. The suggested identification of *ḳʿ* as haematite cannot now be maintained, since it is not in fact the material of the sixteen amulets discussed, but refers to the ten objects mentioned sv. *shrt* (p. 131). The most that can be said of these is that their colour is predominantly green, and there is nothing to support an equation with any particular stone. In default of further evidence therefore, the meaning of *ḳ*ꜥ must remain in doubt.

p. 137. *ṯḥnt*. In discussing *ṯḥnt šꜥd* I overlooked the obvious possibility that it might be a material contained in *ḥnw* vessels, rather than one from which they were made. The most probable translation is then 'chopped' or 'powdered' *ṯḥnt*, and the term may well refer to glaze previously fused and then powdered up ready for application in the manner described by LUCAS, Materials p. 199–200.

*p. 137. *ṯḥnt m*ꜣꜥ is (with *ds*) the material for five amulets,—a scarab, a figure of Thoth, a *ḏd* pillar, a *wꜣḏ* sceptre, and a *wḏꜣt* eye.

*p. 138. *ds* is (with *ṯḥnt*) the material for the five amulets listed in the previous note, while *ds km* (p. 139) refers to two objects,—a pair of feathers and what appears to be a hide.

p. 142. *ꜣwt-ib*. There is in fact some slight evidence for the use of realgar in Egypt, though as yet only in connection with orpiment. SPURRELL, Archaeological Journal 52 p. 232, states that 'realgar . . . was sparingly used being ground together with it' (ie. orpiment), and among the 'fards' examined by BARTHOUX, Congrès international de géographie, 1925, vol. 4 p. 251–67, was one composed of orpiment and realgar. A piece of orpiment streaked with realgar was found at Tanis by PETRIE, Tanis I p. 39, and orpiment and realgar are reputed to occur on St. John's Island, BARTHOUX, op. cit. p. 261.

p. 144. *wꜣḏ*. Wb. Drog. p. 413–4; p. 126, reads *ḫpꜣw n wꜣḏ* for all the Ebers examples, and distinguishes the expression from *ḥsb n wꜣḏ* occurring in other contexts.

p. 149. *sšw*. Wb. Drog. p. 566–7 interprets the summary writings of Ebers, Hearst and P. Med. Berl. as *ṯrw* ('ein Mineral, vielleicht von roter Farbe'), by comparison with the *ṯrw* of P. Ram. V. This is surely correct, and I am now inclined to think that all the other alleged examples of *sšw* are also to be read *ṯrw*, in which case the meaning of the word will require careful reconsideration.

p. 150. *sty*. Grundriß IV (1) p. 46; IV (2) p. 55, shows that the alleged compounds *pḥwy sꜣ sty* (Wb. I. 537.2) and *km n sꜣ ḥt-ꜥwꜣ* (Wb. V. 124.12) do not in fact exist.

p. 153. *ḳnit*. It is possible that orpiment may have been native to Egypt. COUYAT, Bulletin de la Société française de minéralogie 31 p. 348, refers to a specimen said to have come from Kharga, and both orpiment and realgar are alleged to be found on St. John's Island; cf. note on *ꜣwt-ib* above.

p. 167. *biꜣ*. The polished fibrolite axes from the tombs of the Carnac cultural group are called *men-gurun*, 'thunderstones', by the Bretons of Morbihan, and are thought to fall from the sky in lightning.

*p. 170. *bḳs-ꜥnḫ* does not in fact refer to the largest group of twenty-four amulets, but to the sixteen amulets discussed sv. *ḳ*ꜥ (p. 133), where the preponderance of haematite as the material for the plummet, square, and (?) headrest is noted. The inevitable conclusion would seem to be that *bḳs-ꜥnḫ* is the name for haematite (and possibly also magnetite), which is altogether appropriate to the early form *biꜣ ḳsy* and the undoubted connection with *biꜣ*. As suggested above (p. 166f.), the latter was probably a word of fairly general meaning, embracing iron, meteoric agglomerate, certain iron ores including haematite, and perhaps also other substances, and it seems likely that the introduction of *biꜣ ḳsy* as a term for haematite resulted from the realisation that these different materials, though apparently related, were in fact distinct.

p. 173. *mnnn*. There is some evidence for the use of bitumen in embalming before the Graeco-Roman period. Three samples of material from a mummy of Persian date were examined by Zaki and Iskander, Ann. Serv. 42 p. 223f., and found to consist of a mixture of resin and bitumen in the approximate proportion of 3 : 2; cf. also one of the Ptolemaic specimens investigated by Griffiths, Analyst 62 p. 703f. In view of this evidence it is possible that *mnnn* and *mrḥ* may after all refer to bitumen or a mixture containing bitumen, though identification as wood tar or wood pitch seems more probable.

p. 173. *mrḥ*. Cf. note on *mnnn* above.

p. 175. *msdmt*. Recent analyses confirm Lucas' figures. Of a present total of 74 specimens of ancient kohl, including those collected by Lucas, 45 have been found to be galena, and only one an antimony compound.

p. 178. *ḥtm*. Of the 74 specimens of kohl that can now be traced, 10 have been identified as oxide of manganese.

p. 178. *ẖt-ꜥwꜣ*. Cf. note on *sty* above.

p. 179. *sꜣ-wr*. Grundriß IV (1) takes *sꜣ-wr* as a species of resin (Harz), following Dawson, but Wb. Drog. p. 420–1 is less definite and gives due consideration to the evidence of P. Dem. Mag. 3,25.

p. 180. *sꜣ-šmꜥ*. A further example of *sꜣ-šmꜥ* in a similar context occurs in de Wit, Les Inscriptions du temple d'Opet p. 211, 5ᵉ table.

p. 181. *snn*. Grundriß IV (1) takes *snn* as a species of balsam (Balsam), following Ebbell and Lefebvre, though Wb. Drog. p. 448 notes the possibility that it may be a mineral substance.

p. 183. *gsfn*. Grundriß IV (1) tentatively identifies *gsfn* as 'Gummiharz von Asa foetida', following Ebbell and Lefebvre; cf. Wb. Drog. p. 540–1. The evidence suggesting that *gsfn* was a mineral substance used as an alternative to *msdmt* seems to me convincing.

p. 187. *imrw*. Wb. Drog. p. 33–4 agrees with Grundriß IV (1) in tentatively identifying *imrw* as an unknown mineral.

p. 190. *bsn*. Wb. Drog. p. 181–2 translates *bsn* alone and in the compounds as 'Ton', though it is difficult to see what *bsn n ṯḥnt* would then mean.

p. 209. *id*ꜣ. The expression *ḥw*(*i*) *id*ꜣ is doubtful and rests only on the authority of MONTET, Scènes p. 255,—but cf. BIFAO. 9 p. 12. Neither NEWBERRY, Beni Has. II,7, nor CAILLAUD, Recherches sur les arts et métiers pl. 16, saw *ḥ* before the ideogram, which might therefore equally well stand for *sḫt*. Nor does the legend refer to 'des gens qui foulent aux pieds une masse grise', as MONTET states, but to a man who bends over a similar grey mass and (?) smoothes it with his hands. The men trampling the clay are in fact accompanied by the label *sḫt* (or *ḥw*(*i*)) ꜣ*ht*, illegible in NEWBERRY's copy, but clear in CAILLAUD.

p. 217. *itnw*. Grundriß IV (1) p. 306 translates the Ebers passage, '. . . werde bestrichen seine Wände, seine Fußboden an ihren Löchern (*itnw*) mit Wasser'; cf. also IV (2) p. 230.

p. 221. *dḳw*, *ḳ*ꜣ*w*. The problem of the reading of this group is also discussed in Wb. Drog. p. 609–10, where a list is given of the substances with which it occurs.

p. 224. Hathor. The association of Hathor with precious stones may have originated in her being the goddess of the desert and mountain regions outside the Nile Valley, in which the majority of the mines were situated.

LIST OF ABBREVIATIONS

Abyd.
PETRIE, Abydos, 3 vols. London 1902–1904.
Abyd. (Caul.)
CAULFIELD, The Temple of the Kings at Abydos. London 1902.
Abyd. Mar.
MARIETTE, Abydos, 2 vols. Paris 1869, 1880.
Abyd. Rit. Tabl.
in MARIETTE, Abydos I pl. 1f.
Abydos (Calv.)
CALVERLEY, The Temple of King Sethos I at Abydos, 4 vols. London 1933–1958.
Act. Or.
Acta Orientalia.
Admon.
GARDINER, The Admonitions of an Egyptian Sage. Leipzig 1909.
Äg. Inschr. Berl.
Aegyptische Inschriften aus den Königlichen Museen zu Berlin, 2 vols. Leipzig 1913, 1924.
Aeg. Stud.
WIEDEMANN, Aegyptologische Studien (die Augenschminke mestem). Bonn 1889.
Äg. und Nub.
SÄVE-SÖDERBERGH, Ägypten und Nubien. Lund 1941.
Alt. Aeg.
WIEDEMANN, Das alte Ägypten. Heidelberg 1920.
Amada (Wien)
KUENTZ, Deux Stèles d'Amenophis II (Bibliothèque d'Étude, X). Cairo 1925; Urk. IV, 1287f.
Amamu
BIRCH, Egyptian Texts of the Earliest Period from the Coffin of Amamu etc. London 1886.
Amarna
DAVIES, The Rock Tombs of El Amarna, 6 vols. London 1903–1908.
Amenemhet
DAVIES & GARDINER, The Tomb of Amenemhēt. London 1915.
Amenemope
P. B. M. 10474, LANGE, Das Weisheitsbuch des Amenemope. København 1925.
Amonhymn Leid.
P. Leid. 350, ZÄS. 42, 12–42; ZANDEE in Oudheidkundige Mededelingen Nieuwe Reeks 28. Leiden 1947.
Amonrit. Berl.
P. Berl. 3055, Hier. Pap. Berl. I.
Amtliche Berichte
Amtliche Berichte aus den Königlichen (preuß.) Kunstsammlungen. Berlin 1875–1943.
Amulets
PETRIE, Amulets. London 1914.
Anast.
Anastasi Papyri I–IX, Select Papyri in the Hieratic Character from the Collections of the B. M., 2 vols. London 1844, 1860.
Anastasi I in GARDINER, Hier. Texts I pl. 1–40; Anastasi II–VI in GARDINER, LEM.
Anc. Eg.
Ancient Egypt.

Ani
BUDGE, The Book of the Dead, The Papyrus of Ani in the British Museum. London 1895.
Ann. Coll. de France
Annuaire du Collège de France.
Ann. Serv.
Annales du Service des Antiquités de l'Égypte.
Antefoker
DAVIES & GARDINER, The Tomb of Antefoḳer. London 1920.
Apoph.
P. B. M. 10188, 22–32, FAULKNER, The Papyrus Bremner-Rhind. Bibliotheca aegyptiaca III. Bruxelles 1933.
AR.
BREASTED, Ancient Records of Egypt, 5 vols. Chicago 1906–1907.
Arch. Or.
Archiv Orientální.
Ass. Dict.
The Assyrian Dictionary of the Oriental Institute of Chicago.
Atlas
WRESZINSKI, Atlas zur altägyptischen Kulturgeschichte. Leipzig 1923f.
Auswahl
LEPSIUS, Auswahl der wichtigsten Urkunden des aegyptischen Altertums. Leipzig 1842.
Avignon
MORET, Rec. Trav. 32, 137; 34, 87; 34, 182; 35, 48; 35, 193.

Baharia
BALL & BEADNELL, Baharia Oasis, its Topography and Geology. Cairo 1903.
Bau- und Denkmalsteine
SETHE, Die Bau- und Denkmalsteine der alten Ägypter und ihre Namen. Sitz. Berl. Ak. 1933, 864–912.
Bau- und Edelsteine
WENDEL, Über die in altägyptischen Texten erwähnten Bau- und Edelsteine. Leipzig 1888.
Beni Has.
NEWBERRY, Beni Hasan, 4 vols. London 1893–1900.
Berl.
Inscriptions in the Berlin Museum, cf. Äg. Inschr. Berl.
Bersh.
NEWBERRY, El Bersheh, 2 vols. London 1893, 1894.
Bewaffnung
WOLF, Die Bewaffnung des altägyptischen Heeres. Leipzig 1926.
Bib. Ég.
MASPERO, Bibliothèque Égyptologique.
BIFAO.
Bulletin de l'Institut français d'archéologie orientale.
Bijouterie
VERNIER, La Bijouterie et la Joallerie égyptiennes (Mem. Inst. II). Cairo 1907.
Bijoux
VERNIER, Bijoux et Orfèvreries (Cat. Cairo), 2 vols. Cairo 1927.
Bild. Ausdr.
GRAPOW, Die Bildlichen Ausdrücke des Aegyptischen. Leipzig 1924.
B. M.
Inscriptions in the British Museum, cf. Hieroglyphic Texts from Egyptian Stelae etc. in the Brit. Mus., 8 vols. London 1911–1939.
B. M. Guide Sculpture
BUDGE, A Guide to the Egyptian Galleries (Sculpture). B. M. 1909.
Book of Dead
BUDGE, The Book of the Dead, 3 vols. London 1898.

Borch. Baug.
Borchardt, Zur Baugeschichte des Amonstempels von Karnak (Unt. V). Leipzig 1912.
Bub. Port.
Reliefs and Inscriptions at Karnak (III), The Bubastite Portal (Chicago Epigraphic Survey). Chicago 1954.
Bull. Inst. Ég.
Bulletin de l'Institut d'Égypte.
Bull. Inst. Fouad I du Desert
Bulletin de l'Institut Fouad I du Desert.
B. v. Atmen
P. Louvre 3284, Horrack, Le Livre des Respirations etc. Paris 1877.
B. v. Durchw.
P. Berl. 3044, von Bergmann, Das Buch vom Durchwandeln der Ewigkeit. Sitz. Wien. Ak. 1877, 369–412.

Cairo
Inscriptions in the Cairo Museum published in Cat. Cairo.
Cairo Sc. Journ.
Cairo Scientific Journal.
Canop.
Canopus decree, Urk. II, 124–54.
Cat. Cairo
Catalogue général des Antiquités égyptiennes du Musée du Caire.
Cat. des Mon.
De Morgan, Catalogue des Monuments et Inscriptions de l'Égypte antique, 3 vols. Vienne 1894—1909.
Cat. Ryl.
Griffith, Catalogue of the Demotic Papyri in the John Rylands Library Manchester, 3 vols. Manchester 1909.
Cem. Armant
Mond & Myers, Cemeteries of Armant, 2 vols. London 1937.
Cen. Seti I
Frankfort etc., The Cenotaph of Seti I at Abydos, 2 vols. London 1933.
Champ. Mon.
Champollion, Monuments de l'Égypte et de la Nubie.
Chronicle
Caminos, The Chronicle of Prince Osorkon (Analecta Orientalia 37). Rome 1958.
Chronique
Chronique d'Égypte.
Cit. of Akh. (III)
Pendlebury etc., The City of Akhenaten (part III), 2 vols. London 1951.
Cod. Paris
cf. Crum, 713a.
Coff. Texts
De Buck, The Egyptian Coffin Texts, 6 vols. Chicago 1935–1956.
Considérations
Jéquier, Considérations sur les Religions égyptiennes. Neuchatel 1947.
C. R.
Classical Review.
CRAI.
Comptes rendues de l'Académie des inscriptions et belles-lettres.
Crum
Crum, A Coptic Dictionary. Oxford 1939.

Dahchour 1894
De Morgan, Fouilles à Dahchour 1894. Vienne 1895.
Debod Roeder
Roeder, Debod bis Bab Kalabsche (Les Temples immergés de la Nubie), 3 vols. Cairo 1911–1912.

Dec. Roy.
WEILL, Les Décrets royaux de l'Ancien Empire égyptien. Paris 1912.
Deir el Bahari
NAVILLE, The Temple of Deir el Bahari, 6 vols. London 1895–1908.
De Gemmis
BLAKE & DE VIS, Epiphanius de Gemmis. London 1934.
D. el B.
cf. Deir el Bahari.
D. el Geb.
DAVIES, The Rock Tombs of Deir el Gebrawi, 2 vols. London 1902.
Dend. Chass.
CHASSINAT, Le Temple de Dendera, 5 vols. Cairo 1934–1952.
Dend Mar.
MARIETTE, Dendérah, 6 vols. Paris 1870–1880.
Dend. Supp.
MARIETTE, Dendérah, Supplément. Paris 1874.
Dend. ⟨Zim.⟩
cf. SETHE, Bau- und Denkmalsteine p. 9 n. 20.
Dendereh
PETRIE, Dendereh. London 1900.
Denk. Memph.
Ein Denkmal Memphitischer Theologie, in SETHE, Dram. Texte.
Deshasheh
PETRIE, Deshasheh. London 1898.
Destr.
Destruction of mankind, NAVILLE, TSBA. 4, 1–19; MAYSTRE, BIFAO. 40, 53–115.
Die Maasse
EBERS, Papyrus Ebers, Die Maasse und das Kapitel über die Augenkrankheiten. Leipzig 1889.
Dict.
PIEHL, Dictionnaire du Papyrus Harris No. 1. Vienne 1882.
Dict. géog.
BRUGSCH, Dictionnaire géographique de l'ancienne Égypte, 2 vols. Leipzig 1879, 1880.
Dict. géog.
GAUTHIER, Dictionnaire des Noms géographiques contenus dans les textes hiéroglyphiques, 7 vols. Cairo 1925–1931.
Dictionary
CAMPBELL THOMPSON, A Dictionary of Assyrian Chemistry and Geology. Oxford 1936.
Doc. de Fouilles IFAO.
Documents de Fouilles de l'Institut français d'archéologie orientale du Caire.
Dram. Texte
SETHE, Dramatische Texte zu altaegyptischen Mysterienspielen (Unt. X). Leipzig 1928.
Düm. Baug.
DÜMICHEN, Baugeschichte des Denderatempels etc. Straßburg 1877.
Düm. G. I.
DÜMICHEN, Geographische Inschriften altägyptischer Denkmäler, 4 vols. Leipzig 1865–1885.
Düm. H. I.
DÜMICHEN, Historische Inschriften altaegyptischer Denkmaeler, 2 vols. Leipzig 1867, 1869.
Düm. K. I.
DÜMICHEN, Altaegyptische Kalenderinschriften. Leipzig 1866.
Düm. Res.
DÜMICHEN, Resultate der . . . im Sommer 1868 nach Aegypten entsendeten archäologisch-photographischen Expedition. Berlin 1869.
Düm. T. I.
DÜMICHEN, Altägyptische Tempelinschriften, 2 vols. Leipzig 1867.
Dyroff-Pörtner
DYROFF & PÖRTNER, Aegyptische Grabsteine und Denksteine aus süddeutschen Sammlungen: II, München. Strassburg 1904.

Eb.
WRESZINSKI, Der Papyrus Ebers. Leipzig 1913.
Edfou
CHASSINAT, Le Temple d'Edfou (Mem. Miss.), 14 vols. Cairo 1892–1934.
Eg. Hier.
PETRIE (H.), Egyptian Hieroglyphs of the First and Second Dynasties. London 1927.
Eg. Medical Papyri
LEAKE, The Old Egyptian Medical Papyri. Lawrence, Kansas 1952.
Egypt
BUNSEN (ed. BIRCH), Egypt's Place in Universal History, 5 vols. London 1867.
Eisen, Kupfer und Bronze
GSELL, Eisen, Kupfer und Bronze bei den alten Ägyptern. Karlsruhe 1910.
El Arabah
GARSTANG, El Arábah. London 1901.
Elder Pliny
BAILEY, The Elder Pliny's Chapters on Chemical Subjects, 2 vols. London 1929, 1932.
El Kurru
DUNHAM, The Royal Cemeteries of Kush: I, El Kurru. Cambridge Mass. 1950.
Emb. Rit.
SAUNERON, Rituel de l'Embaumement (P. Boulaq III & P. Louvre 5.158). Cairo 1952.
Ét. Ég.
MASPERO, Études Égyptiennes, 2 vols. Paris 1886.
Études[2]
CHABAS, Études sur l'Antiquité historique, (2nd. ed.). Paris 1873.
Ezek.
Ezekiel.

Farbensymbolik
KEES, Farbensymbolik in altägyptischen religiösen Texten (Nachrichten Göttingen 1943 no. 11).
Fayum Pap.
LANZONE, Les Papyrus du Lac Moeris. Turin 1896.
Fellahin
BLACKMAN (W. S.), The Fellahin of Upper Egypt. London 1927.
Fest. Grap.
FIRCHOW (ed.), Ägyptologische Studien (Festschrift für H. Grapow). Berlin 1955.
Five Ram. Pap.
BARNS, Five Ramesseum Papyri. Oxford 1956.
Flor.
Inscriptions in the Florence Museum, cf. SCHIAPARELLI, Museo Archeologico di Firenze, Antichtià Egizie. Roma 1887.
Flore[2]
LORET, La Flore Pharaonique (2nd. ed.). Paris 1892.
Fremdworte
BURCHARDT, Die altkanaanaischen Fremdworte und Eigennamen im Aegyptischen. Leipzig 1909.
Frises
JÉQUIER, Les Frises d'Objets des Sarcophages du Moyen Empire (Mem. Inst. XLVII). Cairo 1921.
FuF.
Forschungen und Fortschritte.

Gartenpflanzen
KEIMER, Die Gartenpflanzen im alten Ägypten (vol. 1 only). Hamburg-Berlin 1924.
Geog. Pap.
PETRIE, Two Hieroglyphic Papyri from Tanis. London 1889.
Geografia
SCHIAPARELLI, La Geografia dell' Africa Orientale. Roma 1916.
Geology
HUME, Geology of Egypt (vol. 1). Cairo 1925.

Gesch.
MEYER, Geschichte des Altertums.
Geschichte
DÜMICHEN & MEYER, Geschichte des alten Aegyptens, in ONCKEN, Allgemeine Geschichte. Berlin 1887.
Giornale
BOTTI & PEET, Il Giornale della Necropoli di Tebe, 2 vols. Turin 1928.
Giza H.
SELIM HASSAN, Excavations at Giza, 8 vols. Cairo 1932–1953.
Giza J.
JUNKER, Bericht über die . . . Grabungen auf dem Friedhof des Alten Reiches etc., 12 vols. Wien 1929–1955.
Glossar
ERMAN, Aegyptisches Glossar. Berlin 1904.
Goshen
NAVILLE, The Shrine of Saft el Henneh and the Land of Goshen. London 1887.
Grammaire (Champollion).
CHAMPOLLION, Grammaire égyptienne . . . Paris 1836.
Grammaire (Lexa).
LEXA, Grammaire démotique. Praha 1940–1950.
Grammar
GARDINER, Egyptian Grammar (3 rd. ed.). London 1957.
Griff. Stud.
Studies Presented to F. Ll. Griffith. London 1932.
Gr. Oase
BRUGSCH, Reise nach der großen Oase El Khargeh in der Libyschen Wüste. Leipzig 1878.
Grosse Pylon
JUNKER, Der große Pylon des Tempels der Isis in Philä. Wien 1958.
Grundriß
GRAPOW, Grundriß der Medizin der alten Ägypter, 9 vols. Berlin 1954f.

Hamm.
cf. Hammamat.
Hammamat
COUYAT & MONTET, Les Inscriptions hiéroglyphiques et hiératiques du Ouâdi Hammâmât (Mem. Inst. XXXIV). Cairo 1912.
Harmhab. Kron.
MASPERO, The Tombs of Harmhabi and Touatankhamanou (pp. 9, 19, 40). London 1912.
Harp.
Mem. Miss. V^2, Neferhotpou pl. 2 (after p. 540). Cairo 1891.
Harr.
ERICHSEN, Papyrus Harris I. Bibliotheca aegyptiaca V. Bruxelles 1933.
Harr. A.
NEWBERRY, The Amherst Papyri pls. VIII–XIV. London 1899.
Hatnub
ANTHES, Die Felseninschriften von Hatnub (Unt. IX). Leipzig 1928.
Hatnub Gr.
cf. Ḥatnub.
Hearst
WRESZINSKI, Der Londoner medizinische Papyrus und der Papyrus Hearst. Leipzig 1912.
Hier. Ostr.
ČERNÝ & GARDINER, Hieratic Ostraca (vol. 1). Oxford 1957.
Hier. Pap.
BIRCH, Facsimile of an Egyptian Hieratic Papyrus of the Reign of Rameses III, now in the Brit. Mus. London 1876.
Hier. Pap. Berl.

Hieratische Papyrus aus den Königlichen Museen zu Berlin, 3 vols. Leipzig 1901–1911.

Hier. Pap. B. M.
Gardiner, Hieratic Papyri in the British Museum, Third Series, 2 vols. London 1935.

Hier. Texts I
Gardiner, Egyptian Hieratic Texts I. Leipzig 1911.

History of Technology
Singer, Holmyard & Hall, A History of Technology (vol. 1). Oxford 1954.

Hood
Gardiner, Onomastica pls. XIV–XV.

Hor & Seth
P. Ch. B. I, Gardiner, Late Egyptian Stories p. 37–60. Bibliotheca aegyptiaca I. Bruxelles 1932.

Hungersnoth
Brugsch, Die biblischen sieben Jahre der Hungersnoth ... einer altägyptischen Felsen-Inschrift. Leipzig 1891.

Huy
Davies & Gardiner, The Tomb of Ḥuy. London 1926.

Hwb.
Spiegelberg, Koptisches Handwörterbuch. Heidelberg 1921.

Inscr. d'Ahmès
De Rougé, Mémoire sur l'Inscription du Tombeau d'Ahmès. Paris 1851.

Inscr. Hier. Char.
Inscriptions in the Hieratic and Demotic Character from the Collections of the British Museum. London 1868.

Is. et Os.
Plutarch, De Iside et Osiride.

Israel
Inscription of Meneptah, ZÄS. 34, 1–25.

JAOS.
Journal of the American Oriental Society.

JEA.
Journal of Egyptian Archaeology.

Jer.
Jeremiah.

Jewelry
Ransom Williams, Gold and Silver Jewelry and Related Objects. New York 1924.

JNES.
Journal of Near Eastern Studies.

Joppa
P. Harris 500 vs. I–III, Gardiner, Late Egyptian Stories p. 82–5. Bibliotheca aegyptiaca I. Bruxelles 1932.

Journ. Man. E & O Soc.
Journal of the Manchester Egyptian and Oriental Society.

JRAS.
Journal of the Royal Asiatic Society.

Kadesh
Kuentz, La Bataille de Qadech (Mem. Inst. LV). Cairo 1928.

Kagemni
Bissing, Die Mastaba des Gem-Ni-Kai, 2 vols. Berlin 1905, 1911.

Kah. Pap.
Griffith, Hieratic Papyri from Kahun and Gurob, 2 vols. London 1898.

Kamose
Chronique 30, 206.

Karlsruhe
WIEDEMANN & PÖRTNER, Ägyptische Grabreliefs aus der Großherzoglichen Altertümer-Sammlung zu Karlsruhe. Straßburg 1906.
Karn.
VARILLE, Karnak I (Doc. de Fouilles IFAO. XIX). Cairo 1943.
Karn. Mar.
MARIETTE, Karnak. Leipzig 1875.
Karn. R. III.
Reliefs and Inscriptions at Karnak, Ramses III's Temple (Chicago Epigraphic Survey), 2 vols. Chicago 1936.
Kawa I.
MACADAM, The Temples of Kawa I, The Inscriptions (Text). London 1949. The inscriptions are cited by number and line.
Kenam.
DAVIES, The Tomb of Ḳen-Amūn at Thebes, 2 vols. New York. 1930.
Kerma
REISNER, Kerma (Harvard African Studies 5–6), 2 vols. Cambridge Mass. 1923.
Koller
GARDINER, Hier. Texts I pl. 41–9; GARDINER, LEM. p. 116–20.
Kommentar
SETHE, Übersetzung und Kommentar zu den altägyptischen Pyramidentexten, 4 vols. Glückstadt 1935–1939.
Koptos
PETRIE, Koptos. London 1896.
Kuban
TRESSON, La Stèle de Koubân (Bibliothèque d'Étude IX). Cairo 1922; SANDER-HANSEN, Historische Inschriften der 19. Dynastie p. 30f. Bibliotheca aegyptiaca IV. Bruxelles 1933.
Kulturgeschichte
KEES, Kulturgeschichte des alten Orients, I, Ägypten. München 1933.

Lansing
ERMAN & LANGE, Papyrus Lansing, eine ägyptische Schulhandschrift der 20. Dynastie. København 1924; GARDINER, LEM. p. 99–116.
LD.
LEPSIUS, Denkmäler aus Ägypten und Äthiopien.
Lebens.
P. Berl. 3024, ERMAN, Gespräch eines Lebensmüden mit seiner Seele. Abh. Berl. Ak. 1896.
Lehre
BRUNNER, Die Lehre des Cheti Sohnes des Duauf. (Ägyptologische Forschungen XIII). Glückstadt-Hamburg 1944.
Leid.
Inscriptions in the Leiden Museum, cf. BOESER, Beschreibung der Ägyptischen Sammlung des Niederländischen Reichsmuseums der Altertümer in Leiden. Haag 1905f.
Leiden Magical Papyrus
MASSART, The Leiden Magical Papyrus (Oudheidkundige Mededelingen Nieuwe Reeks 34 supp.). Leiden 1954.
Leid. Granitnaos
cf. Leid., BOESER vol. VII pls. I–V.
LEM. (Caminos)
CAMINOS, Late Egyptian Miscellanies. Oxford 1954.
LEM. (Gardiner)
GARDINER, Late Egyptian Miscellanies. Bibliotheca aegyptiaca VII. Bruxelles 1937.
Les Mammisis
DAUMAS, Les Mammisis des Temples égyptiens. Paris 1958.
Lesest.
SETHE, Aegyptische Lesestücke (2 nd. ed.). Leipzig 1928.

Lesest. Erl.
Sethe, Erläuterungen zu den Aegyptischen Lesestücken. Leipzig 1927.
Lib. Fun.
Schiaparelli, Il Libro dei Funerali, 3 vols. Torino 1881–1890.
Liebespoesie
Müller (W. M.), Die Liebespoesie der alten Ägypter. Leipzig 1899.
Lit. Frag.
Caminos, Literary Fragments in the Hieratic Script. Oxford 1956.
Louvre
Inscriptions in the Louvre, cf. Pierret, Inscr.; Gayet, Stèles de la XIIe dynastie. Paris 1886.
Louvre Apisstele
Chassinat, Rec. Trav. 21, 56; 22, 9; 22, 163; 23, 76; 25, 50.
Love Beatty
Gardiner, The Chester Beatty Papyri No. I, pls. 16, 22–6, 29–30. London 1931.
Love Harr.
P. Harris 500, Müller, Liebespoesie pl. 3f.; Maspero, Études Égyptiennes I p. 217f.
Love Tur.
P. Tur. PR. 79–82, Maspero, Études Égyptiennes I p. 217f.

Mammisi
Chassinat, Le Mammisi d'Edfou (Mem. Inst. XVI). Cairo 1910/1939.
Man. Mag. Copte
Chassinat, Le Manuscrit magique Copte No 42573 du Musée égyptien du Caire. Cairo 1955.
Manners and Customs
Wilkinson, The Manners and Customs of the Ancient Egyptians (1878 ed. revised by Birch), 3 vols. London 1878.
Mar. Mast.
Mariette, Les Mastabas de l'ancien Empire. Paris 1889.
Mar. Mon. Div.
Mariette, Monuments divers recueillis en Égypte et en Nubie. Paris 1872, 1889.
Marseille
Maspero, Catalogue du Musée Égyptien de Marseille. Paris 1889; Naville, Les Quatre Stèles orientées du Musée de Marseille. Lyon 1880.
Materials
Lucas, Ancient Egyptian Materials and Industries (3 rd. ed. London 1948 unless otherwise stated).
Med. Hab.
Medinet Habu (Chicago Epigraphic Survey), 5 vols. Chicago 1930–1957.
Med. Kent.
Lüring, Die über die medicinischen Kenntnisse der alten Ägypter berichtenden Papyri etc. Leipzig 1888.
Medum
Petrie, Medum. London 1892.
Meir
Blackman, The Rock Tombs of Meir, 6 vols. London 1914–1953.
Mél. Cumont
Mélanges Franz Cumont, 2 vols. Bruxelles 1936.
Mél. d'Arch.
Mélanges d'Archéologie égyptienne et assyrienne, 3 vols. Paris 1873–1876.
Mél. Masp.
Mélanges Maspero (Mem. Inst. LXVI–LXVIII), 3 vols. Cairo 1934–1954.
Mém.
De Rougé, Mémoire sur les six premières dynasties (Bib. Ég. 26). Paris 1918.
Mém.
Dévéria, Mémoires et Fragments (Bib. Ég. 4–5), 2 vols. Paris 1896, 1897.
Mem. Inst.
Mémoires publiés par les membres de l'Institut français d'archéologie orientale du Caire.

Mem. Miss.
Mémoires publiés par les membres de la Mission archéologique française au Caire.
Memphis (V)
PETRIE etc., Tarkhan I and Memphis V. London 1913.
Menkhep.
DAVIES, The Tombs of Menkheperrasonb, Amenmosĕ and Another. London 1933.
Mereruka
The Mastaba of Mereruka (Chicago Sakkarah Expedition), 2 vols. Chicago 1938.
Metallurgy
FORBES, Metallurgy in Antiquity. Leiden 1950.
Métaux
LEPSIUS, Les Métaux dans les Inscriptions Égyptiennes. Paris 1877.
Mett.
GOLÉNISCHEFF, Die Metternichstele. Leipzig 1877.
Mitt. Kairo
Mitteilungen des Deutschen Instituts für Ägyptische Altertumskunde in Kairo.
MMA. Bull.
The Bulletin of the Metropolitan Museum of Art, New York.
Modern Egyptians
LANE, An Account of the Manners and Customs of the Modern Egyptians (Everyman ed.).
Mon. Piot
Fondation Eugène Piot, Monuments et Mémoires.
M. u. K.
ERMAN, Zaubersprüche für Mutter und Kind. Abh. Berl. Ak. 1901.
Müller Res.
MÜLLER (W. M.), Egyptological Researches, 3 vols. Washington 1906–1920.
Münch. Bekh.
BRUGSCH, Thes. 1240–1.
Münch. Glypt.
Cf. Dyroff-Pörtner.
Myrrhe und Stakte
STEUER, Myrrhe und Stakte. Wien 1933.
Mythe d'Horus
NAVILLE, Textes relatifs au Mythe d'Horus recueillis dans le temple d'Edfou. Genève-Bâle 1870.
Mythus
SPIEGELBERG, Der ägyptische Mythus vom Sonnenauge. Sitz. Berl. Ak. 1915, 876–94.

Nastesen
SCHÄFER, Die aethiopische Königsinschrift des Berliner Museums. Regierungsbericht des Königs Nastesen . . . Leipzig 1901.
Naukratis
Naukratis stela, ZÄS. 38, 127.
Nauri
SANDER-HANSEN, Historische Inschriften der 19. Dynastie p. 13–24. Bibliotheca aegyptica IV. Bruxelles 1933.
Neferirkerēᶜ
BORCHARDT, Das Grabdenkmal des Königs Nefer-ỉr-keꜣ-reᶜ. Leipzig 1909.
Neferrenpet
SPELEERS, Le Papyrus de Nefer Renpet. Bruxelles 1917.
New Kingdom Art
ALDRED, New Kingdom Art in Ancient Egypt. London 1951.
Nitocris
ZÄS. 35, 16f.
Not. Descr.
CHAMPOLLION, Monuments de l'Égypte et de la Nubie, Notices descriptives, 2 vols. Paris 1844, 1889.
Nouvelles Inscriptions
GOYON, Nouvelles Inscriptions rupestres du Wadi Hammamat. Paris 1957.

Œuvres Div.
Chabas, Œuvres Diverses (Bib. Ég. 9–13), 5 vols. Paris 1899–1909.
OLZ.
Orientalistische Literatur-Zeitung.
Ombos
De Morgan, Kom Ombos (Cat. des Mon. II–III), 2 vols. Vienne 1895, 1909.
Onom.
Gardiner, Ancient Egyptian Onomastica, 3 vols. Oxford 1947.
Onurislegende
Junker, Die Onurislegende. Wien 1917.
Orb.
Papyrus d'Orbiney, Gardiner, Late Egyptian Stories p. 9–30. Bibliotheca aegyptiaca I. Bruxelles 1932.
Ostraca
Daressy, Ostraca (Cat. Cairo). Cairo 1901.
Ostraca etc.
Hayes, Ostraka and Name Stones from the Tomb of Sen-Mut at Thebes. New York 1942.
Ostr. Cairo
Daressy, Ostraca (Cat. Cairo). Cairo 1901; Černý, Ostraca Hiératiques (Cat. Cairo), 2 vols. Cairo 1935. For 25247 cf. Spiegelberg, OLZ. 5 p. 323f.
Ostr. Cen. Seti
Gunn in Frankfort, The Cenotaph of Seti I at Abydos, pl. 90. London 1933.
Ostr. DM.
Černý, Catalogue des Ostraca Hiératiques non littéraires de Deir el Médineh (Doc. de Fouilles IFAO. 3–7), 5 vols. Cairo 1935–1951.
Ostr. Gard.
Ostraca in the possession of Gardiner, some published in Černý & Gardiner, Hieratic Ostraca. Oxford 1957.
Ostr. J. Č.
Ostraca in the possession of Černý, some in Hier. Ostr.
Ostr. Len. (2973)
Černý, Arch. Or. 3 p. 395 (ostracon in Leningrad).
Ostr. Liv.
Ostraca in the Liverpool Museum, some in Hier. Ostr.
Ostr. MMA. 90.6.3.
Müller, Rec. Trav. 22 p. 103. Revised transcription at end of unpublished Belegstellen.
Ostr. Nash
Ostraca in the possession of Nash (now dispersed), some in Hier. Ostr.
Ostr. Petrie
Ostraca formerly in the possession of Petrie, some in Hier. Ostr.
Ostr. Senmut
Hayes, Ostraka and Name Stones from the Tomb of Sen-Mut at Thebes. New York 1942.
Ostr. Stras.
Ostracon in Strassburg, Iversen, Paints and Pigments p. 35.
Ostr. Tor.
Ostraca in Toronto, Gardiner, in Theban Ostraca (University of Toronto Studies). London 1913.

P. Berl.
Papyri and Ostraca in the Berlin Museum, in Hieratische Papyrus aus den Königlichen Museen zu Berlin, 3 vols. Leipzig 1901–1911.
P. Bib. Nat. 237
cf. Wb. Belegstellen II p. 132.
P. B. M.
Papyri in the British Museum, cf. Peet, Tomb Robberies.
P. Bol.
Gardiner, LEM. p. 1–12.

P. Boul.
MARIETTE, Les Papyrus égyptiens du Musée de Boulaq, 3 vols. Paris 1871–1876.

P. Ch. B.
GARDINER, The Chester Beatty Papyri No. 1. London 1931, and Hieratic Papyri in the British Museum, Third Series, 2 vols. London 1935.

P. Dem. Mag.
GRIFFITH & THOMPSON, The Demotic Magical Papyrus of London and Leiden, 3 vols. London 1904–1909.

P. Gurob
cf. P. Kah.

P. Insinger
LEXA, Papyrus Insinger, 2 vols. Paris 1926.

P. Kah.
GRIFFITH, Hieratic Papyri from Kahun and Gurob, 2 vols. London 1898.

P. Leid.
LEEMANS, Aegyptische Monumenten van het Nederlandsche Museum van Oudheden te Leiden, II pl. 98f. Leiden 1846.

P. Len.
GOLÉNISCHEFF, Les Papyrus hiératiques de l'Ermitage imperial à St. Petersbourg. 1913.

P. Leo. Am.
JEA. 22, 169f.

P. Mag. Gr. Leid.
PARTHEY, Zwei Griechische Zauberpapyri. Berlin 1866.

P. Mag. Harr.
LANGE, Der magische Papyrus Harris. København 1927.

P. Mayer
PEET, The Mayer Papyri A and B. London 1920.

P. Med. Berl.
WRESZINSKI, Der große medizinische Papyrus des Berliner Museums. Leipzig 1909.

P. Méd. Copte
CHASSINAT, Un Papyrus médical Copte (Mem. Inst. XXXII). Cairo 1921.

P. Med. Lond.
WRESZINSKI, Der Londoner medizinische Papyrus und der Papyrus Hearst. Leipzig 1912.

P. Med. Lond. and P. Hearst
cf. P. Med. Lond.

P. Northumberland
BARNS, JEA. 34, 35.

P. Oxy.
The Oxyrhynchus Papyri, 35 vols. to date. London 1898f.

P. Ram.
GARDINER, The Ramesseum Papyri. Oxford 1955; BARNS, Five Ramesseum Papyri. Oxford 1956.

P. Rhind
MÖLLER, Die beiden Totenpapyrus Rhind des Museums zu Edinburg. Leipzig 1913.

P. Ryl.
GRIFFITH, Catalogue of the Demotic Papyri in the John Rylands Library Manchester, 3 vols. Manchester 1909.

P. Salt
(124) CHABAS, Mélanges Égyptologiques III, i; pl. 8–11. Chalon-Paris 1870; ČERNÝ, JEA. 15 pl. 42–6.
(825) BUDGE, Facsimiles of Egyptian Hieratic Papyri in the Brit. Mus., 2nd. Series, pl. 31–40. London 1923.

P. Tur.
Papyri in the Turin Museum, cf. P. Tur. PR., Giornale, RAD., etc.

P. Tur. Leps. Grundplan
LEPSIUS, Grundplan des Grabes König Ramses IV in einem Turiner Papyrus. Berlin 1867; cf. also JEA. 4, 130f.

P. Tur. PR.
PLEYTE & ROSSI, Papyrus de Turin, 2 vols. Leide 1869–1876.

P. Wien
Von Bergmann, Hieratische und hieratische-demotische Texte der Sammlung aegyptischer Alterthümer des . . . Kaiserhauses. Wien 1886.

Paheri
Tylor & Griffith, Ahnas el Medineh and the Tomb of Paheri at El Kab. London 1894.

Paints and Pigments
Iversen, Some Ancient Egyptian Paints and Pigments. København 1955.

Paläographie
Möller, Hieratische Paläographie, 3 vols. with Supplement. Leipzig 1909–1936.

Palermo
Schäfer, Ein Bruchstück altägyptischer Annalen. Abh. Berl. Ak. 1902; Urk. I, 235–49.

Pap. Ani Budge
Budge, The Book of the Dead: Facsimile of the Papyrus of Ani . . . (2 nd. ed.). London 1894.

Pap. Ebers
Ebbell, The Papyrus Ebers, The Greatest Egyptian Medical Document. Copenhagen-London 1937.

Pap. Ebers
Joachim, Papyros Ebers, das älteste Buch über Heilkunde. Berlin 1890.

Pap. Smith
Breasted, The Edwin Smith Surgical Papyrus, 2 vols. Chicago 1930.

Peas.
P. Berl. 3025, P. Berl. 10499 etc., Vogelsang, Die Klagen des Bauern. Berlin 1904.

Per-Nēb
Ransom Williams, The Decoration of the Tomb of Per-Nēb. New York 1932.

Petos.
Lefebvre, Le Tombeau de Petosiris, II, Les Textes. Cairo 1923.

Petubastis
Spiegelberg, Der Sagenkreis des Königs Petubastis, 2 vols. Leipzig 1910.

Peyron, Lexicon
Peyron, Lexicon Linguae Copticae. Turin 1835.

Philensis
Philae inscriptions of Ptolemy Epiphanes, Urk. II, 198–230.

Pianchi
Urk. III, 1–56.

Piehl Inscr.
Piehl, Inscriptions hiéroglyphiques, Série 1–3. Leipzig 1886–1903.

Pierret Inscr.
Pierret, Recueil d'Inscriptions inédites du Musée égyptien du Louvre, 2 vols. Paris 1874–1878.

Porträts
Borchardt, Porträts der Königin Nofret-ete. Leipzig 1923.

Priesterdek.
Spiegelberg, Der demotische Text der Priesterdekrete von Kanopus und Memphis (Rosettana). Heidelberg 1922.

Prices & Wages
Černý, Prices and Wages in Egypt in the Ramesside Period (Cahiers d'Histoire Mondiale vol. 1 no. 4, April. 1954).

Prisse
Dévaud, Les Maximes de Ptahhotep. Fribourg 1916.

PSBA.
Proceedings of the Society of Biblical Archaeology.

Puyem.
Davies, The Tomb of Puyemrê at Thebes, 2 vols. New York 1922, 1923.

Pyr.
Sethe, Die altägyptischen Pyramidentexte, 4 vols. Leipzig 1908–1922.

Pyr. Texts
Mercer, The Pyramid Texts, 4 vols. New York 1952.

Q. Hesy
QUIBELL, Excavations at Saqqara, The Tomb of Hesy. Cairo 1913.
Q. Saqq.
QUIBELL, Excavations at Saqqara, 6 vols. Cairo 1907–1923.

RAD.
GARDINER, Ramesside Administrative Documents. London 1948.
Rec. Champ.
Recueil d'Études égyptologiques dédiées à la Mémoire de Jean-François Champollion. Paris 1922.
Rec. Trav.
Recueil de Travaux relatifs à la Philologie et à l'Archéologie égyptiennes et assyriennes.
Recherches
DE MORGAN, Recherches sur les Origines de l'Égypte, 2 vols. Paris 1896, 1897.
Records
PATON, Early Egyptian Records of Travel, 4 vols. Princeton 1915–1922.
Recueil
BRUGSCH, Recueil de Monuments égyptiens, 2 vols. Leipzig 1862, 1863.
Red.
SANDER-HANSEN, Historische Inschriften der 19. Dynastie p. 25–9. Bibliotheca aegyptica IV. Bruxelles 1933.
Rekhmirē
DAVIES, The Tomb of Rekh-mi-rēᶜ at Thebes, 2 vols. New York 1943.
Reliefs AR. MR. NR.
KLEBS, Die Reliefs (und Malereien) des alten, mittleren, neuen Reiches. Heidelberg 1915; 1922; 1934.
Religion
ERMAN, Die ägyptische Religion. Berlin 1905.
Rev. Arch.
Revue Archéologique.
Rev. d'Ég.
Revue d'Égyptologie.
Rhind
PEET, The Rhind Mathematical Papyrus. Liverpool 1923.
Rit. d'Apis
Vienna Demotic Papyrus 27, VON BERGMANN, pls. 10–18, ZÄS. 56, 1f., references are to column and line of the text.
R. Fun. Guieysse
GUIEYSSE, Rituel funéraire égyptien, Chapitre 64e, quoted by line of text. Paris 1876.
Roman Book
BALL, A Roman Book on Precious Stones. Los Angeles 1950.
Rom. Obelisk
ERMAN, Römische Obelisken. Abh. Berl. Ak. 1917.
Rosetta
Rosetta decree, Urk. II, 166–98.
Rouleau Magique
PLEYTE, Étude sur un Rouleau Magique du Musée de Leide. Leide 1866.

Sahurēᶜ
BORCHARDT, Das Grabdenkmal des Königs Śaꜣhu-rēᶜ, 2 vols. Leipzig 1910, 1913.
Sall.
Sallier Papyri, BUDGE, Facsimiles of Egyptian Hieratic Papyri in the Brit. Mus., 2nd. series. London 1923.
Sammlung
WIEDEMANN, Sammlung altägyptischer Wörter welche von Klassischen Autoren umschrieben oder übersetzt worden. Leipzig 1883.
Sarc.
LACAU, Sarcophages antérieures au Nouvel Empire (Cat. Cairo), 2 vols. Cairo 1904, 1906.

Saqq. Mast.
MURRAY, Saqqara Mastabas, 2 vols. London 1905, 1937.
Scarabs (Hall)
HALL, Catalogue of Egyptian Scarabs etc. (in the Brit. Mus.). London 1913.
Scarabs (Petrie)
PETRIE, Scarabs and Cylinders with names. London 1917.
Scènes
MONTET, Scènes de la Vie privée. Strasbourg 1925.
Scepter
HAYES, The Scepter of Egypt (Part I). New York 1953.
Schriftsystem
JUNKER, Über das Schriftsystem im Tempel der Hathor in Dendera. Berlin 1903.
Sculpture and Painting
SMITH, A History of Egyptian Sculpture and Painting in the Old Kingdom. London 1946.
Sehel
BARGUET, La Stèle de la Famine à Séhel. Cairo 1953.
Sethe
Notebooks of SETHE in the Berlin Academy, quoted by no. of notebook and page.
Sh. S.
P. Len. 1115, BLACKMAN, Middle Egyptian Stories p. 41–8. Bibliotheca aegyptiaca II. Bruxelles 1932.
Sin.
P. Berl. 3022, BLACKMAN, Middle Egyptian Stories p. 1–41. Bibliotheca aegyptiaca II. Bruxelles 1932.
Sinai
GARDINER, PEET & ČERNÝ, The Inscriptions of Sinai, 1, inscriptions, cited by no. and line, 2, commentary, 2 vols. London 1952, 1955.
Sinuhe
GARDINER, Notes on the Story of Sinuhe. Paris 1916.
Sitz. Berl. Ak.
Sitzungsberichte der Preußischen Akademie der Wissenschaften zu Berlin.
Sius.
Khaemwese II, GRIFFITH, Stories of the High Priests of Memphis. Oxford 1900.
Six. Temp.
PETRIE, Six Temples at Thebes. London 1897.
Smith
BREASTED, The Edwin Smith Surgical Papyrus, 2 vols. Chicago 1930.
Sonnenheiligtum
BISSING, Das Re-Heiligtum des Königs Ne-Woser-Re, 3 vols. Berlin 1905–1928.
Sonnenscheibe
BRUGSCH, Die Sage von der geflügelten Sonnenscheibe nach altägyptischen Quellen. Göttingen 1870.
Stat. Taf.
BISSING, Die statistische Tafel von Karnak. Leipzig 1897.
Statues
LEGRAIN, Statues et Statuettes de Rois et de Particuliers (Cat. Cairo), 3 vols. Cairo 1906–1914.
Steinbrüche
FITZLER, Steinbrüche und Bergwerke im Ptolemaischen und Römischen Ägypten. Leipzig 1910.
Stèle de la Famine
BARGUET, La Stèle de la Famine à Séhel. Cairo 1953.
Struggle
MASPERO, The Struggle of the Nations. London 1896.
Studies
FORBES, Studies in Ancient Technology, 6 vols. Leiden 1955–1958.
Sudhoff's Archiv
Archiv für Geschichte der Medizin.

Tanis
PETRIE, Tanis, 2 vols. London 1885, 1888.
Tell el Yah.
NAVILLE, The Mound of the Jew and the City of Onias, and GRIFFITH, The Antiquities of Tell el Yahudiyeh. London 1890.
Temple of Hibis (III)
DAVIES, The Temple of Hibis in el Khargeh Oasis. New York 1953.
Textiles
LUTZ, Textiles and Costumes among the Peoples of the Ancient Near East. Leipzig 1923.
Text. Relig.
LACAU, Textes religieux (Rec. Trav. 26–37).
Theophrastus
CALEY & RICHARDS, Theophrastus on Stones. Columbus, Ohio 1956.
Thes.
BRUGSCH, Thesaurus Inscriptionum Aegyptiacarum, 6 vols. Leipzig 1883–1891.
Ti
STEINDORFF, Das Grab des Ti. Leipzig 1913.
Tomb. Court.
PETRIE, Tombs of the Courtiers at Oxyrhynkhos. London 1925.
Tomb Robberies
PEET, The Great Tomb Robberies of the Twentieth Egyptian Dynasty, 2 vols. Oxford 1930.
Totb. Budge
BUDGE, The Book of the Dead, 3 vols. London 1898.
Totb. Einl.
NAVILLE, Das Aegyptische Todtenbuch, Einleitung. Berlin 1886.
Totb. Leps.
LEPSIUS, Das Totenbuch der Ägypter nach dem hieroglyphischen Papyrus in Turin . . . Leipzig 1842.
Totb. Nav.
NAVILLE, Das Aegyptische Totenbuch der xviii bis xx Dynastie, 2 vols. Berlin 1886.
Totb. (Nu)
cf. BUDGE, The Book of the Dead, 3 vols. London 1898.
Totb. Pl.
PLEYTE, Chapitres supplémentaires du Livre des Morts, 3 vols. Leide 1881.
Totb. Stud.
SETHE, ZÄS. 57f. (Göttinger Totenbuchstudien).
Trésor de Tôd (Cat. Cairo)
BISSON DE LA ROQUE, Trésor de Tôd (Cat. Cairo). Cairo 1950.
Trésor de Tôd (Doc. de Fouilles IFAO.)
BISSON DE LA ROQUE, CONTENAU, CHAPOUTHIER, Le Trésor de Tôd (Doc. de Fouilles IFAO. XI). Cairo 1953.
TSBA.
Transactions of the Society of Biblical Archaeology.

Unt.
SETHE, Untersuchungen zur Geschichte und Altertumskunde Aegyptens.
Unt. Med.
GRAPOW, Untersuchungen über die altägyptischen medizinischen Papyri, 2 vols. Leipzig 1935, 1936.
Urgesch.
SETHE, Urgeschichte und älteste Religion der Ägypter. Leipzig 1930.
Urk.
Urkunden des ägyptischen Altertums (ed. STEINDORFF).
Urk. z. Relig.
ROEDER, Urkunden zur Religion des alten Ägypten. Jena 1915.

V. ä. K.
SCHÄFER, Von ägyptischer Kunst (3rd. ed.). Leipzig 1930.
Voyage
CHABAS, Voyage d'un Égyptien en Syrie en Phénicie en Palestine . . . Chalon-Paris 1866.

Wadi el Hudi
FAKHRY, The Inscriptions of the Amethyst Quarries at Wadi el Hudi. Cairo 1952.
Waffen
BONNET, Die Waffen der Völker des alten Orients. Leipzig 1926.
Wanderung
BRUGSCH, Wanderung nach den Türkis-Minen und der Sinai-Halbinsel. Leipzig 1868.
Wb.
ERMAN & GRAPOW, Wörterbuch der ägyptischen Sprache (cf. WÖRTERBUCH).
Wb.
BRUGSCH, Hieroglyphisch-Demotisches Wörterbuch, 7 vols. Leipzig 1867–1882.
Wb. Erg.
BRUGSCH, ib. Ergänzungsband.
Wen.
GARDINER, Late Egyptian Stories p. 61–76. Bibliotheca aegyptiaca I. Bruxelles 1932.
Westc.
ERMAN, Die Märchen des Papyrus Westcar. Berlin 1890.
Wien.
Inscriptions in the Vienna Museum, WRESZINSKI, Ägyptische Inschriften aus dem Königlichen Hofmuseum in Wien. Leipzig 1906.
Wilbour
GARDINER, The Wilbour Papyrus, 4 vols. Oxford 1941–1952.
Wohlriechende Natron
STEUER, Über das wohlriechende Natron bei den alten Ägyptern. Leiden 1937.
Wolf Opet
WOLF, Das schöne Fest von Opet. Leipzig 1931.
Wortbildung
GRAPOW, Über die Wortbildung mit einem Präfix *m* im Ägyptischen Abh. Berl. Ak. 1914.
WÖRTERBUCH
ERMAN & GRAPOW, Wörterbuch der ägyptischen Sprache, 6 vols. mit Belegstellen zu Bd. I–V. Berlin 1926–1953. 2. unv. Neudruck, Berlin 1957–1959.
Wortforschung
ERMAN, Zur ägyptischen Wortforschung. Sitz. Berl. Ak. 1907, 400–415; 1912, 904–963; 1928, 255–258.

Yuya
NAVILLE, The Funeral Papyrus of Iouiya. London 1908.
Young Hier.
YOUNG, Hieroglyphics collected by the Egyptian Society. London 1823.

Zahlen
SETHE, Von Zahlen und Zahlworten bei den alten Ägyptern. Straßburg 1916.
ZÄS.
Zeitschrift für ägyptische Sprache und Altertumskunde.
Zech.
Zechariah.

INDICES

N. B. In Index I (Egyptian) and Index VII (Minerals etc.) only the substantive references are cited.

I. EGYPTIAN

II. COPTIC

III. AKKADIAN & SUMERIAN

IV. HEBREW ETC.

V. GREEK

VI. LATIN

VII. MINERALS ETC.

VIII. GENERAL

UNPUBLISHED TEXTS

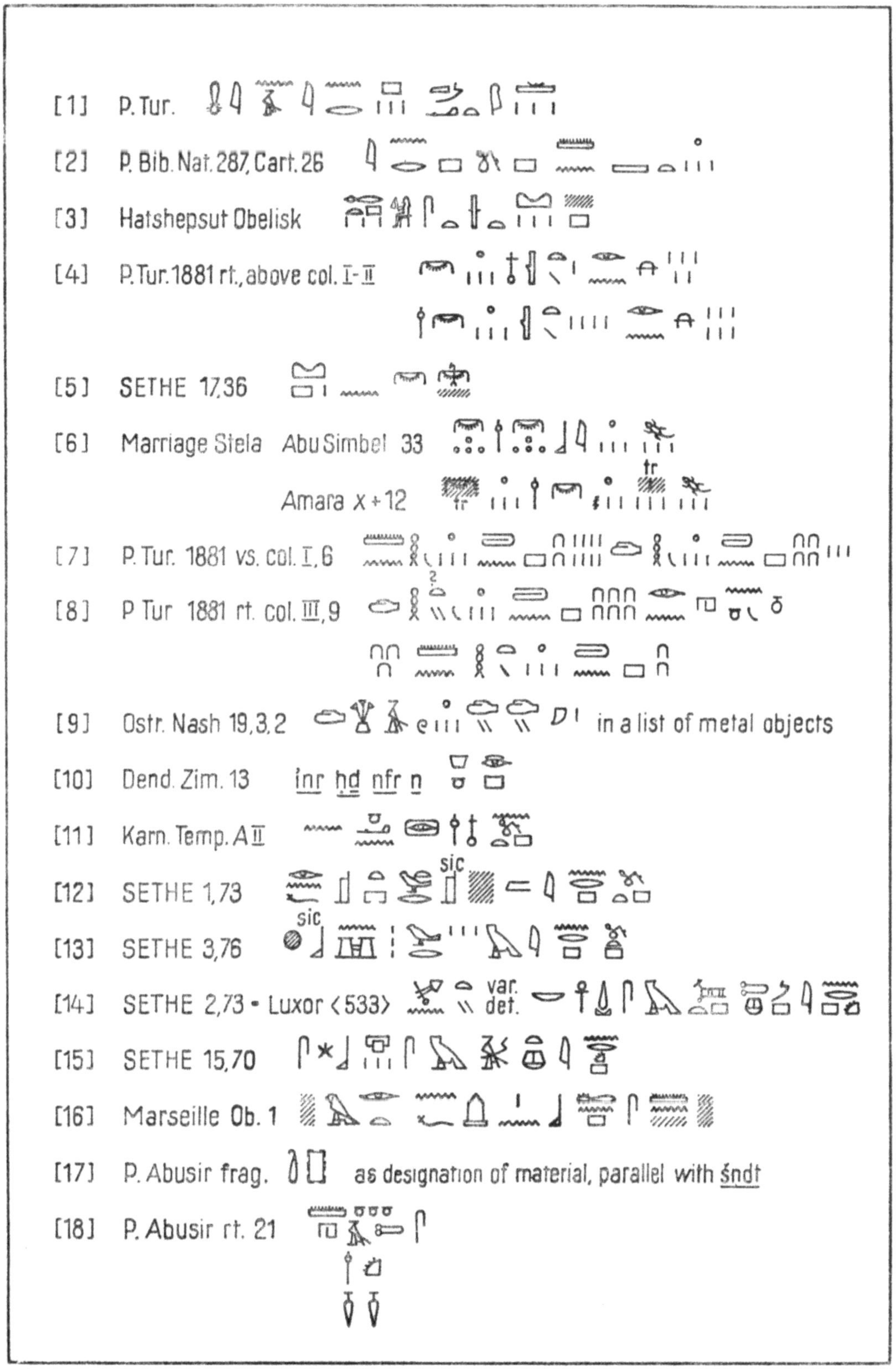
[1] P. Tur.
[2] P. Bib. Nat. 287, Cart. 26
[3] Hatshepsut Obelisk
[4] P. Tur. 1881 rt., above col. I-II
[5] SETHE 17,36
[6] Marriage Stela Abu Simbel 33
tr
Amara x+12
tr
[7] P. Tur. 1881 vs. col. I, 6
[8] P Tur 1881 rt. col. III, 9
[9] Ostr. Nash 19,3,2
in a list of metal objects
[10] Dend. Zim. 13
ỉnr ḥḏ nfr n
[11] Karn. Temp. A II
[12] SETHE 1,73
sic
[13] SETHE 3,76
sic
[14] SETHE 2,73 • Luxor ‹533›
var. det.
[15] SETHE 15,70
[16] Marseille Ob. 1
[17] P. Abusir frag.
as designation of material, parallel with šndt
[18] P. Abusir rt. 21

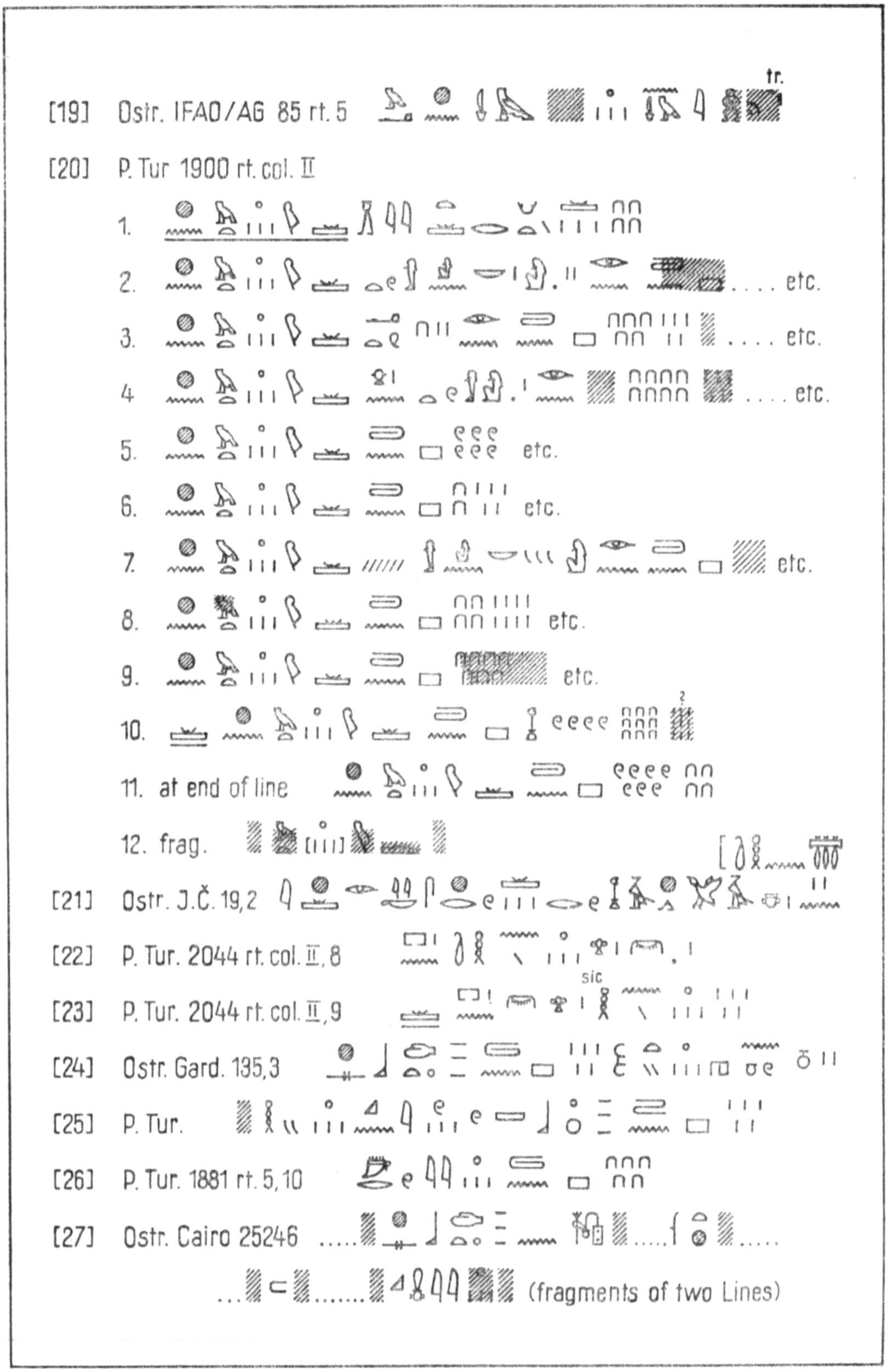
[19] Ostr. IFAO/AG 85 rt. 5
tr.
[20] P. Tur 1900 rt. col. II
1.
2. .... etc.
3. .... etc.
4 .... etc.
5. etc.
6. etc.
7. etc.
8. etc.
9. etc.
10.
11. at end of line
12. frag.
[21] Ostr. J.Č. 19,2
[22] P. Tur. 2044 rt. col. II, 8
[23] P. Tur. 2044 rt. col. II, 9
sic
[24] Ostr. Gard. 135,3
[25] P. Tur.
[26] P. Tur. 1881 rt. 5, 10
[27] Ostr. Cairo 25246
(fragments of two Lines)

[28] Luxor ⟨453⟩, ⟨472⟩ c.f. WB. III 488. 6

[29] P Abusir rt. 19  as designation of material

[30] P. Abusir rt. 21 

Ostr. MMA 90. 6. 3. (1728) : Rec. Trav. 22 p. 103

Revised Transcription

Rt. 1. 

2 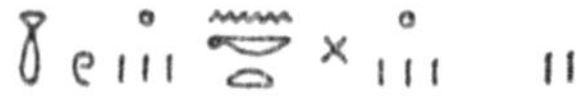

3 

4. 

5. 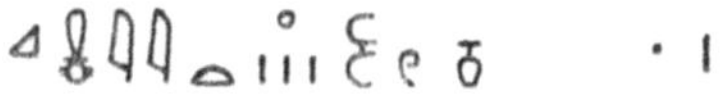

6. 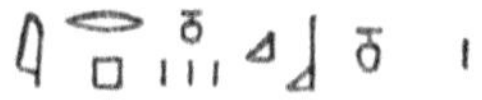

Vs. 1. 

2. 

3 

4. 

Bei Fragen zur Produktsicherheit wenden Sie sich bitte an:
If you have any questions regarding product safety,
please contact:

Walter de Gruyter GmbH
Genthiner Straße 13
10785 Berlin
productsafety@degruyterbrill.com